Springer Series in Solid-State Sciences

Editors: M. Cardona P. Fulde H.-J. Queisser

E. A. Silinsh

Organic Molecular Crystals

Their Electronic States

With 135 Figures

Springer-Verlag Berlin Heidelberg New York 1980

Professor Dr. *Edgar A. Silinsh*

Institute of Physics and Energetics,
Latvian SSR, Academy of Sciences,
Riga-6, USSR

Series Editors:

Professor Dr. Manuel Cardona
Professor Dr. Peter Fulde
Professor Dr. Hans-Joachim Queisser

Max-Planck-Institut für Festkörperforschung, Heisenbergstrasse 1
D-7000 Stuttgart 80, Fed. Rep. of Germany

Revised translation of the Russian edition:
Elektronnye sostoyaniya organicheskikh molekulyarnykh kristallov (Electronic States of
Organic Molecular Crystals)
by E. A. Silinsh
© by Zinatne, Riga (1978)
English translation by J. Eiduss in collaboration with the author.

ISBN-13: 978-3-642-81466-2   e-ISBN-13: 978-3-642-81464-8
DOI: 10.1007/978-3-642-81464-8

Library of Congress Cataloging in Publication Data. Silinsh, E. A. Organic molecular crystals. (Springer series in solid-state sciences ; v. 16) Revised translation of the author's Elektronnye sostoyaniya organicheskikh molekulyarnykh kristallov, 1978 ed. Bibliography: p. Includes index. 1. Molecular crystals. 2. Chemistry, Organic. 3. Crystals–Electric properties. I. Title. II. Series. QD921.S539313 1980   547   80-14967

© by Springer-Verlag Berlin Heidelberg 1980
Softcover reprint of the hardcover 1st edition 1980

Foreword

This book is based on the results of many years of experimental work by
the author and his colleagues, dealing with the electronic properties of
organic crystals. E. Silinsh has played a leading role in pointing out the
importance of the polarization energy by an excess carrier, in determining
not only the character of the carrier mobility in organic crystals, but in
determining the band gap and the nature of the all-important trapping site
in these crystals. The one-electron model of electronic conductivity that
has been so successful in dealing with inorganic semiconductors is singular-
ly unsuccessful in rationalizing the unusual physical properties of organic
crystals. A many-body theory is required, and the experimental manifestation
of this is the central role played by the crystal polarization energies in
transferring the results obtained with the isolated molecule, to the solid.
The careful studies of E. Silinsh in this field have shown in detail how
this polarization energy develops around the excess carrier (and also the
hole-electron pair) sitting on a molecular site in the crystal.

As with all insulators, trapping sites play a dominant role in reducing
the magnitude of the current that can theoretically pass through the organic
crystal. It is usually the case that these trapping sites are energetically
distributed within the forbidden band of the crystal. For many years, an
exponential distribution has shown itself to be useful and reasonably
correct. However, E. Silinsh has pointed out the inadequacies of this assumption, and has demonstrated the greater validity of a Gaussian distribution. Recent work in different parts of the world, particularly with
amorphous materials, has substantiated the point of view of E. Silinsh.

A glance at the Table of Contents of this timely book reveals an unusual-
ly pertinent selection of topics. Not only are charge carrier phenomena
discussed, but also the excitonic phenomena that are exhibited to an unusual
degree in these organic crystals. In all cases, the approach taken by the

author is admirably selective, direct, clear, and positive. He has presented his opinion as to what is correct, and in so doing, has rendered a significant service to the reader who may not be familiar with this field, and who wishes to begin with the best thinking of an outstanding worker in the field. The theoretical treatment is presented as a basis for understanding the experimental results, and the book is filled with excellent figures and tables. These provide a rich source of experimental data. The results from the author's laboratory, dealing with heterocyclic compounds, are not usually treated elsewhere in such a coherent fashion.

The entire area of electronic conductivity in organic materials is expanding at an unbelievable rate. These compounds are being used as photoconductors in electrophotography and as radiation detectors; in the form of polymers, and charge-transfer crystals, it has been possible to develop quasi-one-dimensional crystals with conductivities that approach metals. The book by E. Silinsh provides an excellent starting point for those two wish to learn the important elements that distinguish this group of materials from the more conventional materials. It will also remain useful despite the passage of time because the opinions of the author will retain their validity and superb tutorial quality.

New York, June 1980

Martin Pope
New York University

Preface

In the broad spectrum of solid-state science a new band, overlapping adjacent interdisciplinary areas, has recently emerged—the physics of the organic solid state.

In this new promising field of research organic crystals of aromatic and heterocyclic molecules have become the main objects of scientific interest. This is due to the fact that, in the case of aromatic and a number of heterocyclic organic crystals, the investigator has at his disposal detailed information about their molecular and crystal structure. Such knowledge is an indispensable precondition for successful studies of electron and exciton processes in the crystal. Secondly, the above-mentioned classes of organic crystals open new promising avenues for practical applications. A number of aromatic and, especially, heterocyclic molecular crystals exhibit high quantum efficiency of photoconductivity, interesting photodielectric and photomagnetic properties, pronounced anisotropy of optical and electrical parameters, high electronic polarizability and other exotic characteristics. Thirdly, a systematic study of simple organic molecular crystals is a necessary prerequisite for proper understanding of electronic processes in more complex organic solids, including polymers and molecular biosystems.

At present the investigation of electronic states in organic molecular crystals is regarded as a self-contained problem of solid-state physics. This is partially caused by the fact that, for organic molecular crystals, a number of traditional concepts of solid-state physics are not valid, e.g., band theory of one-electron approximation.

In molecular crystals, due to the weakness of Van der Waals interaction, the main electronic characteristics of molecules undergo only slight changes in the process of solid phase formation. The domination of molecular properties over the crystalline ones leads to a marked tendency of localization of charge carriers and excitons on individual molecules of the crystal. As a result, a number of qualitatively new properties emerge in organic mole-

cular crystals, such as electronic polarization of the lattice by a localized
charge carrier, vibronic relaxation of molecules in the process of the formation of genuine ionic states, formation of local states of polarization
origin in the regions of structural defects of the crystal, and other phenomena, unfamiliar to covalent and ionic crystals.

The main purpose of the book is to investigate the present status and
future trends in the field of energy structure studies in molecular crystals.
The basic emphasis is laid on stationary conduction levels of thermalized
charge carriers, as well as on local trapping states. Dynamic aspects of
charge carrier generation and transport are discussed only fragmentarily, in
context with the description of stationary states of the crystal.

The phenomenological models and basic concepts of energy structure of
molecular crystals are based on investigations of the author and his collaborators, as well as on data reported by other workers. Although more than
500 references appear in the course of the discussion, the book is not meant
to be a review. The results of other authors have been mainly used for interpreting or confirming theoretical principles and phenomenological models,
or for illustrating methods of investigation.

This book was stimulated by the desire to formulate a scientific basis
for experimental investigation of molecular crystal energy structure. Since
this new field is still being developed, we discuss both recognized physical
models and theories, as well as problematic, controversial questions which
have, in our view, considerable heuristic value in stimulating experimental
and theoretical research in organic solid-state physics.

The physics of the organic solid state occupies a borderline between a
number of different disciplines—molecular spectroscopy, quantum chemistry,
solid-state physics and electronics, crystal chemistry, organic and physical
chemistry, and even molecular biology. Therefore the present book may be
interesting and useful for specialists working in different fields.

The present work could hardly have been possible without the assistance
and cooperation on the part of a large number of colleagues, near and far.

The author is grateful to Springer-Verlag and, in particular, to Dr. H.
Lotsch for the opportunity of publishing his work in English

The author greatly appreciates the help of his collaborators Dr. L.F.
Taure, Dr. A.K. Gailis, D.R. Balode, I.S. Kaulach, V.A. Kolesnikov, I.J.
Muzikante, V.A. Skudra and the Photoemission group of the Laboratory,
headed by Dr. A.I. Belkind, for providing new experimental data for the book,
as well as the fruitful collaboration with A.J. Jurgis in electronic polariz-

ation and local state energy spectra calculations, and with Dr. S. Nešpurek in the study of SCLC theory for Gaussian trap distribution.

For translation of the original Russian book, as well as of the revised and newly added portions of text the author is much indebted to Dr. J. Eiduss.

All those colleagues who have helped with valuable comments in private discussions, conferences and correspondence are gratefully acknowledged.

In addition to the acknowledgements expressed in the preface to the original Russian edition of the book [Zinatne, Riga 1978], the author wishes to thank in particular Professor E.L. Frankevich, Dr. M.V. Kurik, Professor T.J. Lewis, I.B. Rips, Dr. J. Sworakowski, Dr. L.F. Taure, Dr. A.T. Vartanyan, Dr. J.O. Williams for useful comments and suggestions during the preparatory period of the manuscript of this English edition.

The author is much indebted to D.R. Balode, I.J. Muzikante, I.E. Rozīte, V.A. Skudra and L.F. Taure for meticulous and patient technical assistance at all stages of this work. This does not, however, free the author from full responsibility for any shortcomings or possible errors in the text.

Riga, January 1980   *Edgar A. Silinsh*

Acknowledgments

The author is indebted to the following publishers for permission to reproduce their figures in the text:

The Royal Society of London:
Fig. 3.17.

The American Institute of Physics:
Figs. 1.1a,5,18; 4.4-6; 5.21.

The Institute of Physics (London):
Figs. 5.29,41,42,46.

The Chemical Society (London):
Figs. 1.14; 3.12,19,25; 5.24.

The North-Holland Publishing Co.:
Figs. 1.15; 3.14; 4.7-9,11; 5.43.

Wiley-Interscience: Figs. 3.6,9b,11,
30; 5.10.

McGraw-Hill Book Co.: Figs. 3.4,10.

Taylor and Francis Ltd: Fig. 3.27.

Blackwell Scientific Publications
Ltd.: Fig. 3.29.

Academic Press Inc.: Fig. 3.9a.

Akademie Verlag, Berlin: Figs. 2.8,
14; 3.1,2,16; 5.25-34,40,45,50.

Wroclaw Technical University Press:
Figs. 3.23,24; 5.35,36,39,47,48.

Gordon and Breach Science Publishers
Ltd.: Figs. 3.8,28,31; 4.1-3;
5.4,6.

Pergamon Press Ltd., Oxford: Figs. 3.3,
7,13;5.29b,50.

Nauka, Moscow: Figs. 1.1b,4;2.1;5.5,12

We also wish to thank these authors for their kind consent to reproduce
the following figures:

J.M. Thomas, F.R.S.: Figs. 1.15;
3.3,12,13,17-19,23-25,28; 4.7-9;
5.24,41,42.

A.S. Davydov: Fig. 2.1.

J.O. Williams: Figs. 3.27,29.

N. Karl: Fig. 5.50

J. Sworakowski: Figs. 3.8; 5.4,6,40,
44.

L. Schein: Fig. 1.19.

H. Baessler: Figs. 5.29,43.

V.A. Lisovenko: Figs. 3.14; 4.10,11.

A.I. Kitaigorodsky: Figs. 1.1b,4,17;5.5

Contents

1. Introduction:
Characteristic Features of Organic Molecular Crystals

Science is not just a collection of laws,
a catalogue of unrelated facts.
It is a creation of the human mind,
with its freely invented ideas and concepts.

Albert Einstein

Organic molecules in condensed solid phase form molecular crystals which differ considerably in their mechanical, optical, and electronic properties from such traditional solid-state objects as covalent or ionic crystals. This is due to weak intermolecular interaction forces of the Van der Waals type, with bonding energy considerably lower than that of covalent or ionic bonds in atomic crystals. Lattice energy is accordingly low in molecular crystals, which means low melting and sublimation temperatures, low mechanical strength, and high compressibility. The heat of sublimation H_s of anthracene-type molecular crystals (H_s = 70-130 kJ$\cdot$mol^{-1} [1.1]) lies between that of noble-gas crystals (H_s = 3-16 kJ$\cdot$mol^{-1} [1.2,3]) and lattice energies of typical ionic crystals such as alkali halides (600-1000 kJ$\cdot$mol^{-1} [1.3]).

Van der Waals forces of attraction are caused by specific nonvalent weak interaction between electrically neutral particles. The most general form of this kind of interaction is known as dispersion forces of attraction between nonpolar, electrically neutral molecules. It is just these forces which are most essential in the case of crystals of aromatic hydrocarbons, such as anthracene, consisting of nonpolar molecules.

This chapter gives a brief description of Van der Waals interaction forces determining the electronic properties of a molecular crystal. It also contains data about molecular and crystal structure and properties of aromatic hydrocarbons - the main model compounds of organic molecular crystals. Specific properties of electronic states and contemporary concepts of the nature of electronic conductivity in organic crystals are briefly discussed.

Chapter 2 is devoted to the investigation of electronic states of a perfect molecular crystal. The main theoretical principles and experimental methods for ionized state studies in molecular crystals are dealt with.

The Lyons model of ionized states and its modified and extended variants describing phenomenologically the many-electron effects of electronic polarization and vibronic relaxation on the self-energy of charge carriers have been treated in some detail, as well as the specific role of charge transfer (CT) states in photogeneration and recombination processes. Energy diagrams for a number of aromatic and heterocyclic molecular crystals are presented and analyzed.

Chapter 3 contains an analysis of the role of crystal irregularities and structural defects in the formation of local electronic states of polarization origin. Energetical and configurational properties of various specific structural defects in molecular crystals are considered and experimental techniques for their investigation described.

Chapter 4 presents a brief description of local exciton trapping states of structural origin and data on their experimental investigation in anthracene-type crystals.

Chapter 5 deals with local charge carrier trapping states. The formation of local electronic states of polarization origin in structural defects of a molecular crystal is discussed. Different randomizing factors are analyzed, assumed to be responsible for the statistical Gaussian distribution of local state energy spectra. A phenomenological space charge limited current (SCLC) theory, developed for Gaussian trap distribution is discussed in detail and illustrative examples of its application for the investigation of energy spectra of local states in a number of aromatic and heterocyclic molecular crystals are given. The problem of validity of the Gaussian model for description of local trapping states of structural origin in molecular crystals is approached on the basis of experimental data obtained by independent methods. Correlations between distribution parameters of local trapping states and crystalline structure of the samples are analyzed.

1.1 Interaction Forces in Molecular Crystals

A quantum-mechanical treatment of dispersion forces was given by LONDON [1.4]. It is based on the concept of electrical interaction between fluctuating multipole moments of molecules [1.2]. A static approach yields zero interaction between neutral nonpolar molecules. However, even in atoms and molecules with zero mean value of multipole moments there exist dynamically fluctuating multipole moments depending on momentary states of electron motion. The momentary electric field created by these moments induces multipole moments in adjacent atoms and molecules. Interaction between the electric moment of the initial particle and induced moments of the adjacent ones, averaged over the total set of particles, produces forces of attraction between them.

As was shown by LONDON [1.4], this effect is primarily due to external, more weakly bonded electrons such as valent σ and π electrons of aromatic molecules. These electrons are also responsible for the dispersion of light. That is why LONDON called this phenomenon of attraction between neutral molecules dispersion forces, and this term has become generally accepted.

Using quantum-mechanical second-order perturbation theory LONDON [1.4] derived an approximate, but rather adequate expression for dispersion interaction energy $U_{dis}(r)$ as a function of distance r between two molecules. In this approach the interacting molecules are treated as quasi-elastic oscillators possesssing zero energy in the ground state. The potential energy of the interacting molecule pair is expressed in the form of a multipole series, of which only the dipole-dipole term is taken into account. This yields an expression of interaction energy $U_{dis}(r)$, averaged over various orientations of the dipoles with respect to the vector $\underline{r}$, connecting both molecules [1.2]

$$U_{dis}(r) = -\frac{A}{r^6} \quad , \tag{1.1}$$

where A is a constant, equal to

$$A = \frac{2}{3} \sum_{i'} \sum_{j'} \frac{|<i|\mu_1|i'>|^2 |<j|\mu_2|j'>|^2}{E_{i'} + E_{j'} - E_i - E_j}$$

where i,j denote ground states of nonperturbed molecules; i',j' excited states of the molecules; E_i, E_j and E_i', E_j', are energies of molecules in the ground and excited states, respectively; $<i|\mu_1|i'>$ is the matrix element of dipole moment between states i and i' for molecule 1; $<j|\mu_2|j'>$ is the matrix element of dipole moment between states j and j' for molecule 2.

According to (1.1) $U_{dis}(r)$ is negative and describes the Van der Waals attractive interaction between neutral molecules. Since (1.1) is not expedient for practical calculations, LONDON proposed the following approximated formula, expressed in terms of the principal molecular parameters — ionization energy I and molecular polarizability α [1.4]:

$$U_{dis}(r) = -\frac{3}{2} \alpha_1 \alpha_2 \frac{I_1 I_2}{I_1 + I_2} \cdot \frac{1}{r^6} \quad , \tag{1.2}$$

α_1, α_2 denoting the polarizability of molecules 1 and 2, respectively, and I_1, I_2 their ionization energies.

London's theory of dispersion forces provides a satisfactory description of molecular interaction at distances of the order of Van der Waals radii. At closer intermolecular distances considerable overlapping of molecular orbitals takes place, leading to strong mutual repulsion between the molecules. Quantum-mechanical treatment of repulsion potential U_{rep} yields a rather complicated dependence of U_{rep} on distance r. For this reason $U_{rep}(r)$ is usually described by means of empirical formulae. Most frequently an exponential approximation of the following type is used

$$U_{rep}(r) = B \exp(-Cr) \quad , \tag{1.3}$$

where B and C are empirical constants.

At a certain distance r_0 attractive interaction U_{dis} and repulsive interaction U_{rep} become equal, and correspond to the potential energy minimum of the system.

The full potential curve of intermolecular interaction can be described by the empirical BUCKINGHAM formula [1.5,6]

$$\varphi(r) = -\frac{A}{r^6} + B \exp(-Cr) \quad , \tag{1.4}$$

The potential $\varphi(r)$, in (1.4), is often called the (6-exp) Buckingham potential.

Another empirical formula, proposed by LENNARD-JONES, is also frequently used for $\varphi(r)$ calculations [1.7]

$$\varphi(r) = -\frac{A}{r^6} + \frac{B'}{r^{12}} \quad , \tag{1.5}$$

B' also being an empirical constant.

Potential $\varphi(r)$, in (1.5) is usually known as the (6-12) Lennard-Jones potential.

In the case of polar molecules other kinds of interaction, apart from dispersion forces, can take place, viz. orientational and inductive interaction, both of electrostatic origin.

The energy of orientational or dipole-dipole interaction between molecules a and b, with dipole moments μ_a and μ_b, respectively, equals

$$U_{d-d} = -\frac{2}{3kT} \cdot \frac{\mu_a^2 \mu_b^2}{r^6} \quad , \tag{1.6}$$

If we have molecule a with dipole moment μ_a interacting with molecule b, possessing quadrupole moment Q_b, then the interaction energy equals

$$U_{d-Q} = -\frac{1}{kT} \cdot \frac{\mu_a^2 Q_b}{r^8} \quad , \tag{1.7}$$

Induction forces appear, if molecule a with constant dipole moment μ_a polarizes a nonpolar molecule b, thus inducing a dipole moment in the latter. The mean energy of induction interaction is in this case

$$U_{ind} = -\frac{2\mu_a^2 \alpha_b}{r^6} \quad , \tag{1.8}$$

where α_b is the polarizability of molecule b.

The mean energies of dipole-dipole and induction interaction are of the same order as that of dispersion interaction.

1.2 The Atom-Atom Potential Method

The theory of London-type Van der Waals forces, as well as the corresponding formulae for intermolecular interaction potential (1.4) and (1.5), are based on the assumption of a field of central forces. Hence, quantitative calculations of intermolecular interaction energy, using the Buckingham or the Lennard-Jones formula, were first carried out on the simplest kind of molecular crystals, formed by noble-gas atoms. The high symmetry of these crystals and the spherical nature of the atoms excellently suit the main requirements of London's theory. As regards more sophisticated systems, such as crystals of aromatic hydrocarbons consisting of large polyatomic molecules, direct interaction energy calculation using potential functions of types (1.4) and (1.5) is not feasible. A more complex empirical method of calculation has to be employed in this case, known as the atom-atom potential method, proposed and developed by KITAIGORODSKY et al. [1.2,8-12]. According to the atom-atom potential approach the interaction energy of molecules is obtained as the sum of interaction energies between atoms of neighboring molecules. In such an approximation the forces of interaction between atoms can be considered as central ones conforming with the basic concepts of London's theory and thus permitting the use of interaction potentials of type (1.4) or (1.5).

The method is based on the following assumptions, the validity of which has been confirmed empirically for a large number of aromatic and aliphatic molecular crystals [1.2,12].

1) The energy U of molecular interaction is calculated as the additive sum of interaction energies φ_{ij} of atoms constituting the molecule

$$U = \sum_{ij} \varphi_{ij} \quad , \tag{1.9}$$

where i,j are the symbols for all types of nonvalent interaction between atoms of neighboring molecules (e.g., interaction between atoms C...C, C...H and H...H in hydrocarbons).

2) It is assumed that the nonvalent interaction forces between atoms may be regarded as central.

3) The interaction potential φ_{ij} between two atoms i and j, separated by distance r_{ij}, can be expressed either via the Buckingham (6-exp) potential (1.4) or the Lennard-Jones (6-12) potential (1.5).

Parameter A for the attraction potential in (1.4,5) can, in principle, be calculated theoretically [cf. (1.1)]. In practice, however, it is usually chosen from empirical data. As to parameters B, B' and C for the repulsion potential, there are no adequate theoretical methods for their evaluation, and they can only be found empirically [1.2].

4) Potentials φ_{ij} are assumed to be universal, depending only on the kind of interacting atoms and not on the kind of molecule of which they are part, or on their valency state in the molecule.

This assumption of the universal nature of φ_{ij} is one of the principal advantages of the atom-atom potential method in calculating interaction energy between polyatomic organic molecules. It permits one to choose interatomic potential from experimental data for a few model compounds and then to use the interatomic potentials thus obtained for calculating interaction energies for other analogues of the same series. Thus, knowing the interaction potentials for atom pairs C...C, C...H, and H...H, it becomes possible to evaluate the interaction energies for a large variety of different hydrocarbons [1.2,13,14].

Table 1.1. Empirical values of parameters A, B, and C for interatomic Buckingham potentials [cf. (1.4)] in hydrocarbon crystals

Inter-acting atoms	A [kJ·mol^{-1} Å^{-6}] according to			B [kJ·mol^{-1}] according to			C [Å^{-1}] according to		
	[1.2]	[1.14]	[1.17]	[1.2]	[1.14]	[1.17]	[1.2]	[1.14]	[1.17]
C...C	1500	2380	913	17.6×10^4	35×10^4	728×10^4	3.58	3.60	4.32
C...H	645	524	370	17.6×10^4	3.7×10^4	9.7×10^4	4.12	3.67	4.20
H...H	239	114	151	17.6×10^4	1.7×10^4	2.03×10^4	4.86	3.74	4.08

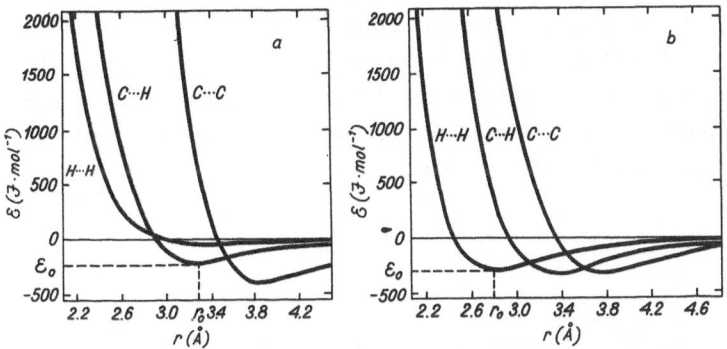

Fig. 1.1. Curves of nonvalent interaction potentials between H...H, C...H and C...C atoms in crystals of aromatic hydrocarbons: (a) according to WILLIAMS [1.14], (b) according to KITAIGORODSKY [1.2]

KITAIGODORSKY et al. [1.2,8-12] have convincingly demonstrated the validity of the atom-atom potential method for calculating crystal structures, dynamic and energy parameters of the lattice, and other data for a number of hydrocarbon crystals. In these studies the optimum values of parameters A, B and C, giving best agreement with experimental data, were also proposed. The search for optimum values of interaction potentials between atom pairs C...C, C...H, and H...H in hydrocarbon crystals and further extension of the method was continued by WILLIAMS [1.13-15], MOMANY et al. [1.16],

LIQUORI et al. [1.17], WARSHEL and LIFSON [1.18], MIRSKAYA et al. [1.19], TADDEI et al. [1.20] and others. For comparison, parameter A, B, and C values, used by KITAIGORODSKY [1.2] WILLIAMS [1.14] and LIQUORI et al. [1.17] are presented in Table 1.1. The corresponding curves of nonvalent interaction potentials between H...H, C...H, and C...C atoms, according to [1.2,14], are shown in Fig.1.1. As may be seen from Table 1.1 and Fig.1.1, the "optimum" values of parameters A, B, and C as well as potential curves of atom-atom interaction, as proposed by various authors, differ considerably from each other. It has, however, been demonstrated in practical application of the method, that these differences in parameters do not affect significantly the final results of calculation, since these parameters are chosen purely empirically and are based on comparison between calculation and empirical data. Since the choice of parameters depends only on the error in the final results of calculation, as compared with experimental data, a good agreement can obviously be obtained with various sets of empirical parameters A, B, and C [1.2].

Parameters A, B, and C in the (6-exp) potential of interaction between neutral atoms [cf. (1.4)] do not possess any obvious physical meaning. They can, however, be expressed by means of the following three physical parameters of the potential curves (Fig.1.1): equilibrium distance r_0, depth of potential well ε_0, and value of the second derivative D at minimum point $(r = r_0)$ [1.2].

The values of r_0 and ε_0 can be easily substituted into (1.4) and we obtain [1.2]

$$\varphi(r) = \frac{\varepsilon_0}{6 - Cr_0} \left\{ -\frac{Cr_0^7}{r^6} + 6 \exp[-C \, (r - r_0)] \right\} \quad , \tag{1.10}$$

Constant C is connected with parameters r_0, ε_0, and D in the following way [1.2]:

$$\frac{Dr_0^2}{\varepsilon_0} = \frac{6(Cr_0)^2 - 42Cr_0}{6 - Cr_0} \tag{1.11}$$

The values of parameters r_0, ε_0 [cf.Fig.1.1], and the product Cr_0 for interaction between atoms C...C, C...H and H...H in hydrocarbon crystals are given in Table 1.2.

Attractive as well as repulsive energies between atoms decrease rapidly with distance [cf. (1.4)]. Therefore summation according to (1.9) need not be carried out in practical calculations beyond a distance of 10-20 Å. The error in this case does not exceed 1-2 % [1.9]. For estimating purposes it is sufficient to consider the interaction only between neighboring molecules.

Table 1.2. Values of parameters of interatomic equilibrium distance r_0, potential well depth ε_0, and product Cr_0 in hydrocarbon crystals

Inter-acting atoms	r_0 [Å] according to		ε_0 [J·mol^{-1}] according to		Cr_0 according to	
	[1.2,10]	[1.14]	[1.2]	[1.14]	[1.2]	[1.14]
C...C	3.8	3.88	250	398	13.60	13.96
C...H	3.3	3.31	250	201	13.59	12.15
H...H	2.8	3.37	250	42	13.60	12.60

Recently SWORAKOWSKY [1.21], SILINSH and JURGIS [1.22], MIRSKY and COHEN [1.23,24], and RAMDAS et al. [1.25] have demonstrated the applicability of the atom-atom potential method for computational approach in the studies of extended structural defects, such as stacking faults in anthracene-type crystals. General principles, methodological problems and limitations of such an approach will be further discussed in some detail (see Sect.3.9.3).

Finally, it is instructive to compare typical intermolecular (dispersion) and intramolecular (covalent) interaction energies between atoms in aromatic hydrocarbon crsytals. As may be seen from Table 1.3 the dispersion interaction energy constitutes only ca. 0.1 % of the corresponding covalent interaction energy. Hence, intermolecular interaction forces may well be qualified as "very weak", as compared to covalent intramolecular ones. A similar comparison between typical interaction forces in molecular and covalent crystals leads to the conclusion that such a pronounced difference of three orders of magnitude in interaction energy values must necessarily imply also considerable qualitative differences in the properties of molecular and covalent crystals. This is indeed the case, as will be amply demonstrated in the following sections.

Table 1.3. Intermolecular (dispersion) and intramolecular (covalent) interaction energies between atoms in aromatic hydrocarbon crystals

Dispersion interaction [1.2]			Covalent interaction [1.26]		
Type of interaction	Mean energy of interaction		Type of bond	Mean energy of covalent bond	
	$[J \cdot mol^{-1}]$	[eV]		$[J \cdot mol^{-1}]$	[eV]
C....C	398	4.1×10^{-3}	C–C	347×10^3	3.6
C...H	201	2.1×10^{-3}	C–H	422×10^3	4.4
H...H	42	0.4×10^{-3}	H–H	435×10^3	4.5

1.3 Aromatic Hydrocarbons — Model Compounds of Organic Molecular Crystals

Aromatic hydrocarbons, especially linear polyacenes, have served during the last decades as the basic model compounds for organic molecular crystal studies.

1.3.1 Anthracene

Anthracene as Model Compound

The most unique in this respect is anthrancene which has become a rather favored object in experimental, as well as in theoretical studies in a large number of laboratories working in the field of organic solid-state chemistry and physics.

This, in our opinion, is mainly due to the following circumstances.

First, the crystal and molecular structure of anthracene has been determined with high accuracy [1.27-34], a factor of first-rate importance for theoretical studies of the crystal, such as band structure calculations, electronic polarization energy determination, etc.

Secondly, the most effective methods of purification and techniques of obtaining ultra-pure organic single crystals of relatively high perfection have been developed mainly for anthracene [1.35-37]. The latter circumstance appears to have been decisive in choosing anthracene as a basic model system in many experimental studies of electronic properties and energy structure of neutral and ionized states [1.38-51]. In this respect anthracene has played a role similar to that of germanium in covalent and NaCl in ionic crystal investigations.

Historically anthracene has served as model crystal for various theoretical calculations. Thus, anthracene is at present the only polyatomic organic crystal for which a full calculation of exciton band structure has
been performed [1.52]. Anthracene and its analogues have been widely used
for experimental testing of the basic concepts of molecular exciton theory
[1.43,53].

The most elaborate attempts of calculating the structure of conductivity
bands in one-electron approximation have also been carried out on anthracene [1.54-58]. Subsequently anthracene has served as model crystal for
demonstrating the validity limits of such an approximation [1.58].

Anthracene has also been used as model system in a large munber of electronic polarization energy calculations [1.38,59-64]. Historically, it also
served as a basis for the development of Lyons' phenomenological model
[1.38,65] and its modifications [1.42,62-64,66-69].

A large amount of experimental work has been carried out on anthracene
in studying local trapping centers for excitons (see, e.g., [1.70-74]), and
charge carriers (see, e.g., [1.75,76]).

Finally, the most detailed studies of various linear and planar structural defects have also been performed on anthracene and other crystals
belonging to the same space group [1.77-81].

It is also anthracene that has been used as the main model crystal in
the present work for detailed analysis of experimental data and a number of
calculations (electronic polarization of a perfect crystal, energy of structural defects, energy spectra of local states of polarization origin, etc.).

We therefore consider it appropriate to give in this introductory chapter the basic data concerning the crystal and molecular structure of anthracene (in some detail) and of other polyacene analogues. These data will be
repeatedly referred to in further discussions.

Molecular Structure

Anthracene ($C_{14}H_{10}$) (Fig.1.2) belongs to the series of linear polyacenes
(benzologues). Molecular weight M = 178.24 [1.3]. The symmetry of the molecule is D_{2h}. The planar condensed ring structure of the anthracene molecule was conclusively established by ROBERTSON's [1.27] early applications
of X-ray crystal structure analysis. This planarity of the molecule is due
to sp^2 hybridization of three of the valent electrons of the carbon atom,
the three hybridized electron orbitals being located in the xy plane of the

12

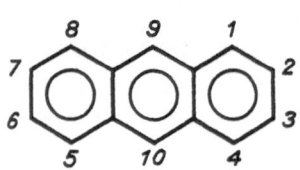

Fig. 1.2. Anthracene (Ac)

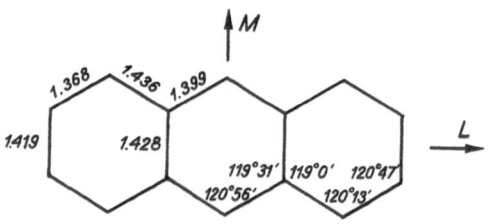

Fig. 1.3. Anthracene. Mean bond lengths [Å] and bond angles. L and M are molecular axes

molecule and forming a planar trigonal configuration. The fourth of the carbon valency electrons retains $2p_z$ symmetry, viz. its orbital is directed at right angles to the xy plane of the molecule. The aromatic nature of anthracene is due to 14 coplanar $2p_z$ electron orbitals of carbon atoms, forming, as a result of conjugation, delocalized π-electron orbitals (cf. Fig.1.20). The individual C-C bond lengths and bond angles of the anthracene molecule have been experimentally determined by MATHIESON et al. [1.28], and SINCLAIR et al. [1.29]. A detailed refinement of these data was later performed by CRUICKSHANK [1.30,31], and by CRUICKSHANK and SPARKS [1.33].

Corrected mean bond lengths and bond angles of anthracene, based on combinations of several independent refinements of three-dimensional X-ray data sets, according to [1.33], are presented in Fig.1.3. The final averaged bond lengths and bond angles, as shown in Fig.1.3, have, according to [1.33], estimated standard deviations (e.s.d.) of about 0.004 Å and 12', respectively. The latest X-ray results are in remarkably good agreement with gas phase electron diffraction data; the root-mean-square (r.m.s.) difference is only 0.004 Å [1.34]. Recent X-ray results for bond lengths in anthracene also agree to about 0.01 Å with the calculated ones [1.34].

X-ray analysis confirms that the anthracene molecule is practically planar. Small deviations of individual carbon atom coordinates from the mean molecular plane do not exceed 0.012 Å and are caused by close intermolecular approaches of some atoms in the lattice [1.30]. As may be seen from Fig.1.3., the bond angles differ only slightly from 120° value of ideal trigonal sp^2 hybrid orbitals.

Mean C-H bond length is 1.084 ± 0.006 Å. Effective Van der Waals thickness of the anthracene molecule, i.e., the thickness of the π-electron cloud is ca. 3.4 Å.

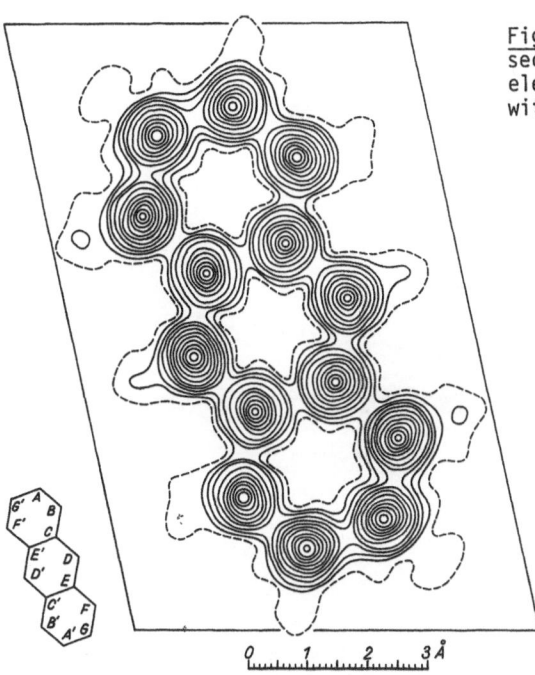

Fig. 1.4. Anthracene. A cross section of the three-dimensional electron density map, coinciding with the molecular plane [1.2]

A cross section of the three-dimensional electron density map of anthracene along the plane of the molecule is shown in Fig.1.4. Every contour line corresponds to an increase in electron density of ca. 0.5 electrons/$\AA^3$. The first contour shows marked bulges due to hydrogen atoms.

Basic Molecular Parameters

First ionization potential values I_G of the anthracene molecule, as estimated by different experimental methods, are presented in Table 2.11. The most plausible experimental values of electron affinity A_G of the anthracene molecule are given in Table 2.12. Experimental and calculated values of mean isotropic polarizability α of anthracene, as well as values of the main components b_i of the molecular polarizability tensor are presented in Table 2.2.

Crystal Structure

The crystallographic data on anthracene have been studied *in extenso* by ROBERTSON et al. [1.27-29,32] and CRUICKSHANK [1.30,31].

14

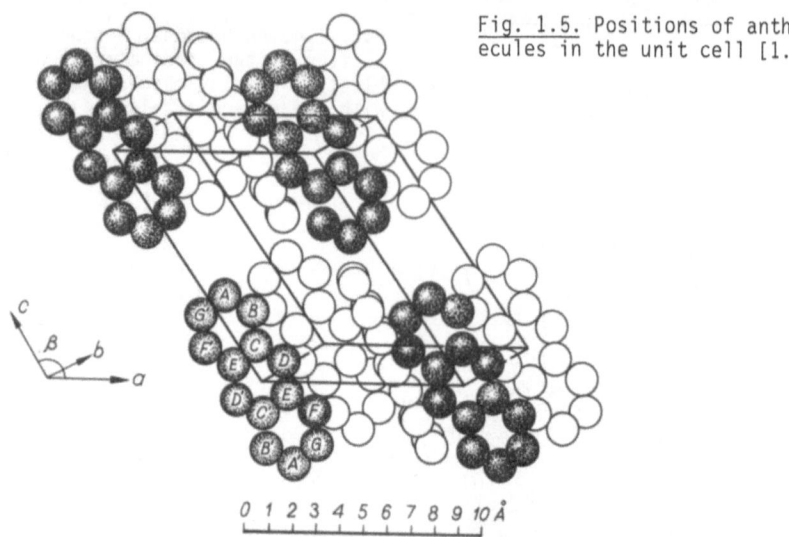

Fig. 1.5. Positions of anthracene molecules in the unit cell [1.32]

0 1 2 3 4 5 6 7 8 9 10 Å

The anthracene crystal is monoclinic, it belongs to space group $P2_{1/a}$
with two molecules in a unit cell (Z=2).

Positions of the anthracene molecules in the unit cell are shown in
Fig.1.5, the hydrogen atoms being omitted for clarity. The structure is ap-
proximately close packed, for each molecule has 12 near neighbors. Molecule
I of the unit cell which is situated at the center of symmetry (0,0,0) is
transformed by the glide plane operation into molecule II at (1/2,1/2,0).
Crystallographic data of anthracene, according to [1.28], are given in
Table 1.4.

The elementary cell volume of anthracene equals V = 474 $Å^3$; molecular
volume V_{mol} = 170.2 $Å^3$; packing coefficient k = 0.718, according to [1.2],
k = 0.733 at 0 K, according to [1.82]. Reported measured density d_{meas}
equals 1.283_4^{25} g·cm^{-3}. Melting point is at 216.2-0.4, boiling point at ca.
340°C [1.3]; heat of sublimation H_s = 101.8 kJ·mol^{-1} [1.1].

The orientation of the molecules in the crystal is given in Table 1.5.
Refined values of carbon atom coordinates and closest intermolecular approaches, calculated from carbon atom coordinates and assumed hydrogen atom
coordinates, have been given by CRUICKSHANK [1.30].

Table 1.4. Crystallographic data on naphtalene (Nph), anthracene (Ac), tetracene (Tc), pentacene (Pc), and hexacene (Hc)

	Nph $C_{10}H_8$ [1.93]	Ac $C_{14}H_{10}$ [1.28]	Tc $C_{18}H_{12}$ [1.107]	Pc $C_{22}H_{14}$ [1.107]	Hc $C_{26}H_{16}$ [1.107]
Crystal system	mono-clinic	mono-clinic	tri-clinic	tri-clinic	tri-clinic
Space group	$P2_{1/a}$	$P2_{1/a}$	P$\bar{1}$	P$\bar{1}$	P$\bar{1}$
a [Å]	8.24	8.56	7.90	7.90	7.9
b [Å]	6.00	6.04	6.03	6.06	6.1
c [Å]	8.66	11.16	13.53	16.01	18.4
α [°]	90.0	90.0	100.3	101.9	102.7
β [°]	122.9	124.7	113.2	112.6	112.3
γ [°]	90.0	90.0	86.3	85.8	83.6
V [Å³]	360	474	583	692	800
Z	2	2	2	2	2
d_{cal}	1.17	1.24	1.29	1.33	1.35
d_{meas}	1.15	1.25	1.29	1.32	1.34

Table 1.5. Orientation of molecules in naphthalene (Nph), anthracene (Ac), tetracene (Tc) and pentacene (Pc) crystals

Orientation angles	Nph [1.93]	Ac [1.29]	Tc [1.107] molecule I	Tc [1.107] molecule II	Pc [1.107] molecule I	Pc [1.107] molecule II
χ_L	115.8°	119.7°	105.8°	105.5°	104.4°	104.3°
ψ_L	102.6	97.0	105.5	103.6	106.4	104.0
ω_L	29.0	30.6	22.5	20.8	22.1	20.3
χ_M	71.2	71.3	69.2	115.4	67.9	118.5
ψ_M	29.45	26.6	30.1	26.0	30.7	29.2
ω_M	68.2	71.8	69.2	85.0	69.8	84.0
χ_N	32.8	36.2	26.6	30.6	26.6	32.4
ψ_N	116.3	115.5	115.2	67.8	115.7	65.0
ω_N	71.9	66.2	81.9	70.1	83.8	70.7

Note: χ_L, ψ_L, ω_L; χ_M, ψ_M, ω_M; χ_N, ψ_N and ω_N are the angles which the molecular axes L, M (see Figs. 1, 3, 7, 10, 11) and the molecular plane normals N make with orthogonal axes a, b and c of the crystal.

The r.m.s. amplitudes of translational oscillation of molecules in the directions of molecular axes L, M and N, determined by X-ray anisotropic thermal motion analysis, are 0.20, 0.17, and 0.16 Å, respectively [1.30]. The corresponding r.m.s. amplitudes of angular oscillations are $3.9°$, $2.2°$, and $3.0°$. The amplitudes of angular oscillations may be interpreted in terms of average frequencies of the rotational branches of the normal lattice vibrations [1.30]. Such analysis yields mean frequencies at 81, 65 and 43 cm^{-1} for the branches corresponding to oscillations about each of the molecular axes. Considering the e.s.d. values of the amplitudes of oscillation, these results agree satisfactorily with optical lattice frequencies determined spectroscopically.

Elastic and Optical Properties

Elastic constants c_{ij} and main components of mechanical compliance tensor s_{ij} of an anthracene crystal are given in Table 3.3. For the anisotropic coefficients of linear compressibility β of the crystal see Table 5.2.

The values of anisotropic shear modulus μ and Poisson coefficient ν, as well as that of shear modulus μ_V averaged according to Voigt, are presented in Table 3.4.

The main anisotropic refraction coefficients n_i and optical dielectric permeability $\bar{\varepsilon}$ are given in Table 2.4.

Reported optical lattice vibration frequencies of anthracene are classified according to CRUICKSHANK [1.83] as rotational modes (librons) at 120, 68, and 48 cm^{-1} and translational mode at 59 cm^{-1}. DOWS et al. [1.83a] have compiled the reported and their own data on anthracene lattice frequencies, have compared them with calculated values and give the following assignments of lattice mode symmetry: Raman active modes A_g at 121, 70 and 39 cm^{-1} and B_g at 125, 65 and 45 cm^{-1}; IR active modes A_u at 101 and 44 cm^{-1} and B_u at 60 cm^{-1}. HADNI et al. [1.84] have identified three translational modes at 120, 101 and 61 cm^{-1}.

Metastable Phases in Anthracene

Recently RAMDAS et al. [1.85] and PARKINSON et al. [1.86] have reported the discovery of a new metastable phase of crystalline anthracene. It can be readily produced at room temperature by application of a compressive force perpendicular to the (001) (i.e., basal) planes of single crystals of the thermodynamically stable ($P2_{1/a}$) form of anthracene. On the basis of the approximate unit cell parameters derived from electron diffraction analysis [1.85] a refined crystallographic structure of the new phase was computed using the atom-atom potential method [1.86]. It was found that the space group is P1 with a = 8.31, b = 6.31, c = 11.035 Å; $\alpha = 123.7°$, $\beta = 101.3°$, and $\gamma = 89.5°$; V = 468.4 Å^3; Z = 2. This structure is estimated to possess a lattice energy of only 5.5 kJ·mol^{-1} less than that of the monoclinic $P2_{1/a}$ form [1.86] .

Electron microscopic observations indicate that moderate stress produces crystallites of the new phase in coherent contact with the parent crystal matrix in one specific orientation.

Subsequent observations, following UV irradiation of the stressed specimens, revealed that these crystallites act as nuclei for photochemical dimerization. It is of considerable significance, in this case, that there is a diminution in $C(9)...C(9')$ distance from 4.5 Å in the $P2_{1/a}$ phase to 4.2 Å in the new P1 structure. Of equal significance is the fact that in the triclinic phase there is enough room for rotation of neighboring molecules, so as to bring them into overlapped registry necessary for easy production of the di-paradimer [1.86].

Earlier CRAIG et al. [1.87], using the atom-atom potential method, showed that in an anthracene crystal at least three metastable crystal structures may exist: two monoclinic of $P2_{1/c}$ and $P2_{1/n}$ symmetry, and one triclinic of $P\bar{1}$ symmetry, close in energy to the stable $P2_{1/a}$ phase.

These metastable structures have the following calculated lattice parameters at 0 K:

$P2_{1/c}$: a = 4.30 Å, b = 5.86 Å, c = 22.03 Å, β = 124.4°;

$P2_{1/n}$: a = 8.46 Å, b = 5.85 Å, c = 22.63 Å, β = 122.7°;

$P\bar{1}$: a = 4.35 Å, b = 5.67 Å, c = 11.37 Å, α = 88.6°;

β = 122.9°, γ = 91.9°.

These calculated metastable forms possess a lattice energy of about 7 to 9 kJ·mol^{-1} less than the stable monoclinic structure. CRAIG et al. [1.87] suggested that in compressed regions of dislocation cores the local lattice energies are much larger than the energy differences between calculated metastable phases and the stable $P2_{1/a}$ phase. Consequently, metastable phase inclusions can arise in stressed defect regions of an anthracene crystal.

Thus, experimental observations [1.86], as well as the computational approach [1.87] supply direct evidence towards the suggestion that hybrid (biphase) structures may exist even in anthracene-type crystals for which previously only one stable form was known. Similar biphasic structures were earlier observed also in substituted anthracene [1.88,89]. One may therefore expect that such hybrid bi- or even multiphase structures might well exist in the case of heterocyclic molecular crystals which are (e.g., acridine or phenazene) polymorphic by nature.

1.3.2 Naphthalene

Naphthalene has been widely used as model compound for the investigation of optical and elastic properties, as well as for various theoretical calculations (cf. [1.2,38,61,64,90,91]). It has, however, not been a favored object for studies of electronic properties. This may be due to its very high electrical resistivity, low photosensitivity, and to various methodological problems of sample preparation. Accordingly, data on the ionized state energy structure of naphthalene are poor, in comparison with those on anthracene. There is, for instance, no reliable reported value of the energy gap for naphthalene (cf. [1.92]).

18

We shall use naphthalene in our further discussion mainly as a simple
model crystal for different calculation purposes (such as electronic polarization energy calculations, evaluation of dislocation energies, etc.).

Molecular Structure

Naphthalene ($C_{10}H_8$) (Fig.1.6) is the first benzologue of the linear polyacene series. Molecular weight M = 128.19 [1.3]. The molecule is planar of
D_{2h} symmetry.

Molecular structure of naphthalene was first established by ABRAHAMS et
al. [1.93]. Its detailed refinement was carried out later by CRUICKSHANK
[1.33,34,94]. The corrected mean bond lengths and bond angles are given in
Fig.1.7. The e.s.d. values for bond lengths are ca. 0.005 Å. X-ray analysis
confirms the planar structure of the naphthalene molecule. The deviations
of carbon atom coordinates from the mean molecular plane do not exceed
0.007 Å [1.94].

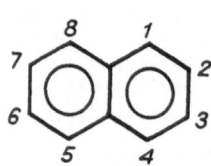

Fig. 1.6. Naphthalene (Nph)

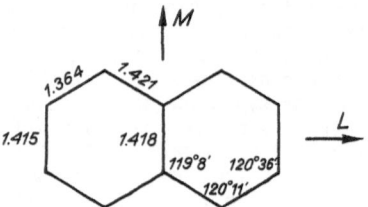

Fig. 1.7. Naphthalene. Mean bond
lengths [Å] and bond angles

Basic Molecular Parameters

The mean ionization potential value of the naphthalene molecule equals
I_G = 8.12 eV [1.38,96]. The most plausible experimental value of electron
affinity A_G of the naphthalene molecule is A_G = 0.15 eV [1.95] [some authors even give negative values of A_G (cf. [1.38,92])]. Experimental and
calculated molecular polarizability α and b_i values are given in Table 2.2.

Crystal Structure

The crystallographic data on naphthalene have been determined by ABRAHAMS
et al. [1.93] and refined by CRUICKSHANK [1.94]. Like anthracene, the naphthalene crystal is monoclinic and belongs to space group $P2_{1/a}$, with two

molecules in a unit cell (Z=2). The positions of the molecules in the unit cell are analogous to those in anthracene (cf. Fig.1.5).

Crystallographic data on naphthalene are presented in Table 1.4. The packing coefficient k equals 0.721 at 0 K, according to [1.82]. Reported measured density equals 1.0253_4^{20} g·cm^{-3} [1.3]. Melting point is at 80.55°C, boiling point at ca. 218°C [1.3]; sublimates in vacuo at ca. 87°C [1.3]; heat of sublimation H_s = 72.99 kJ·mol^{-1} [1.1].

The orientation of the molecules in the crystal are presented in Table 1.5.

Refined values of carbon atom coordinates and closest intermolecular approaches are given by CRUICKSHANK [1.94].

Elastic and Optical Properties

Elasticity parameters c_{ij} and s_{ij} of the naphthalene crystal are given in Table 3.3. The values of shear moduli μ and μ_V, as well as that of the Poisson coefficient ν are presented in Table 3.4. The values of the main anisotropic refraction coefficients n_i and optical dielectric permeability $\bar{\varepsilon}$ are given in Table 2.4.

Reported optical lattice vibration frequencies of naphtalene are classified by SUZUKI et al. [1.91] as rotation modes (librons) at 125; 109; 74; 71; 51 and 46 cm^{-1}, and translational modes at 98; 73 and 39 cm^{-1}. DOWS et al. [1.83a] have compiled the reported and their own data on naphtalene lattice frequencies, have compared them with calculated values, and give following assignments of lattice mode symmetry: Raman active modes A_g at 109; 74 and 51 cm^{-1} and B_g at 125; 71 and 46 cm^{-1}; IR active modes A_u at 102 and 49 cm^{-1} and B_u at 65 cm^{-1}.

1.3.3 Higher Linear Polyacenes

Crystals of tetracene and pentacene, the higher linear benzologues of the naphthalene-anthracene series, have become the most favored objects for studies of electronic states and energy structure in recent years [1.67,68, 97-104].

This is partially due to the high photosensitivity of tetracene and, especially, of pentacene [1.99], as well as to methodological advantages in preparing evaporated layers of tetracene and pentacene with different crystalline structure [1.97,98].

Owing to these circumstances tetracene and pentacene have lately served as the main model compounds for studying the influence of crystalline structure on the energy structure of ionized states in molecular crystals [1.67, 104]. Our further discussion in this work will also largely be based on our own research, as well as on reported data on tetracene and pentacene.

Tetracene and Pentacene

Molecular Structure. Tetracene (Tc) ($C_{18}H_{12}$) (Fig.1.8) and pentacene (Pc) ($C_{22}H_{14}$) (Fig.1.9) are higher benzologues of the linear polyacene series. Their molecules are planar, of D_{2h} symmetry. Molecular weight of Tc is M = 228.3 [1.3], that of Pc M = 278.36 [1.3]. Molecular structure of Tc and Pc was determined by ROBERTSON et al. [1.105], and CAMPBELL et al. [1.106,107].

Corrected bond lengths and bond angles for Tc and Pc are given in Figs. 1.10 and 11, respectively, according to [1.107]. Both Tc and Pc molecules are completely planar within the limits of experimental error of X-ray analysis [1.105,106].

Basic Molecular Parameters. Mean I_G and A_G values for Tc and Pc are given in the energy structure diagram in Fig.2.16. Molecular polarizability α and b_i values for Tc are given in Table 2.2.

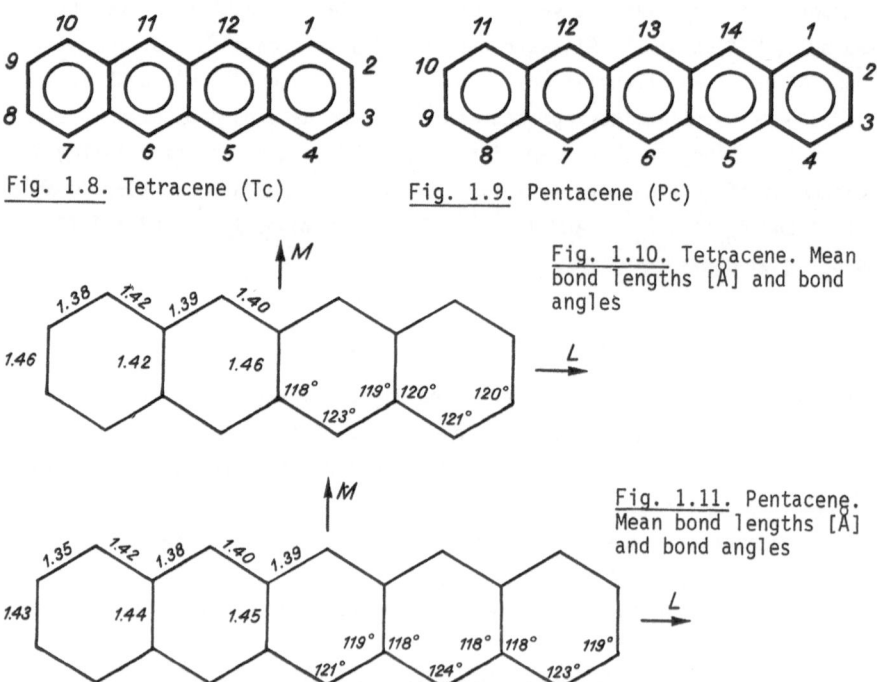

Fig. 1.8. Tetracene (Tc)

Fig. 1.9. Pentacene (Pc)

Fig. 1.10. Tetracene. Mean bond lengths [Å] and bond angles

Fig. 1.11. Pentacene. Mean bond lengths [Å] and bond angles

Crystal Structure. The crystallographic structure of Tc and Pc has been established and later refined by ROBERTSON et al. [1.105], and CAMPBELL et al. [1.106,107].

Tc and Pc crystals are triclinic and belong to space group $P\bar{1}$ with two molecules in a unit cell (Z=2). Their crystallographic parameters are given in Table 1.4. The orientation of the two unequivalent molecules in the unit cell of Tc and Pc crystals are presented in Table 1.5. The latter shows that molecules I and II in the triclinic unit cell of Tc and Pc are not directly related by symmetry elements. However, the symmetry centers of both mole-cules are situated at sites (0,0,0) and (1/2,1/2,0), forming an arrangement in the unit cell, which bears a close resemblance to the anthracene structure (cf. Fig.1.5). This structure belongs to space group $P\bar{1}$ with pseudo-$P2_{1/a}$ symmetry [1.105].

Slight deviations from monoclinic symmetry in Tc and Pc crystals (cf. Table 1.5) result in closer packing of molecules in the lattice, and a cor-responding increase in density (ca. 2%). We wish to emphasize the close similarity of crystal structure of all the members of the polyacene series, from naphthalene to pentacene. This circumstance is of high importance for the comparison of electronic properties of polyacene crystals. Recently KOLENDRITSKII et al. [1.107a] reported on phase transition in Tc single crystals at 160 K. Their conclusions were based on the studies of peculiari-ties in reflection spectra measured in polarized light at temperatures from 300 to 4.2 K and confirm earlier more indirect observations. The authors suggested that the phase transition observed may be connected with crystal symmetry change from triclinic to monoclinic owing to internal stress in the crystal with lowering of the temperature. On the other hand, JANKOWIAK et al. [1.107b] have reported the existence of two distinct phase transi-tions in Tc single crystals at 182 and 144 K, based on temperature-depen-dence studies of Raman-active lattice phonon spectra.

There are indications [1.107b] that vacuum evaporated layers of Tc may exist in monoclinic form already at room temperatures. According to JAN-KOWIAK et al. [1.107b] the Raman-active lattice frequencies of Tc are at 44, 49, 61, 94, 120 and 132 cm^{-1} at 296 K and confirm the earlier reported data by TOMKIEWICZ et al. [1.107c].

The melting point of the Tc is ca. $357^{\circ}C$ [1.3]; it sublimates in vacuo at $140-160^{\circ}C$; heat of sublimation H_s = 124.9 $kJ \cdot mol^{-1}$ [1.1].

Pc decomposes at temperature $\gtrsim 300^{\circ}C$; it sublimates in vacuo at $220-240^{\circ}C$; heat of sublimation H_s = 125.7 $kJ \cdot mol^{-1}$.

Carbon and hydrogen atom coordinates are given in [1.105] for Tc, and in [1.106] for Pc crystals.

22

Hexacene

The next member in the linear polyacene series is hexacene ($C_{26}H_{16}$) (Fig. 1.12)

   This is a promising virgin compound, the electronic and optical properties of which are practically unknown. This is due to decreasing stability and increasing ease of oxidation, which preclude the preparation of hexacene, as well as higher benzologue crystals (cf. [1.108]). However, an extrapolation of electronic properties from tetracene and pentacene to higher benzologues shows that hexacene should possess even higher photosensitivity than pentacene, higher polarizability and other exotic qualities.

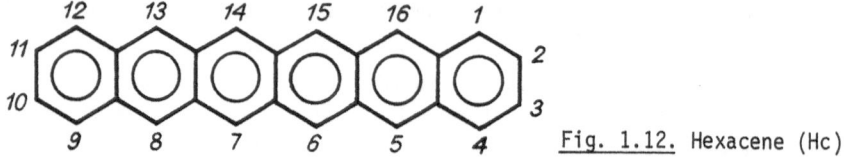

Fig. 1.12. Hexacene (Hc)

   Crystallographic data on hexacene were obtained by CAMPBELL et al. [1.107] (cf. Table 1.4). A comparison indicates that hexacene has crystal structure very similar to that of tetracene and pentacene, the only significant difference in cell dimensions being a further increase of 2.4 Å in the c axis to accommodate the extra ring.

1.3.4 Other Model Aromatic Compounds

Anthracene-type crystals with two nonparallel near-neighbour molecules in a unit cell (cf. Fig.1.5) belong, according to STEVENS' classification [1.109] to lattice type A of aromatic crystals. There is another more exotic lattice type B — with two pairs of parallel adjacent molecules in a unit cell to which such aromatic compounds as pyrene and perylene belong. Owing to such dimer structure of closely spaced parallel molecules in the lattice pyrene and perylene crystals exhibit strong excimer fluorescence and other unusual optical properties. They have therefore been for almost two decades the most frequently used crystals for optical studies (we recommend for details and references BIRKS' reviews [1.41,46]) (see also [1.110-117]). The binding energies of pair-states in an α-perylene crystal and their manifestation in optical properties have been recently treated theoretically by COHEN et al. [1.118]. On the other hand, reported data on electronic properties of pyrene and perylene crystals are rather fragmentary. ALEKSANDROV et al. [1.119] have recently studied energy structure of ionized states in perylene crystals in comparison with those of tetracene and pentacene, with the purpose of evaluating the influence of molecular structure on the energy structure of the given aromatic crystals (see Sect.2.11).

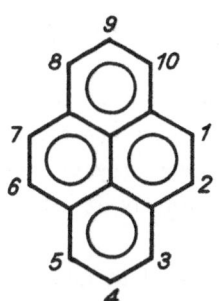

Fig. 1.13. Pyrene (Py)

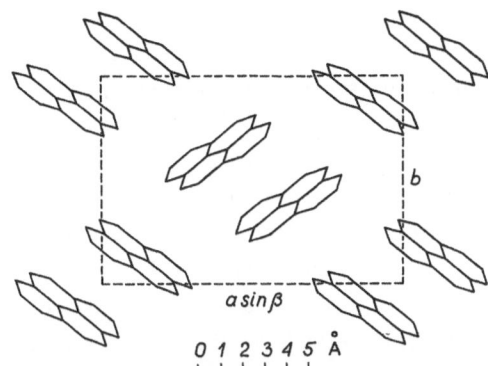

Fig. 1.14. Schematic crystallographic structure of pyrene: projection of molecules in ab plane [1.120]

Table 1.6. Crystallographic data on pyrene (Py) and perylene (Pl) crystals

	Py $C_{16}H_{10}$	Pl $C_{20}H_{12}$	
	[1.121]	α form [1.123]	β form [1.110]
crystal system	monoclinic	monoclinic	monoclinic
space group	$P2_{1/a}$	$P2_{1/a}$	$P2_{1/a}$
a [Å]	13.65	11.35	11.27
b [Å]	9.26	10.87	5.88
c [Å]	8.47	10.31	9.65
β [°]	100.28	100.8	92.1
V [Å³]	1052.9	1249	394.3
Z	4	4	2

Considering the promising prospects of pyrene and perylene as model compounds of type B lattice class for future investigations of energy structure, we give here a brief summary of their crystallographic structure.

Pyrene ($C_{16}H_{10}$). Molecular weight M = 202.2 [1.3], molecular symmetry D_{2h} (Fig.1.13). The crystallographic structure of pyrene has been determined by ROBERTSON and WHITHE [1.120] and later refined by CAMERMAN and TROTTER [1.121]. The crystallographic data are presented in Table 1.6.

The packing of molecules in the pyrene lattice is unusual: the role of the "crystallographic molecule" is played by molecular pairs having a dimer

structure (Fig.1.14). The adjacent parallel molecules of dimers are relatively closely spaced, the separation distance being 3.53 Å. Such configuration causes considerable overlap of π orbitals (cf. Fig.1.15), and consequently, considerable interaction.

In other aspects the crystal structure is similar to that of anthracene:
it is monoclinic, belongs to space group $P2_{1/a}$, except that there are four
molecules in a unit cell (Z = 4). Calculated density of a pyrene crystal is
d_{cal} = 1.288; the measured value is d_{meas} = 1.27. Melting point is at 150°C;
heat of sublimation is H_s = 94.5 kJ $\cdot$ mol^{-1} [1.122].

The orientations of the molecule in the crystal are presented in Table 1.7.

In recent studies JONES et al. [1.122] have estimated the existence of
two phases of pyrene crystals: the well-known phase at room temperature —
pyrene I, stable above 120 K; and pyrene II which is formed in a phase transition below 120 K. Pyrene II retains the space group and structure of pyrene I, with slightly changed lattice parameters. Pyrene II exhibits closer
packing of molecules (e.g., the separation of molecular dimer pairs (cf.
Fig.1.15) decreases from 3.53 Å to 3.44 Å). Accordingly, the lattice energy value is increased by about 4 kJ $\cdot$ mol^{-1}.

Table 1.7. Orientation of molecules in pyrene (Py) and α-perylene (Pl)
crystals

Orientation angles	Py [1.121]	α-Pl [1.123]
χ_L	60.9°	83.3°
ψ_L	77.1	89.2
ω_L	32.5	6.8
χ_M	53.3	55.4
ψ_M	51.1	35.0
ω_M	120.0	94.5
χ_N	130.0	144.5
ψ_N	41.7	55.0
ω_N	80.0	84.9

Note: χ_L, ψ_L, ω_L, χ_M, ψ_M, ω_M and χ_N, ψ_N and ω_N are the angles which the
molecular axes L, M and the molecular plane normals N make with the
orthogonal axes a, b, c' of the crystal.

Perylene ($C_{20}H_{12}$). Molecular weight M = 252.32 [1.3]; molecular symmetry
D_{2h} (Fig.1.16). The crystal structure of perylene has been determined by
DONALDSON et al. [1.123] and by TANAKA [1.110].

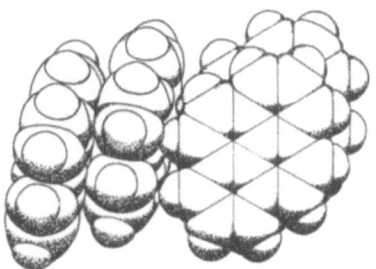

Fig. 1.15. Projection drawing of
molecular pairs in pyrene crystals
[1.122]

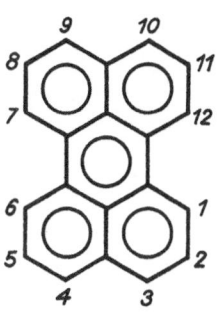

Fig. 1.16. Perylene (Pl)

Perylene exists in two crystallographic forms: α perylene, having dimer
pyrene-like structure with four molecules in a unit cell (B-type of crystal
structure) [1.123], and β perylene, having an anthracene- or A-type of
crystal structure with two molecules in a unit cell [1.110]. The crystallographic data for both forms are given in Table 1.6.

The formation of α or β structure depends on the conditions of crystal
growth [1.110]. β perylene usually forms rectangular yellow crystals of
monoclinic symmetry. The melting point of α perylene is at ca. 266-268°C.
β perylene usually forms small hexagonal green-yellow prisms, also of monoclinic syngony [1.110]. Both forms belong to the $P2_1/a$ space group.
β perylene changes into the α form at about 140°C. At and below room
temperature both forms are stable, and no transition is observed at lower
temperatures [1.110].

Perylene may thus be regarded as a "translational" compound between the
A and B types of aromatic crystal structure. From this point of view it can
be regarded as an excellent model compound.

The orientation of the molecules in an α perylene crystal are presented
in Table 1.7.

1.4 Specific Properties of Electronic States in a Molecular Crystal

Weak intermolecular interaction forces (cf. Table 1.3) produce only slight
changes in the electronic structure of molecules upon formation of the solid phase, and molecules retain their individuality.

Thus, X-ray analysis of electron density distribution shows that electronic configuration of the molecules remains practically unchanged in a
crystal. Figure 1.17 illustrates the aforesaid, presenting a picture of
electron density in coronene, which is a seven-ring benzologue of the aromatic hydrocarbon crystal. Maximum electron density can be seen to concen-

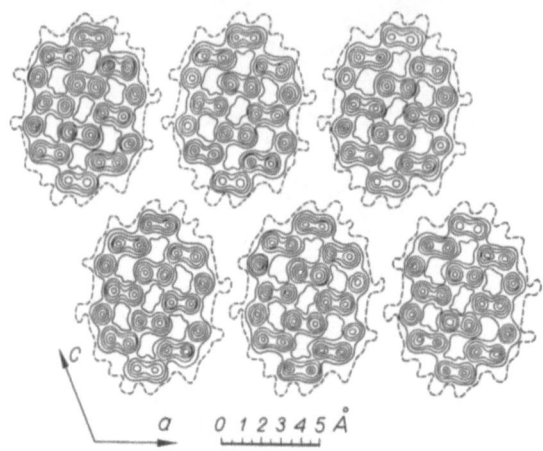

Fig. 1.17. Coronene. Electron density projection on ac plane of the crystal [1.124]

trate around carbon atoms, dropping to practically zero value in the inter-molecular space. The aromatic molecules form in the crystal something like an archipelago of molecular "islands" with delocalized π electrons, separated by "channels" of electronless space.

From this aspect molecular crystals are closer to an "oriented molecular gas" than to traditional solids formed by covalent or ionic crystals with a strongly bonded rigid atomic or ionic lattice characterized by complete loss of individual properties of the particles in the crystal.

Such electronic structure determines the basic specific features of optical and electronic properties of molecular crystals, as well as the peculiarities of the energy structure of neutral and ionized states.

The optical spectra of an isolated molecule and a molecular crystal are pretty similar. The crystal spectrum completely retains the spectral features of individual molecules, including their electronic-vibrational structure.

On the other hand, certain new optical and electronic properties appear in molecular crystals, due to collective molecular interaction. For instance, such specific phenomena in the spectra of crystals as Davydov splitting, the peculiar shape of exciton bands and its temperature dependence, etc., are caused by resonance and dispersion interaction of the excited molecule with its crystalline surroundings, and by lattice vibrations on the structure of electronic-vibrational bands of the molecule spectrum.

The self-energy value of a free charge carrier in a molecular crystal also strongly depends on its interaction with surrounding molecules of the lattice. On the other hand, in the formation of the energy structure of ionized states in a crystal an equal part is played by the individual properties of the molecule, such as electron affinity, and electronic and vibronic polarizability. These determine, as we shall see later, the localization of the charge carrier on individual molecules.

Such "dualism" is a specific feature of molecular crystals which does not permit one to apply traditional concepts of solid-state physics in their study and requires a search for new approaches (see, e.g. [1.38,42, 58,60-64,67,68]).

It is therefore essential to combine "molecular" and "solid-state" aspects in the treatment of optical and electronic properties of molecular crystals. Energy structure and electronic properties of molecular crystals are determined both by crystalline and molecular structure. Thus, for instance, detectable electrical conductivity and photosensitivity can be found mainly in molecular crystals containing molecules with polyconjugated bond systems, such as aromatic hydrocarbons and conjugated heterocyclic compounds. These molecules contain delocalized, weakly bonded conjugated π-electron systems, as well as heteroatoms with lone pairs of n electrons. It is just the π and n electrons, which are the potential sources of free charge carriers formed through action of light or temperature. Owing to its specific electronic properties a large series of molecular crystals consisting of conjugated organic molecules has been named "organic semiconductors".

Since molecular crystals possess a number of specific features making them essentially different from covalent and ionic crystals, the study of their energy structure and electronic properties has become an independent scientific problem in solid-state physics.

1.5 Basic Characteristics of Electronic Conduction States in Molecular Crystals

The question about the nature of electronic conduction states and possible conductivity mechanisms has been the most controversial problem of organic solid-state physics during the last two decades. To a considerable extent it has been caused by deliberate or unconscious disregard of the characteristic dualism of molecular and solid-state approaches to molecular crystals (cf. Sect.1.4). Scholars frequently tend to overemphasize only one single aspect of the complementary problem - either the solid state or the molecular one.

Thus, for more than a decade organic molecular crystal physics has been a battle-field between two apparently contradicting approaches to the problem of electronic conductivity states: the *hopping model* approach which emphasizes the molecular character of conduction states, describing them in terms of energy eigenvalues of charge carriers localized on individual molecules; and the *band model* approach which, on the other hand, emphasizes the solid-state character of conduction states, describing them in terms of traditional conductivity bands of delocalized charge carriers.

It was assumed, for a long time, that both approaches logically exclude one another and only one of them can be feasible. Recently, however, it has been demonstrated that both approaches are actually complementary and may be valid, each in its own temperature or charge carrier energy range (cf. Sect.1.5.3).

Historically, the hopping model was handicapped with respect to the band theory approach by a number of circumstances. In the case of charge carrier hopping there is no single unequivocal solution to the problem, and several possible models may be suggested. At least a dozen different hopping models have been proposed, based on different physical principles and approximations. Unfortunately, the majority of them give only a qualitative description without the possibility of computational treatment. Or, vice versa, they are too sophisticated for such a treatment (cf. Sect.1.5.2).

The band model approach has been in a much more favorable position. In this case the problem may be treated within the framework of simple one-electron approximation for which an elegant and well-developed mathematical formalism exists, opening possibilities for quantitative results. And, most important, band theory had also purely psychological advantages over other theories, having so successfully served as the most powerful tool in solving energy structure problems in traditional solid-state physics.

Apparently, because of this a number of leading theoreticians working in the field of organic solid-state physics made, in the sixties, the most elaborate efforts of calculating the band structure for excess charge carriers in anthracene-type crystals. The results, however, as we shall see in the following section, were more than controversial and demonstrated that the applicability of simple band theory is rather dubious for molecular crystals.

It is curious to observe that for more than a decade the band theory served as a Procrustian bed into which, figuratively speaking, such tender "victims" as organic crystals were forced to fit the rigid "bed" of germanium or silicon lattice.

1.5.1 Band Theory Approach

It is generally believed that the presence of translational symmetry in a
crystal is a sufficient condition for applying band theory; however, this
condition is necessary, but not sufficient. The other important condition
is that the interaction of particles in a solid should be sufficiently
strong. If the periodicity of particle structure permits the use of Bloch-
like wave functions for describing a delocalized electron, the character of
interaction forces between particles determines the formation of band structure, effective band width, etc., thus setting natural limits for applicability of the band theory.

Weak interaction forces in molecular crystals lead to strong localization of excess charge carriers. Under conditions of such pronounced localization other types of interaction between charge carriers and the lattice
may take place, e.g., electronic polarization of surrounding polarizable
molecules. This forbids, as we shall see, the application of one-electron
approximation on which simple band theory is based.

The first band-theory calculations for an anthracene crystal were performed by LE BLANC [1.54]. He used tight-binding approximation in order to
describe more or less adequately the character of weak intermolecular interaction forces which cause strong localization of charge carriers on individual molecules. One-electron crystal wave functions $\psi_k(r)$ are constructed
from linear combinations of one-electron molecular wave functions χ_n

$$\psi_k(r) = N^{-1/2} \sum_{n=1}^{N} \exp(i\underline{k} \cdot \underline{r}_n) \chi_n(\underline{r} - \underline{r}_n) \quad . \tag{1.12}$$

Here $\underline{r}_n$ locates the geometric center of molecule n, and the sum extends
over the N molecules in the crystal. The molecular wave functions χ_n, in
turn, are constructed, according to molecular orbital (MO) theory, as a
linear combination of Slater-type $2p_z$ atomic orbitals (cf. Sect.2.5.2),
using for the electron and hole band, respectively, Hückel coefficients for
the lowest antibonding and the highest bonding π orbitals.

The final result of LE BLANC's calculations [1.54] was more than controversial. The calculated effective bandwidths δE of electron and hole bands
in anthracene were extremely small — ca. 0.015 eV, equalling, in particular,
0.6 kT for electrons and 0.5 kT for holes at room temperature.

THAXTON et al. [1.55] used a similar approach in band calculations for
naphthalene, tetracene, and pentacene crystals. They obtained values of the

same order for effective bandwidth δE for electrons and holes, viz. $\delta E =$ (0.2-0.3) kT for naphthalene, and $\delta E = $ (0.5-0.6) kT for pentacene crystals at romm temperature.

Band theory concepts are in principle not valid for such narrow bands. LE BLANC himself was rather cautious about the applicability of band theory for anthracene-type crystals and pointed out that the narrowness of the bands indicates that it would be preferable to use localized rather than band approach (cf. [Ref.1.38, p.228]).

It was, however, suggested that such narrowness of conduction bands may be due to rather rough approximations used by LE BLANC, and THAXTON et al. in their calculations.

At a later stage KATZ et al. [1.56] improved LE BLANC's calculation procedure introducing self-consistent field (SCF) atomic orbitals instead of Slater orbitals in the framework of MO SCF theory (cf. Table 2.5). He also improved the computational accuracy. SILBEY et al. [1.57] still further refined the calculation procedure, including effects of intermolecular electron exchange and vibronic coupling in the weak coupling scheme. The refined calculations by KATZ et al. and by SILBEY et al. yielded larger bandwidths with δE ca. 0.05 to 0.2 eV.

In this case, however, other internal inconsistencies of band theory approximation became evident. An estimate of the mean free path $\bar{\ell}$ of the charge carrier in anthracene gave a value of 3 to 4 Å, thus being below the elementary cell parameter a_0 value, i.e., $\bar{\ell} < a_0$. These results clearly showed that charge carriers are scattered on every molecule of the lattice, in other words, they are actually strongly localized. Such localization, naturally, excludes traditional band theory concepts, since electron motion cannot be adequately described by a wave vector $\underline{k}$.

GLAESER and BERRY [1.58] performed a comprehensive analysis of the limits of band theory applicability for anthracene-type crystals. They made a critical comparison of band and hopping models. The band model was presented in terms of simple band theory in which a delocalized, tight-binding approximation was used. The hopping model was treated in terms of simple hopping theory in which one-site localized representation was applied. At first the authors carried out band theory calculations for an anthracene crystal introducing still further refinements of calculation procedure in comparison with KATZ et al. [1.56], and SILBEY et al. [1.57]. In this case the effects of electron exchange (shown to be very important by SILBEY et al.) as well as the effects of polarization by the excess charge carrier of the electrons on neighboring molecules were taken into consideration.

According to GLAESER and BERRY [1.58] the crystal wave function ψ_ℓ in localized representation may be constructed as an antisymmetrized product of molecular wave functions, in which one molecule is either a negative or a positive ion, and the remainder are perturbed (polarized) by the ionic molecule. Symbolically the wave function corresponding to the electron or hole sitting on molecule i is [1.58]

$$\psi_\ell = A \; \psi_i \; (2a \pm 1) \prod_{j \neq i} \psi_j^{(i)} \; (2a) \quad , \tag{1.13}$$

a being the number of filled orbitals in the neutral molecule, $\psi_i(2a \pm 1)$ denotes the wave function of the appropriate molecular ion, and $\psi_j^{(i)}$ the wave function of neutral molecule j in the field of this ion.

In the case of Bloch representation the tight-binding wave functions are simply linear combinations of localized crystal functions

$$\psi_\pm(\underline{k}) = \sum_\ell (\pm 1)^\ell \; \exp(i\underline{k}\underline{r}_\ell) \; \psi_\ell \quad . \tag{1.14}$$

It should be emphasized that in this case the approximate wave functions chosen are in fact many-electron functions, since electronic polarization effect is taken into account. Thus, the authors of [1.58] actually extended simple band theory of one-electron approximation to many-electron approximation.

According to [1.58] charge carrier self-energy value dependences on wave vector $\underline{k}$ in the conductivity band $\varepsilon_k^\pm(\underline{k})$ can be described by the following expression:

$$\varepsilon_k^\pm(\underline{k}) = \sum_j \exp(i\underline{k}\underline{r}_j) \; J_{ij} \quad , \tag{1.15}$$

where J_{ij} is the charge carrier transfer integral.

J_{ij} can be expressed in the following general form:

$$J_{ij} = (E_R - \Delta E_R - E_{ES}) \; S \quad , \tag{1.16}$$

where E_R is the term of resonance (exchange) interaction between molecules i and j; ΔE_R is the correction to the resonance term which results from the use of polarized orbitals; E_{ES} is the term of long-range electrostatic interaction between the excess charge and the induced dipoles on neighboring

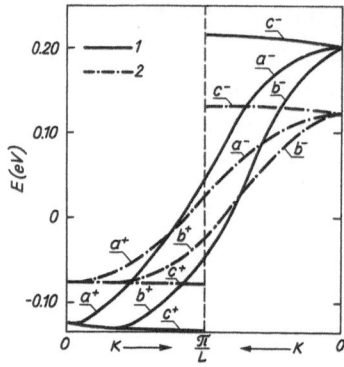

Fig. 1.18. Band structure for excess electron in an anthracene crystal, showing the dependence of energy along the a^{-1}, b^{-1} and c^{-1} directions of the lattice. 1: without electronic polarization; 2: including electronic polarization [1.58]

molecules; S is the overlap factor of polarized electron orbitals of neighboring molecules.

The calculated band structures for excess electrons in anthracene (without and including electronic polarization effects) are illustrated in Fig.1.18.

As may be seen from Fig.1.18, electronic polarization effects reduce the effective electron bandwidth δE_e by about 40 to 50 %. Physically it amounts to increased charge carrier localization.

Subsequently PETELENZ [1.125] evaluated by independent approach the effect of polarization on the charge carrier transfer integrals (in anthracene model dimers) and found that polarization actually lowers the transfer integral value by about 50 %. These data are in good agreement with GLAESER and BERRY's [1.58] calculations.

Recently ČAPEK [1.126] analyzed the influence of electronic polarization on excess electron transport in molecular solids. The author treats electronic polarization self-consistently with respect to electronic motion and shows that in realistic situations the correction factor ΔW of band narrowing might be of the order of typical half-bandwidth in molecular solids, namely 0.03-0.1 eV.

This shows that the excess charge carrier possesses pronounced tendency towards localization due to the formation of the polarization cloud.

There are two computational checks on the validity of the band approximation. First, we can ask whether the computed mean free path of charge carriers $\bar{\ell}$ is significantly larger than the lattice parameters a_0, and, in a similar way, whether the computed free (relaxation) time τ_0 is significantly longer than the lifetime τ associated with the quantum mechanical principle $\tau = \hbar / \delta E$.

These data for testing self-consistency of the band model in an anthracene crystal are presented, according to GLAESER and BERRY [1.58], in Table 1.8 (In order to be nearer the real situation we present in Table 1.8. only such calculation data which include polarization effects).

As may be seen from Table 1.8, the evaluated mean free path $\bar{\ell}$ of a charge carrier is smaller than the molecule-to-molecule spacing, and the free time τ_0 is by only about a factor of two or three larger (in the a^{-1} and b^{-1} directions) than the lifetime τ required by the uncertainty principle.

The authors of [1.58] concluded that these results (as well as those reported earlier by KATZ et al. [1.56] and SILBEY et al. [1.57]) convincingly demonstrate the inconsistency of the band model for anthracene-type crystals.

Taken literally, these results say that elastic scattering with lattice phonons is so strongly coupled to charge carrier motion that the wave momentum $\underline{k}$ is not at all conserved [1.58]. In other words, charge carrier motion is "randomized" by scattering at each lattice site and, apparently, a model of stochastic hopping suits the real physical situation better.

It should be mentioned that additional inelastic interaction processes between charge carrier and lattice should strengthen the coupling still further and increase its localization, as demonstrated in the case of electronic polarization effects (cf. Fig.1.18).

Table 1.8. Data for testing self-consistency of band model in anthracene crystal according to GLAESER and BERRY [1.58]

Charge carrier	Band direction	δE [eV]	$\bar{\ell}$ [Å]	a_0 [Å]	τ_0 $[10^{-15}s]$	$\tau = \hbar/\delta E$ $[10^{-15}s]$
Electron	a^{-1}	0.20	2.6	8.56	7.6	3.3
	b^{-1}	0.20	3.6	6.04	11.1	3.3
	c^{-1}	0.01	49.9	11.60	485.0	96.2
Hole	a^{-1}	0.11	4.3	8.56	12.6	6.1
	b^{-1}	0.23	2.6	6.04	7.0	2.9
	c^{-1}	0.10	9.8	11.60	37.9	6.8

Notation: δE is the calculated effective bandwidth; $\bar{\ell}$ is the evaluated mean free path of charge carrier; τ_0 is its free (relaxation) time; $\tau = \hbar/\delta E$ is the lifetime of a charge carrier, according to the uncertainty principle; a_0 is the lattice paramter. Calculations were performed including electron polarization effect (oscillator strength f = 2.0).

1.5.2 Hopping Versus Band Model

As we have already pointed out hopping models may be formulated in a multi-
tude of ways and, indeed, those reported show a great variety in physical
principles, complexity and limits of applicability. This is understandable,
since the approach in this case is an inductive, phenomenological one.

It is not our purpose to give a detailed review of reported hopping models here. We shall discuss the charge carrier transport problem only in
general terms and only as far as it is connected with the studies of stationary electronic states in a crystal, which is the main object of the present work.

Reported hopping models may be divided into three main groups. Simple
tunneling models (first suggested in most general terms by ELEY et al.
[1.127]), describe electron transfer in a crystal by quantum-mechanical
tunneling through intermolecular potential barriers. Since tunneling probability depends on the thickness and form of the barrier, a variety of possible "static" types of intermolecular barriers may be suggested. As a matter of fact, such simple models give an oversimplified, idealized picture
of the real situation. First, intermolecular barriers should not be static.
Actually they are dynamic, fluctuating. Secondly, quantum-mechanical tunneling through the intermolecular barrier does not require activation energy and therefore should be temperature independent. Actually, in real molecular crystals we observe more or less pronounced temperature dependence
of charge carrier mobility.

According to the second simple hopping model the charge carrier is
"trapped" at each molecular site and moves over the intermolecular barrier
via a thermally activated process. Such a phonon-activated hopping mechanism cannot, however, be applied to anthracene-type crystals, since it
requires exponentially increased mobility with temperature which is not observed for this type of crystal.

An intermediate between these two limiting cases is the resonance transfer model proposed by GLAESER and BERRY [1.58] According to this model the
primary mechanism is "tunneling" or "resonance transfer" of an excess charge
carrier between neighboring molecules. The probability of this process is
considered as a sensitive function of the vibrational states of the lattice.
For example, in an excited vibrational state two molecules approach each other more closely than they do in the vibrational ground state. Since the
transfer integral J_{ij} is essentially exponential in the intermolecular
spacing, the integral will be increased in magnitude, as the vibrational

energy of the crystal increases. This type of phonon-assisted tunneling could be formally described using a probability function in which not only the distribution of neighboring sites of a localized charge carrier, but also intermolecular vibrational quantum numbers should be specified [1.58]. To consider actually such probability distribution would, however, require a detailed specification of the phonon spectrum, as well as an estimate of how the various transfer integrals would be affected.

Owing to a lack of sufficient quantitative data, the authors of [1.58] used discrete probability distribution based on the transfer integrals J_{ij} evaluated at equilibrium configuration of the crystal. The problem is thus reduced to a simple hopping theory of resonance transfer, resulting in random hopping of the charge carrier from one localized state to another.

The hopping model proposed in [1.58] is based on the assumption that each molecule-charge carrier encounter is equivalent to a very strong scattering event. This assumption follows logically from the band theory computational inconsistency (cf. Table 1.8) . Consequently, the charge carrier motion may reasonably be regarded as "randomized" at each jump. This means that each jump is actually independent of the preceding one, and the "hopping" process may be discussed in the spirit of a Markovian model.

The transfer integrals J_{ij} [cf. (1.16)] may be directly related in this case to the jump probabilities, and to the mean time interval between separate jumps, i.e., to the mean hopping time τ_h.

The hopping time τ_h is inversely proportional to transfer integral J_{ij}, in accordance with the quantum-mechanical uncertainty principle

$$\tau_h = \frac{\hbar}{J_{ij}} \quad , \tag{1.17}$$

where $\hbar$ is the Planck constant, $\hbar = 0.66 \cdot 10^{-15}$ eV $\cdot$ s.

Calculated transfer integral J_{ij} values and mean hopping times τ_h for electron and hole in an anthracene crystal are given, according to [1.58] in Table 1.9. This table shows that the transfer integrals in anthracene possess considerable anisotropy. Thus, for electrons the highest probability of hopping lies between the molecules in crystal sites (0,0,0) and (1/2,1/2,0) in the unit cell, whereas for holes the hopping probability is highest between the molecules at crystal sites (0,0,0) and (0,1,0) in the direction of the b axis of the crystal.

As may be seen from Table 1.9, the calculated hopping time τ_h values are in good agreement with those evaluated from experimental data.

Table 1.9. Values of transfer integral J_{ij} and mean hopping time for charge carriers between molecules i and j, according to the hopping model in an anthracene crystal [1.58]

Coordinates of molecule i	Coordinates of molecule j	Electron			Hole		
		J_{ij}, [eV]	τ_h [s]	Mean experimental estimate τ_h [s]	J_{ij}, [eV]	τ_h [s]	Mean experimental estimate τ_h [s]
(0,0,0)	(1/2,1/2,0)	0.025	2.6×10^{-14}	1.89×10^{-14}	0.020	3.4×10^{-14}	1.85×10^{-14}
(0,0,0)	(0,1,0)	0.013	5.0×10^{-14}		0.026	2.6×10^{-14}	
(0,0,0)	(1,0,0)	0.8×10^{-4}	8.0×10^{-12}		0.7×10^{-4}	9.7×10^{-12}	
(0,0,0)	(0,0,1)	1.9×10^{-4}	3.5×10^{-12}		1.4×10^{-4}	4.6×10^{-12}	

A rather remarkable feature is the extremely small value of transfer integral J_{ij} which generally lies below 0.03 eV (cf. Table 1.9). This can be considered as a direct manifestation of very weak intermolecular interaction as the most characteristic peculiarity of molecular crystals. The latter circumstance, in its turn, results, according to (1.17), in comparatively large (for typical electronic processes) time intervals τ_h between separate jumps: $\tau_h \gg 10^{-14}$ s.

Such comparatively "slow" transfer of charge carriers in the crystal leads to a qualitatively new effect of interaction between the charge carrier and the surrounding lattice, namely electronic polarization of the crystal by the charge carrier (cf. Sect.1.5.1), the main consequences of which on charge carrier self-energy will be discussed briefly in Sect.1.5.4 and, in some detail, in Sect.2.3.

Contraria sunt complementa

Niels Bohr

1.5.3 Band-to-Hopping Transition

The inconsistency of the band model for stationary charge carriers in anthracene-type crystals has been proved beyond doubt for room temperature. At the same time the question about the character of charge carrier motion at very low temperatures remained open.

Thus cyclotron resonance of holes observed by BURLAND [1.128], and BURLAND and KONZELMANN [1.129] in anthracene crystals at 2 K might well be interpreted in terms of band theory. The mean hole scattering time obtained from these experiments was $7 \cdot 10^{-11}$ to $4 \cdot 10^{-10}$ s, thus showing that scattering at such low temperature is negligible and charge carriers are sufficiently delocalized to orbit in the a̱ḇ plane of the crystal and provide a measurable cyclotron resonance signal.

These experimental observations, as well as general considerations on temperature dependences of charge carrier motion confirmed the point of view that band and hopping models are not exclusive, but mutually complementary for different temperature regions, and that there may exist experimentally observable transition regions from band-to-hopping motion in organic molecular solids.

Actually such band-to-hopping transition has been anticipated for nearly twenty years in the lattice-polaron literature [1.130-132], but has been only recently observed experimentally by SCHEIN et al. [1.133,134]. The authors of [1.133] have used the transient photoconductivity method for measuring electron drift mobility μ_e in the c' direction of a naphthalene single crystal in an extended temperature interval from 54 to 324 K.

As may be seen from Fig.1.19, three regions can be distinguished in the temperature dependence of electron mobility $\mu_e(T)$. In the first (I) region, from 150 to 324 K the mobility μ_e is essentially independent of temperature, similarly to the behavior, as reported by KARL et al. [1.45,135].

In the second (II) region, below 150 K, the mobility slightly decreases, by about 10%.

Finally, in the third (III) region, below 100 K, the mobility μ_e increases rapidly with decreasing temperature, and the $\mu_e(T)$ dependence can be described by $\exp(E/kT)$ where $E \approx 5.8$ meV (47 cm^{-1}) (see inset of Fig. 1.19). These data were later confirmed independently by KARL [1.136].

Recently a similar mobility $\mu_e(T)$ rise with decreasing temperature below 100 K has been observed by SCHEIN and McGHIE [1.136a] also along the b direction of naphthalene indicating a band-hopping mobility transition at 100 K. In this case the E value equals 10.9 meV (88 cm^{-1}). On the other

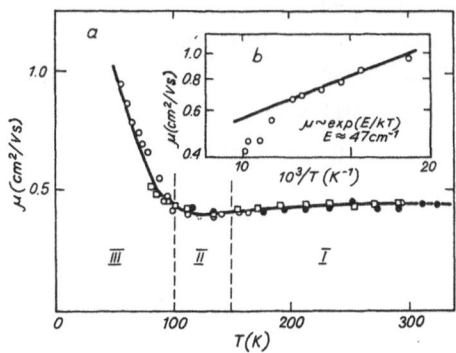

Fig. 1.19a,b. The temperature dependence of electron mobility μ_e (T) in c' direction of naphthalene single crystal (a). I: hopping motion region; II: band-to-hopping transition region; III: band motion region. The lowest temperature (III) region data in semilogarithimic coordinates: $\mu_e = f(1/T)$ (b) [1.133,134]

hand, the $\mu_e(T)$ dependence along the c' direction of deuterated naphthalene is identical within experimental error (±15%) to that of naphthalene (see Fig.1.19).

The authors of [1.133,134] interpreted the sharp rise of μ_e in region III, below 100 K, evident in Fig.1.19, as being due to the onset of band motion. The observed temperature dependence in this low-temperature region, which fits the exp(E/kT) rule, is consistent with electron scattering by lattice modes, this temperature dependence reflecting the concentration of optical phonons. Specifically, the value of $E = 47$ cm^{-1} (or, corrected according to [1.136a], $E = 53$ cm^{-1}) corresponds to the energy of one of the lowest libronic, i.e., rotational lattice modes (cf. Sect.1.3.2). The authors suggested that many lattice modes and molecular vibrations scatter electrons at higher temperatures. Only after they are frozen out at sufficiently low temperatures, the electron scattering becomes weak enough for band motion to take place [1.133,134]. The observed value of E is simply characteristic of the specific lattice vibration which limits electron mobility at T $\lesssim$ 100 K. Obviously, at still lower temperatures scattering by acoustic lattice modes and crystalline defects can become important in limiting μ_e value [1.134].

The authors of [1.133,134] further concluded that mobility μ_e above 150 K (region I, cf. Fig.1.19) is of the hopping type, the intermediate region II being a transition region in which both band and hopping motion occur.

However, as has been pointed out by SCHEIN [1.134], the hopping motion region (I) cannot be described in terms of simple hopping models (cf. Sect. 1.5.2). As it was earlier anticipated by GLAESER and BERRY [1.58] more sophisticated dynamic models should be developed for the description of real situations in organic molecular solids. Specifically, in this case, the hopping model should account for essentially temperature-independent electron mobility over a wide temperature range [1.45,133-136,136a] which has been observed, to be electric field independent up to very high field intensities (1.6 × 10^5 V/cm in anthracene at 100 K and 1.7 × 10^5 V/cm in naphthalene at 160 K) [1.136b]. These data provide additional evidence against the applicability of small lattice polaron models to anthracene-type molecular crystals [1.136a,b] (see Sect.1.5.5).

Recently MADHUKAR and POST [1.137] extended the model proposed earlier by HAKEN et al. [1.138,139] for describing noncoherent motion of excitons (cf. Sect.2.1), including also the problem of noncoherent transfer of charge carriers in molecular crystals. In their approach the authors of [1.137] specifically demonstrated that lattice-vibration fluctuations modulate the hopping integrals, as well as site energies of the electron. An exact solution for the diffusion of a particle in a medium with site and off-diagonal dynamic disorder, presented by the authors of [1.137], leads subsequently to very weak temperature dependence for charge carrier mobility in such a medium.

Another interesting approach proposed recently by SUMI [1.140] describes the temperature-independent electron mobility in the c' direction of anthracene-type crystals in the framework of an anisotropic phonon-assisted hopping model (see also [1.141]). This approach, however, fails to explain reported data, first that band-hopping mobility transition in naphthalene occurs at 100 K both along the c' as well as b directions [1.136a]; secondly, that hopping mobility is field independent [1.136b]. Finally, a more general approach has been proposed by EFRIMA and METIU [1.140a]. The authors analyzed a Hamiltonian in which were included both linear and quadratic phonon coupling to the site energy (cf. Sect.2.1) as well as anharmonic phonon effects and found that the temperature-independent mobility could be obtained if anharmonic effects were included, and the minimum of $\mu_e(T)$ dependence near the transition (see Fig.1.19a) could be produced if site quadratic coupling were considered.

It should be mentioned that the whole situation is complicated by the fact that on other occasions the temperature dependence of charge carrier mobility in the hopping region may have a different character. Thus, for example, the temperature dependence of hole mobility $\mu_h(T)$ in anthracene can be described as $\mu_h \sim T^{-n}$, with n = 1.3 over a temperature interval from ca. 150 to 300 K [1.136]. Similar dependence has been observed for electrons in TCNQ crystals with n ∼ 1 [1.142], and for holes and electrons in a number of other organic molecular crystals.

Thus, it seems that pertinent, rather sophisticated micromodels for charge carrier hopping phenomena should be proposed in the near future not only for different species of organic molecular crystals, but also separately for hole and electron transport in one and the same crystal, e.g., in an anthracene crystal.

It is curious to observe that in the case of charge carrier transport problems in organic molecular solids theoreticians are compelled to turn away from accustomed deductive thinking of traditional solid-state physics to laborious inductive puzzle solving. There is also additional challenge for experimental physicists. In order to develop new pertinent hopping models accurate measurements of $\mu(T)$ values within a wide temperature range and in different crystallographic directions should be provided. Such $\mu(T)$

dependence studies in some model molecular crystals were recently carried out by KARL [1.136] and by SCHEIN et al. [1.133,134,136a,b,142]. This, however, is only the beginning of a difficult and laborious task in the future. Up to now the band-hopping mobility transition has been more or less reliably observed only in naphthalene. For anthracene the transition is anticipated to occur below 77 K but has not yet been observed [1.136a].

At present the dynamic "hopping region" of noncoherent charge carrier motion in organic solids may be regarded as a virgin field where, according to Schein's expression " a comprehensive theory is required" [1.134]. There does not exist, at present, any appropriate microscopic model for quantitative treatment of hopping motion. Phenomenological approach is therefore frequently used in practice.

At present the charge carrier transport problems are often treated in terms of classical diffusion equations, using corpuscular representation. This approach has shown to be fruitful for describing such processes as hot charge carrier thermalization, free charge carrier generation and recombination according to Onsager's model, etc. (cf. Sect.2.8.3).

Specific charge carrier hopping transport problems may be analyzed by Monte Carlo simulation methods (see, e.g., recent studies by SILVER and RESCO [1.143]).

A different kind of band-to-hopping transition may occur in the thermalization processes of "hot" electrons at room temperature. As shown in photoelectron and secondary emission studies from organic solids, the mean free path $\bar{\ell}$ of hot electrons between two scattering events may considerably exceed the $\bar{\ell}$ value of stationary thermalized charge carriers [1.144]. Electron emission studies yield an important emission parameter, namely electron escape depth L which may be defined as the distance at which an electron may pass without scattering with a probability equal to 1/e. It can be shown that the mean free path $\bar{\ell}$ of an electron is connected with escape depth L by the simple relation $\bar{\ell} \approx L/2$ [1.144].

The value of $\bar{\ell}$ for hot electrons may, accordingly, be estimated from experimentally determined L data.

Thus, photoemission data for pentacene crystals show that hot electrons having excess energy E_f ca. $\lesssim$ 3 eV may have a mean free path $\bar{\ell} \lesssim 37 \pm 5$ Å [1.144-147]. A similar value of $\bar{\ell}$ was obtained from photoemission data for tetracene crystals [1.144,147]. These results are in good agreement with the value of $\bar{\ell}$ deduced from secondary emission studies [1.148], viz. $\bar{\ell} =$ 12-40 Å for hot electrons with $E_f \lesssim 3$ eV.

It is characteristic for organic molecular crystals that for hot electrons, having excess energies E_f greater than the excitation energy of
first singlet or triplet excitons, the dominant inelastic scattering mechanism is electron-electron interaction resulting in exciton generation.
Therefore the hot electron generated by autoionization of a photoexcited
neutral molecular state may be regarded as delocalized in the initial phase
and be described by a wave vector $\underline{k}$. After a few electron-electron scattering events the hot electron rapidly loses its excess energy, becomes damped
and localized. When the excess energy of the hot electron decreases below
the energy of triplet exciton formation energy (viz. 1.0-1.5 eV in aromatic
crystals) electron-phonon scattering becomes dominant and the mean free
path $\bar{\ell}$ of an electron, apparently, decreases from 30-40 Å to the value of
intermolecular spacing, i.e., 5-10 Å. This process may be regarded as a
band-to-hopping motion transition of a hot charge carrier in the process of
thermalization, which corresponds to a continuous change in the nature of
motion from a purely coherent to a purely diffusional one. This type of
band-to-hopping transition for charge carriers is analogous to a similar
transfer from coherent to incoherent motion of excitons, caused by site diagonal and off-diagonal dynamic disorder, first anticipated and studied by
HAKEN et al. [1.138,139] and later discussed in detail in a number of recent papers (see, e.g. [1.137]) (cf. Sect.2.1).

From this point of view it is viable to consider the initial phase of
hot electron motion in a time interval Δt as coherent. Such an approach has
been, for example, recently used by BELKIND et al. [1.144,149] in photoelectron emission theory based on the concept of coherent quasi-free emitted
electrons which may be described by wave vector $\underline{k}$.

One should, however, be cautious using definite, coherent, or noncoherent representation for the dynamic transition interval of electron motion.
Besides, in a number of organic molecular crystals the initial mean free
path $\bar{\ell}$ of hot electrons may be considerably smaller than in aromatic crystals. Thus, e.g., in phthalocyanine crystals the initial mean free path $\bar{\ell}$
for hot electrons of $E_f \lesssim 3$ eV has a value of 5 to 7 Å [1.150,151], i.e.
$\bar{\ell} \approx a_0$. In such a situation coherence of electron motion cannot be pre-
served, even in the initial phase.

1.5.4 Electronic Polarization and Charge Carrier Self-Energy

We discussed briefly the influence of electronic polarization effects on excess charge carrier transfer processes in organic molecular crystals (see

Sect.1.5.1). Much more important, however, is the role of electronic polarization in the process of formation of stationary states in organic solids consisting of highly polarizable molecules, such as aromatic and heterocyclic crystals.

Electronic polarization of the crystal by a quasi-localized charge carrier, i.e., its interaction with valency π and σ electrons of surrounding
molecules determines the self-energy of the given charge carrier, and, consequently, the position of its conductivity level in the energy diagram of
ionized states of the crystal.

The details of the electronic polarization process are rather subtle
and will be discussed later (see Sect.2.3), but the very essence of the
phenomenon is simple. During the localization time of the charge carrier on a
definite molecule, i.e., during the hopping time τ_h (τ_h being $> 10^{-14}$ s for
anthracene-type crystals), a localized charge carrier manages to polarize
the electron orbitals, mainly highly polarizable π orbitals, of surrounding
neutral molecules in the crystal.

The electronic polarization process of an aromatic crystal in the field
of a localized positive charge carrier is schematically shown in Fig.1.20.
Here electronic polarization is "visualized" in the form of displacement of
π orbitals of neighboring neutral molecules in the field of the charge
carrier. As a result of such displacement dipole moments μ are induced
around the localized charge. The total energy of charge-induced dipole in-

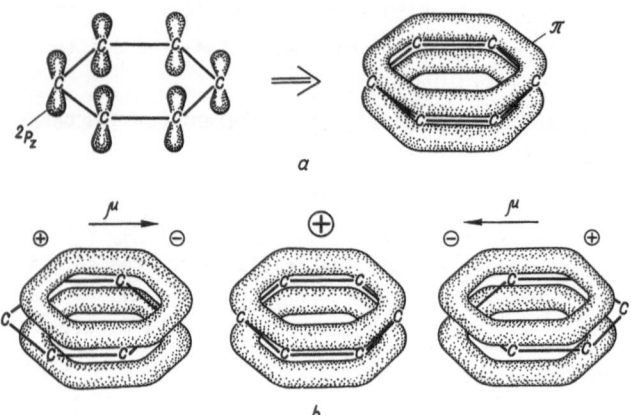

Fig. 1.20 a,b. Schematic picture of electronic polarization of a molecular
crystal by a localized charge carrier: (a) formation of a molecular π orbital
from atomic carbon 2 p_z orbitals in benzene; (b) process of electronic
polarization - formation of induced dipoles μ on neutral molecules of the
crystal in the field of a localized positive charge carrier

teraction is considerable in aromatic and heterocyclic molecular crystals. Thus, typical electronic polarization energy P amounts to about 1.5 to 2.0 eV in this case.

Electronic polarization is a very fast process. The time necessary for displacement of the π orbitals of neighboring molecules, i.e., the time needed for induced dipole formation is only 10^{-16} to 10^{-15} s. Consequently, the typical charge carrier localization time τ_h in anthracene-type crystals exceeds the electronic polarization time τ_e by several orders of magnitude (i.e., $\tau_h \gg \tau_e$, as may be seen from Fig.1.22). Because of this it is appropriate to regard the localized charge carrier as "static" relative to the very fast process of electronic polarization. Electronic polarization in molecular crystals can, accordingly, be described in the framework of microelectrostatic approximation as charge-induced dipole interaction (cf. Sect.2.3). Subsequently stationary charge carriers move together with their electronic polarization "cloud" by means of noncoherent hopping in molecular crystals. A schematic picture of the formation of an electronic polarization "cloud" around a localized charge carrier in a hypothetic molecular crystal consisting of benzene-type molecules is shown in Fig.1.21.

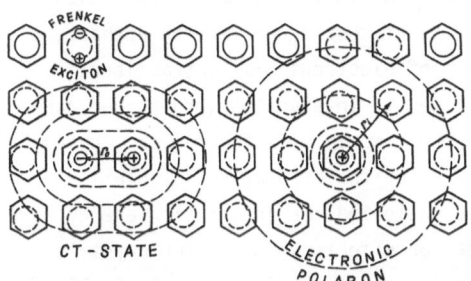

Fig. 1.21. Schematic picture of formation of a Frenkel exciton, charge transfer (CT) state and electronic polaron in a hypothetic molecular crystal consisting of benzene-type molecules

Such a quasi-static formation may be conditionally called an "electronic polaron". Actually, this term was first used by TOYOZAWA [1.152] for describing a dynamic state of a Bloch-like delocalized electronic polarization wave in a crystal. Now we know, however, from previous discussions (cf. Sect.1.5.3) that such delocalized charge carrier states may be formed in molecular crystals either at very low temperatures, or in the case of hot charge carriers for which the effective hopping time τ_h may be of the same order as the time τ_e of relaxation for electronic polarization (see Fig.1.22). Dynamic electronic polaron states, as we shall see later, are of great importance in fast electronic relaxation processes.

On the whole, electronic polarization is one of the most fundamental electronic processes determining the self-energy of charge carriers and

real energy structure of ionized states in molecular crystals. It also
plays a first-rate role, as will be shwon in Chaps.3 and 5, in the formation of local states of structural origin in defective molecular crystals.
For this reason a number of sections in Chap.2 will be devoted to a treatment of electronic polarization phenomena, as well as to methods of its experimental and theoretical investigation.

At this stage we only wish to emphasize that electronic polarization is
essentially a many-electron phenomenon. For this reason electronic conduction states in organic crystals cannot, in principle, be approached in
terms of one-electron approximation. Consequently such traditional solid-
state concepts as simple band theory are not applicable in this case.

In organic molecular crystals the many-electron phenomenon of electronic
polarization has been taken into account in a phenomenological model first
proposed by LYONS [1.65] and later modified in our work [1.61,63,64,66,69]
(cf. Sect.2.2).

1.5.5 Other Types of Interaction

The table in Fig.1.22 summarizes the various types of interaction, relaxation time scales and interaction energies of charge carriers in anthracene-
type crystals. Typical relaxation times of electronic τ_e, intramolecular
(vibronic) τ_v and intermolecular (lattice) τ_ℓ interaction with charge carriers are compared with a typical hopping time τ_h of a thermalized charge
carrier in stationary conductivity states of the crystal (cf. Table 1.9).
Figure 1.22 shows the interaction time scale and the interaction energy
"hierarchy" of charge carrier polarization effects in molecular crystals.

1) The electronic polarization relaxation time τ_e is extremely small compared with hopping time τ_h, i.e., $\tau_h \gg \tau_e$. This means that quasi-localized
charge carriers jump from one molecular site to another one, surrounded by
a steady-state polarization "cloud" (cf. Fig.1.21). Polarization is therefore the dominant, primary-order type of interaction of an excess charge
carrier with the surrounding molecular lattice, having typical energy of ca.
1.5-2.0 eV. Consequently, electronic polarization is the most important factor determining the charge carrier's self-energy and, correspondingly, the
position of electronic conduction states in the energy diagram of the crystal.

2) Intramolecular (vibronic) relaxation time τ_v is of the same order as
hopping time τ_h, i.e., $\tau_v \sim \tau_h$. Two limiting cases are possible here. If τ_h
$\lesssim \tau_v$, dynamic interaction of a quasi-localized charge carrier with intra-

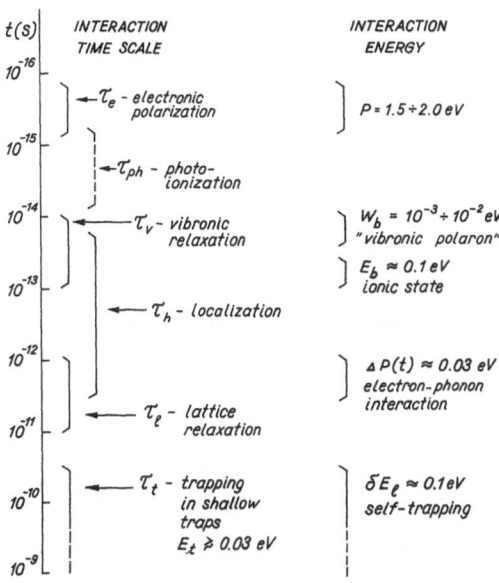

CHARGE CARRIERS
in ANTHRACENE-TYPE CRYSTALS

Fig. 1.22. Various types of interaction, typical relaxation time scales and interaction energies of a charge carrier in an anthracene-type crystal

molecular vibronic modes may take place. As a result of such interaction a dynamic state of a "vibronic polaron" may be formed, according to a model proposed by SIEBRAND and MUNN [1.153-156]. The estimated energy of such interaction in an anthracene crystal is ca. 10^{-3}-10^{-2} eV [1.69], and it is one of the factors determining the effective width of electronic conduction levels (cf. Sect.2.7.2). In the second limiting case, when $\tau_h > \tau_v$, a stationary molecular ion state may be formed during the localization time of the charge carrier (cf. Sect.2.7.2). For this extreme case interaction energy equals bonding energy E_b of the charge carrier by the molecular ion [1.69]. The value of E_b for anthracene-type crystals is ca. 0.1 eV (cf. Sect.2.7.2), and it determines the virtual position of ionic conduction states in the energy diagram of the crystal (see Fig.2.7).

3) For anthracene-type crystals having relatively large molecules the lattice relaxation time τ_ℓ is usually greater than the typical hopping time τ_h, i.e., $\tau_\ell \gg \tau_h$. In such a case the quasi-localized charge carrier may not manage to polarize the surrounding lattice during the localization time τ_h. This agrees with calculated estimates of VILFAN [1.157] that small lattice polaron formation in organic molecular crystals is not likely. In addition, the temperature dependences of charge carrier mobilities $\mu(T)$ in molecular crystals do not speak in favor of the small polaron model [1.45,133,

134,136]. As shown by GOSAR and CHOI [1.158] and VILFAN [1.157], electron-phonon interaction manifests itself mainly via modulation of electronic polarization energy $\Delta P(t)$. This modulation, having a value of ca. 0.03 eV at room temperature [1.157], is also one of the factors determining the effective width of electronic conduction levels.

4) In real crystals charge carriers may get trapped in shallow structural traps (cf. Sect.5.15). The trapping time of such shallow traps τ_t for trap depth $E_t \gtrsim 0.03$ eV is greater than $5 \cdot 10^{-10}$ s. Consequently, in such a case $\tau_t > \tau_\ell$, and local lattice polarization, i.e., lattice deformation around the trapped charge carrier, caused by charge carrier induced dipole interaction may occur. Such local "self-trapping" of charge carriers deepens the existing trap depth by about 0.1 eV [1.21,159].

This brief introductory review shows that the interaction time "spectrum" of charge carriers in molecular crystals covers a wide and varied range and is rather complex. This complexity is further increased by anisotropy of the carrier hopping time τ_h in different directions of the lattice (cf. Table 1.9).

Apart from organic molecular crystals DUKE [1.160,161] has recently made an analytical survey of the charge carrier localization effects in other typical Van der Waals solids, such as organic polymers, orthorhombic sulfur, monoclinic selenium, etc. The author demonstrated that localization in these molecular insulators is caused, similarly as in molecular crystals, by different types of electronic and atomic interactions with excess charge carriers.

2. Electronic States of an Ideal Molecular Crystal

*It seems that the human mind has first to con-
struct forms independently, before we can find
them in things ... knowledge cannot spring from
experience alone but only from the comparison
of the inventions of the intellect with observed fact.*

Albert Einstein

Excited electronic states may be classified as neutral or ionized. In the
present work we shall focus our attention on the problems of formation, behavior and energetics of ionized states. But since the creation of ionized
states in organic molecular crystals proceeds mainly, as will be seen later,
via neutral states, we devote the next section to a brief description of
excited neutral states, their energetics and behavior.

2.1 Neutral Excited States in a Molecular Crystal

Neutral excited states in a molecular crystal can be described in terms of
molecular exciton theory developed by DAVYDOV (see [1.53,2.1,2] and references therein).

DAVYDOV extended the model of the small-radius exciton, proposed earlier
by FRENKEL [2.3] to anthracene-type molecular crystals and developed a consistent theory of light absorption, exciton formation and coherent motion
in such crystals [1.53,2.1,2].

The Frenkel exciton model proved to be the most appropriate for describing the behavior of neutral excited states in organic molecular crystals
characterized by weak intermolecular interaction forces.

The validity of the molecular exciton theory was experimentally verified
and proved in the work of PRIHODKO et al., McCLURE, WOLF, LYONS, TANAKA,
BROUDE et al., and many other investigators (see [1.53,2.2,4,5] and references therein, as well as review articles [2.6,7]). This theory was further
extended and developed by AGRANOVICH [2.4,5], CRAIG [2.8], RICE and JORTNER
[2.9], HAKEN et al. [1.138,139,2.10-12], KENKRE and KNOX [2.13], KENKRE
[2.14,15], GROVER and SILBEY [2.16], MUNN and SILBEY [2.17], SUMI [2.18]
and others.

According to molecular exciton theory [1.53,2.2], in a crystal containing several molecules in a unit cell, several branches of excited states of the crystal correspond to each excited electronic state of a molecule.

In the case of anthracene-type crystals containing two molecules in a unit cell, two bands of excited states correspond to each molecular term. These form the so-called Davydov doublet [1.43,53,2.2]. Davydov splitting is due to symmetry properties of the crystal, viz. to the non-equivalence of the positions of two identical molecules in the unit cell (cf. Fig.1.5).

The excitation energy of these states can be expressed in the following way [1.43,2.2]:

$$E_\mu(\underline{k}) = \Delta\varepsilon + D + \varepsilon_\mu(\underline{k}) \quad , \quad (\mu = 1,2)$$

where $\Delta\varepsilon$ is the excitation energy of an isolated molecule; D is the change in energy due to dispersion interaction of the excited molecule with the crystalline environment; $\varepsilon_\mu(\underline{k})$ is a term determining the energy dispersion of the exciton in the μth band (term of resonance interaction between molecules).

The effect of Davydov splitting provides direct evidence of the presence of collective excitation of the crystal which may be described in terms of a delocalized exciton.

In a perfectly rigid crystal (at helium temperatures) behavior of delocalized excitations can be described by Bloch waves and correspondingly be characterized by a wave vector $\underline{k}$ and energy dispersion $\varepsilon(\underline{k})$ in the exciton band [1.53,139,2.2].

Exciton band structure calculations for the first singlet excited state in an anthracene crystal were performed by DAVYDOV and SHEKA [1.52]. The calculated exciton energy $\varepsilon(\underline{k})$ dependences for both exciton bands ($\mu = 1,2$) in three crystalline directions (c,a,b) of the wave vector $\underline{k}$ are presented in Fig.2.1.

As may be seen from Fig.2.1, the singlet exciton bands in anthracene are extremely narrow, the bandwidth Δ being ca. 100-300 cm^{-1} (0.01-0.04 eV), and showing similarly expressed anisotropy as in the case of charge carrier bands (cf. Fig.1.18).

The close analogy between the behavior of excitons and charge carriers in molecular crystals goes still further.

For charge carriers the temperature-dependent band-to-hopping transition was observed and discussed only recently (cf. Sect.1.5.3). Yet similar

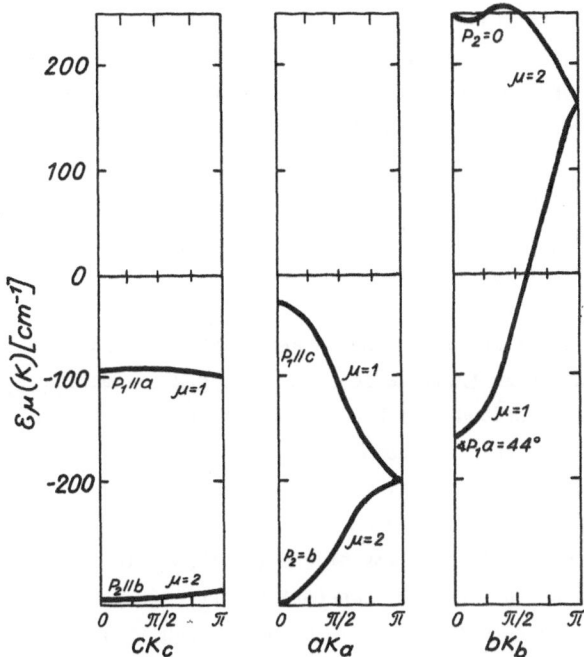

Fig. 2.1. Energy dependence $\varepsilon_\mu(\underline{k})$ of the first two singlet exciton bands $(\mu = 1,2)$ in an anthracene crystal [2.2]

phenomena in exciton physics have been an object of extensive studies for a number of years.

It has been shown that exciton transport in molecular solids depends strongly on exciton-phonon coupling. Already at temperatures above 30-40 K considerable interaction between the exciton and lattice vibrations starts taking place. As a result the exciton wave is repeatedly scattered, its initial amplitude decreases exponentially with time, and the excited state gets localized on individual molecular sites [1.138,139]. This corresponds to transition from coherent wave-like motion to incoherent diffusion of localized excitons by site-to-site hopping in the lattice.

In close similarity to charge carriers (cf. Sect.1.5.3) excitons in anthracene-type crystals interact most effectively, as shown by KURIK and TSIKORA [2.19] with low-frequency optical lattice phonons (librons). The scattering on optical phonons which steadily increases with temperature [2.19] is obviously one of the main factors causing exciton localization.

It has been experimentally established that in the high-temperature range (at ca. T $\gtrsim$ 70 K for singlet excitons in anthracene) the exciton transport may be well described in the framework of macroscopic diffusion theory regarding exciton migration as a form of random walk (see, e.g. [2.20] and references therein, as well as [2.5]).

This phenomenological approach yields such experimental transport parameters as diffusion coefficient D, mean diffusion length L_E, mean exciton lifetime τ_E, etc., which are so important for semi-quantitative description of exciton motion in a crystal (see [2.4,5,20]).

A theoretical treatment of the problem by HAKEN and REINEKER [1.138], HAKEN [1.139] and GROVER and SILBEY [2.16] also shows that exciton motion in molecular crystals can be described as coherent only at very low temperatures, whereas at room temperature it is completely incoherent. Migration of molecular excitons is found to be diffusion-like in a long-time scale for all temperatures, as long as exciton-phonon coupling is present. Thus one of the most complex and difficult problems of exciton theory at present [similarly as in the case of charge carrier transport theory (cf. Sect. 1.5.3)] is the dynamics of temperature- and time-dependent transition from coherent to incoherent motion.

The time interval Δt of exciton localization depends on the nature of the excited state (singlet or triplet), on properties of the crystal lattice, on temperature, and other factors. Thus, for instance, the transfer of a triplet exciton is usually treated in terms of the hopping model, owing to extremely narrow bandwidth $\Delta \lesssim 10$ cm^{-1} [2.5,8,21], whilst a singlet exciton may be considerably delocalized at the initial period Δt of its existence. It may lose, however, its coherency very quickly, in the course of the first picoseconds, due to interaction with vibrational modes [2.22].

In dealing with these and other above-mentioned problems of exciton dynamics, the stochastic approach, according to HAKEN et al. [1.138,139], may be regarded as the most appropriate. Actually, the first stochastic model of exciton-phonon interaction in molecular crystals was proposed by HAKEN and STROBL [2.10]. The authors treated the motion of a Frenkel exciton using a Hamiltonian which comprises a completely coherent part and a fluctuating part which describes both fluctuations of energy of the localized excitons and fluctuations of the transition matrix elements between different lattice sites [2.11]. The model has been further developed and the range of its applicability analyzed by HAKEN et al. [1.138,2.11,12], ERN et al. [2.21], REINEKER [2.23], OVCHINNIKOV and ERIKHMAN [2.24], RIPS [2.25] and others.

It has been shown that the Haken-Reineker-Strobl model is applicable in the case of extremely narrow excitonic bands having bandwidth $\Delta \ll kT$ for triplet excitons in anthracene-type crystals in the temperature range 20 K < T < 150 K (see [1.138,2.5,11,12,21]). The stochastic model has been recently modified by SUMI [2.18] making it more appropriate for the description of singlet exciton motion. A different statistical approach to exciton transfer in molecular crystals, based on a generalized master equation, has been proposed by KENKRE and KNOX [2.13,15] (see also [2.26]). The model yields a unified description of coherent motion at short times and diffusive transport at long times.

KENKRE [2.14] has analyzed the relations existing between the three above-mentioned exciton transfer theories and has established an equivalence between the HAKEN-REINEKER-STROBL [1.138,2.10,11] and the GROVER-SILBEY [2.16,27] formalisms on the one side, and the KENKRE-KNOX [2.13,15] formalism on the other.

In the approach of GROVER and SILBEY [2.16], as well as in other above-mentioned theories, linear exciton-phonon coupling has been assumed, similar in spirit to the work on the small lattice polaron (see, e.g. [1.132, 2.28].

In contrast to this MUNN and SIEBRAND [1.154,156] in their earlier treatment of exciton interaction with "high frequency" intramolecular vibrational modes, considered quadratic coupling in vibrational coordinates, accompanied by frequency changes in the interacting modes.

In general, it may be shown [2.17] that if the given mode is a symmetric intramolecular one, linear coupling usually dominates. On the other hand, for modes of certain symmetries (e.g., out-of-plane bending modes in aromatic hydrocarbons) the linear term vanishes and quadratic coupling dominates. Certain aspects of the earlier Munn-Siebrand treatment were criticized by DRUGER [2.29] because they had not been derived from the Hamiltonian directly, but instead were assumed "heuristically".

In a recent paper by MUNN and SILBEY [2.17] the exciton transport has been investigated for a model Hamiltonian with both linear and quadratic coupling. The main results are broadly consistent with earlier treatments [1.138,154,2.13-16] and with experimental data [2.17]. The extended model proposed in [2.17] should be valid for triplet excitons in aromatic hydrocarbon crystals. The authors of [2.17] plan to extend the validity range of the theory, including in future studies also the "bound" exciton states caused by strong exciton-vibron coupling. Thus we see that molecular exciton

theory has made substantial advances in recent years. Taking into account
the above-mentioned analogies and parallelisms of exciton and charge carrier
behavior in molecular crystals one should expect a considerable impact of
the present stochastic exciton transfer theories on the future development
of similar models for charge carrier transport (cf. Sect.1.5.3). One should,
however, bear in mind that the variety of charge carrier interaction types
within a molecular crystal is wider than in the case of excitons: apart from
inter- and intramolecular coupling we should primarily take into account also the influence of electronic polarization energy fluctuations, and charge
carrier self-trapping effects (cf. Fig.1.22).

This concludes our brief account of exciton dynamics. In our further discussions we shall mainly single out energetical aspects of neutral excited
states in close context with ionized ones.

It follows from the localized molecular exciton concept that neutral excited electronic states of a molecular crystal can be represented by discrete energy levels completely analogous to the energy levels of an isolated
molecule (cf. Fig.2.2-4). One has only to allow for a level shift towards
lower energies owing to dispersion and resonance interaction between molecules in a crystal medium. In the energy diagram of neutral excited states
of a crystal one usually presents, in addition to pure electronic transitions (0-0 transitions), also the vibronic sublevels and Davydov splitting
components (cf. Figs.2.2-4).

Finally, one should notice that the above-mentioned molecular solid-
state dualism (cf. Sect.1.4) clearly manifests itself also in the case of
excited neutral electron states. Owing to the pronounced tendency towards
final localization of molecular excitons on individual molecules, the neutral excited states in a molecular crystal retain to a considerable extent
their molecular nature. In processes of exciton formation, motion and localization the role of "molecular" properties of the crystal clearly prevails over that of the "solid state".

2.2 Ionized States in a Molecular Crystal

A phenomenological model of ionized state energy structure, accounting for
electronic polarization effects of many-electron interaction in organic
molecular crystals (cf. Fig.1.21) was first proposed by LYONS in 1957

[1.65] (see also [1.38,39]) and is generally known as the Lyons model [1.40, 42,45]. We will first discuss the Lyons model in its original form and then focus our attention in following sections on its various modifications (cf. Sect.2.2.2).

In this connection one should bear in mind that a physicist applying phenomeonological models is often encountering a dilemma ingeniously formulated by McCREA [2.30]: "One cannot make a model until one knows what one is talking about; one cannot say what one is talking about without a model".

The evolution of the Lyons model presents a vivid illustration of such a dilemma and the ways of its stepwise solution.

2.2.1 The Lyons Model of Ionized States

Being himself a molecular spectroscopist, LYONS was first to draw attention [1.65] to a very essential experimental fact characteristic of organic molecular crystals, namely, that the ionization energy I_C of a crystal, such as anthracene, is by some 1.5 eV lower than the corresponding value of ionization energy I_G of an isolated molecule (see Fig.2.2).

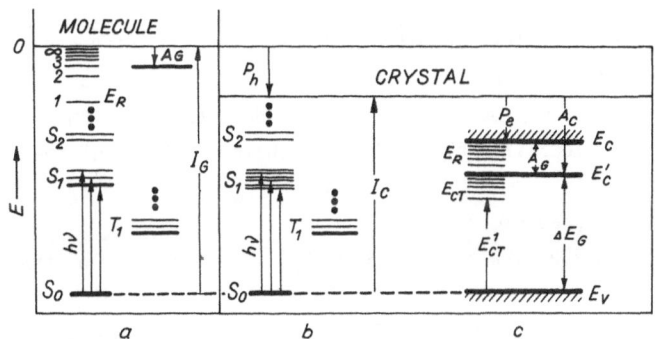

Fig. 2.2. Energy level diagram of neutral and ionized states of an ideal anthracene-type crystal: (a) energy levels of the molecule; (b) neutral and (c) ionized states of the crystal, according to the initial LYONS model (cf. [1.38-40,42,65])

Since the electronic structure of a molecule changes only very slightly upon formation of a molecular crystal (cf. Sect.1.4), it is in principle unlikely that these small changes cause such considerable difference between the I_G and I_C values. LYONS therefore assumed [1.65,38,39] that ionization of a crystal requires less energy than that of a molecule owing to the energy gain due to electronic polarization of the crystal by the positive charge (hole) created in the ionization process. This assumption forms the basic postulate of the LYONS model [1.65,38,39] and may be described by a simple expression

$$P_h = I_G - I_C \quad .$$

(2.1)

where P_h is the energy of electronic polarization of the crystal by a hole (cf. Fig.2.2).

Already the first rough estimates of the value of P_h, made by LYONS and co-workers, using electrostatic approximation [1.38,65,2.31,32], yielded a P_h value of ca. 1.7-2.0 eV, thus giving satisfactory support to the basic postulate (2.1) of the model. Similarly, the self-energy of an excess electron is also determined, according to LYONS, by its electronic polarization energy P_e of the crystal. Consequently, the electron conduction band E_C is situated below the crystal ionization threshold level by the value of P_e (see Fig.2.2). LYONS further assumed that, according to traditional band theory concepts, the excess electron is delocalized forming a "wide" conduction band E_C.

Since, however, it was known that aromatic hydrocarbon molecules possess positive electron affinity A_G of the order of 0.5-1 eV [1.38], LYONS assumed that an excess electron may be "bound" by individual molecules and thus become localized. The localized electrons form their own "narrow" conduction band E_C' lying below the "wide" conduction band E_C of delocalized electrons by a value of A_G (see Fig.2.2).

As may be seen from Fig.2.2, the "narrow" band E_C' is situated below the crystal ionization threshold level by the value A_C of electron affinity of the crystal, A_C being the sum of electron affinity A_G of the molecule and of polarization energy P_e of the crystal by the electron

$$A_C = A_G + P_e \quad .$$

(2.2)

Expression (2.2) constitutes the second main postulate of the Lyons model.

Since there is not an obvious corresponding counterpart of electron affinity, viz. a hole affinity of a molecule, it logically follows that in the framework of the Lyons model there should be only one "wide" conduction band E_V of delocalized holes analogous to the valency band in covalent crystals (see Fig.2.2).

Thus, according to Lyons' model, one should accept an internal asymmetry of the ionized state energy structure in molecular crystals, viz. the existence of "wide" E_C and "narrow" E_C' bands for delocalized and localized electrons, respectively, and only one "wide" band E_V for delocalized holes.

The eclectic nature of the Lyons model is rather conspicuous. It attempts to combine concepts of many-electron interaction (electronic polarization) with concepts and terminology of traditional band theory of one-electron approximation.

This may be one of the main reasons why the Lyons phenomenological model did not gain particular support, especially among solid-state physicists. Only after the inconsistency of the one-electron band model for molecular crystals became evident towards the end of the sixties (cf. Sect.1.5.1), interest was focused once more on the many-electron aspects of the Lyons model. This interest was enhanced by results obtained in 1968, pointing towards the possible existence of two conduction bands for electrons in an anthracene crystal, in agreement with the Lyons model. VAUBEL and BAESSLER [2.33] found, applying the photo-injection method, that the energy gap ΔE_G in anthracene crystal equals 3.72 ± 0.03 eV, and that a second conduction band E_C is most probably situated above the first narrow band E_C' at a distance of ca. 0.6 eV, which approximately corresponds to the value of A_G for anthracene.

The validity of the basic concepts of the Lyons model was confirmed in extended studies by POPE and co-workers, BAESSLER et al., as well as by LYONS and his group and others. The model became thus generally accepted in the seventies as a basis for interpretation of energy structure of ionized states in anthracene-type crystals (cf., e.g. [1.38-40,42,45,62,2.34]).

At the same time, however, certain criticism was voiced, pointing out some internal contradictions and inconsistencies of the model.

Thus already POPE [1.42] drew attention to the above-mentioned controversial problem about the asymmetry of the ionized state energy structure. The difficulties were enhanced in this case by the fact that there should be an essential difference between the bands responsible for stationary conductivity of the crystal: the electrons in the "narrow" E_C' band are localized, whereas holes in the "wide" E_V band are delocalized. This means that electron and hole transfer parameters, such as carrier mobility, must differ considerably. This is not, however, observed experimentally. Mobility values μ_e and μ_h are approximately of the same order along the basic axes of an anthracene crystal (see, e.g. [1.38,40,45]).

On the other hand, DRESNER's experiments [2.35] produced indirect evidence of the existence of two symmetrical bands both for electrons and holes, most probably separated by an interval not exceeding 0.05 eV which is far below A_G value.

Finally, since the hopping mechanism of charge carrier transfer dominates in anthracene-type crystals at room temperatures (cf. Sect.1.5.3) conduction band concepts are altogether inadequate.

The main point, however, was brought out by BAESSLER and KILLESREITER [2.34] who showed that the value of the energy gap for anthracene, as determined earlier in [2.33], namely ΔE_G = 3.72 eV, was incorrect. It was determined using the photo-injection method without accounting for charge-transfer (CT) state formation at the metal-anthracene crystal interface. Allowing for activation energy of CT state decay, a value of ΔE_G = 4.1 ± 0.1 eV was obtained for anthracene.

This disproved the basic and only report on the experimentally ascertained two conduction bands for electrons in anthracene, separated by 0.6 eV, i.e, by a gap of the order of A_G.

In the Lyons model polar nonconducting states are also considered (cf. Fig.2.2):

a) presumed Rydberg states in the crystal which should be situated below the "wide" conduction band E_C for delocalized electrons (cf. Fig.2.2). These hydrogen-like states may be regarded as analogues of Wannier-Mott excitons (cf. [2.4,36]). Their existence, although vigorously searched for, has not been experimentally established in molecular crystals (see, e.g. [1.40, 2.37]). This is understandable, since the effect of strong localization of charge carriers and their transfer by stochastic hopping a priori excludes the probability of the existence of such exciton states (cf. Sects.1.5.3 and 2.8);

b) charge transfer (CT) states situated below the "narrow" conduction band E_C' of localized electrons (cf. Fig.2.2). The CT state is a nonconducting polar state formed by an electron-hole pair bound by Coulomb interaction. The energy E_{CT} of such a state is lower than the width of the energy gap, i.e., $E_{CT} < \Delta E_G$ (cf. Fig.2.2).

CT states in a molecular crystal are not exciton states in the traditional meaning. Owing to localization the Coulomb-field bound electron and hole migrate by stochastic hopping and are usually treated in terms of the Onsager model (cf. Sect.2.8). CT states play a rather considerable role in molecular crystal electronic microprocesses (cf. Sects.2.8.3,4). Their existence has been confirmed by direct and indirect experimental methods [1.42, 2.34,38] as well as by theoretical calculations [1.38,60,61,64] (cf. Sect. 2.8.2).

Thus, summing up, we see that the Lyons model contains internal inconsistencies and a number of assertions which have not been experimentally confirmed. It may not, accordingly, be applied, without modifications, for an adequate description of ionized states of molecular crystals.

2.2.2 A Modified Lyons Model

In an alternative modified model of ionized states proposed by SILINSH et al. [1.62,63,69,2.39] the basic postulates of Lyons' model have been retained: electronic polarization is still considered as the main factor which determines the self-energy of charge carriers in the conductivity states. The most important feature of the modified model is, in our opinion, that it allows avoidance of the unsubstantiated internal asymmetry of the original Lyons model. This asymmetry, as will be seen further, was caused by inconsistent interpretation of the role of the empiric molecular parameter A_G in the formation of ionized conductivity states for electrons.

Secondly, in the modified model one-electron concepts and terminology, typical of traditional band theory, are completely avoided as alien and inapplicable for the description of many-electron conductivity levels.

Finally, the separation of charge carrier interactions with electrons (electronic polarization) and intramolecular vibrations (vibronic coupling), based on different interaction time scales, as compared to mean charge carrier hopping time τ_h (cf. Fig.1.22) allows one to single out, first, purely electronic conduction states (ionized states) of the crystal in order to introduce later real ionic states, when strong vibronic coupling is taken into account (cf. Sects. 2.7.2,3).

Thus, according to our modified model [1.63,69,2.39] the ionized states in an ideal molecular crystal are primarily considered as purely electronic conduction states, characterized by full conservation of equilibrium configuration of the neutral molecule during the time τ_h of charge carrier localization.

In the given case the expression (2.1) should be replaced by a more accurate one

$$P_h = I_G^V - I_C^0 \quad , \tag{2.3}$$

where I_G^V is the experimental vertical ionization energy of the molecule; I_C^0 is the experimental crystal ionization energy determined either from the intrinsic photoemission threshold value $I_{C,Y}$ of the crystal, or from the energy distribution of emitted electrons $I_{C,E}$ according to Einstein's law (cf. Sect.2.9).

As shown by SILINSH [1.63,69], the empirical molecular constant A_G, i.e., the electron affinity of the molecule, as used in the Lyons model, actually consists of an electronic and a vibronic component of affinity

$$A_G = A_G^e + E_b^e \quad , \qquad\qquad (2.4)$$

where A_G^e is the electronic component of the molecule's electron affinity, and E_b^e is the electron bonding energy in the process of formation of equilibrium configuration of a real ionic state (cf. Sect.2.7.2).

As will be shown further (see Sect.2.7.1) the A_G^e value is determined by intrinsic polarization of the electron orbitals of the molecule on which the electron is localized. A_G^e thus actually forms part of the total energy A_C^e of electronic polarization of the crystal by a localized electron, determining the electron affinity of the crystal

$$A_C^e = P_e + A_G^e \quad . \qquad\qquad (2.5)$$

Expression (2.5) shows that the electron affinity A_C^e of the crystal is determined by the sum of electronic polarization P_e of the surrounding molecules and of the intrinsic electronic polarization of the molecule on which the electron is localized.

This means that the self-energy of an electron in a molecular crystal is thus determined by the total energy of interaction between the electron and the valency electrons of the surrounding molecules, including valency electrons of the molecule, with the localized electron. This self-energy of the electron determines the position of the conduction level E_e of the electron in the energy diagram of the crystal (cf. Fig.2.3). As may be seen from Fig. 2.3, inclusion of the parameter A_G^e does not require formation of any additional conduction level E_C' and an energy gap $\Delta E = E_C - E_C'$, as assumed in the original Lyons model [1.65] (cf. Fig.2.2). A_G^e is a component part of the total electronic polarization of the crystal and has a relaxation time of the same order as P_e, viz. $\tau_e \approx 10^{-16} - 10^{-15}$ s which is typical for electronic processes (cf. Fig.1.22). The magnitudes A_G^e and P_e are actually

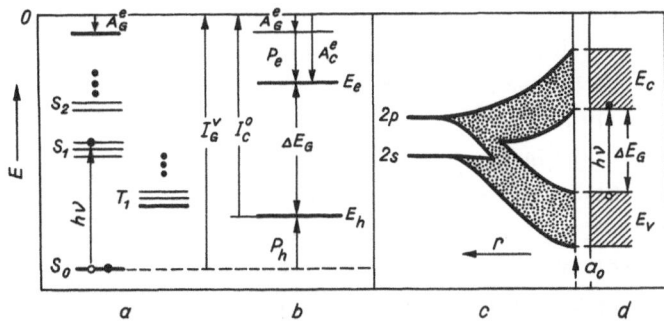

Fig.2.3. Energy level diagram of neutral (a) and ionized (b) states of an ideal anthracene-type crystal, according to the modified Lyons model [1.63, 69,2.39]; (c) formation of energy bands with approach of carbon atoms in a diamond crystal; (d) valency band (E_V) and conduction band (E_C) in a diamond crystal

separated for purely methodological reasons, owing to different methods used in their evaluation (cf. Sect.2.7.1).

A comparison of Fig.2.2 and 2.3 shows that the physical meaning of A_G in the Lyons model radically differs from that of A_G^e in the modified model proposed by us.

As regard the self-energy of a hole in a molecular crystal, it is determined by the polarization energy P_h of interaction between the hole and valency electrons of the surrounding molecules. If the experimentally determined I_G^V value is used in the construction of the energy diagram in Fig.2.3, then the polarization self-energy of the molecule by the localized hole is automatically accounted for, as will be shown further (see Sect.2.7.1). The value of P_h thus determines the position of the conduction level E_h of the hole in the energy diagram in Fig.2.3.

The aforesaid means that the conduction levels E_e of an electron and E_h of a hole in the energy diagram of a molecular crystal (Fig.2.3) are actually self-energy levels of quasi-free charge carriers determined by their polarizational, many-electron interaction with the crystal. Consequently, these many-electron levels have nothing in common with the one-electron conduction bands of traditional solid-state physics. This becomes evident if we compare energy diagrams of ionized states of a molecular crystal (Fig. 2.3b) with the band diagram of a covalent crystal of the diamond type (cf. Fig.2.3d).

In the case of a diamond crystal the discrete atomic carbon levels 2s and 2p split and interpenetrate as the atoms approach each other and, finally, as interatomic distance becomes equal to the lattic parameter a_0, form two bands of quasi-continuous states: the valency band E_v, completely occupied by electrons, and the unoccupied conductivity band E_c.

The picture is radically different in the case of ionized states of a molecular crystal (Fig.2.3b). The conductivity level E_h of holes is not an analogue of the valency band E_v of a covalent crystal, because it is not being formed as a result of splitting of the highest occupied level S_0 of the molecule. The E_h level is, as already stated, the actual self-energy value of a hole in the crystal, and its position in the energy diagram is determined by many-electron interaction effects. Quite similar is also the nature of the conductivity level for electrons E_e. Consequently, there cannot occur direct optical transition between levels E_e and E_h, similar to band-to-band transition in a covalent crystal (cf. Fig.2.3d). In molecular crystals of anthracene type primary processes of optical excitation are, as a rule, of molecular nature (cf. Fig.2.3a). Free charge carriers are usually formed in secondary processes — mainly through auto-ionization of an excited molecular state (cf. Sect.2.8.3).

We have therefore, in order to avoid any possible misunderstandings, resigned from the traditional symbols E_v and E_c of band theory used also in the Lyons model [1.38,40,42,45]. We have introduced, in our model [1.62,63, 69,2.39], the symbols E_h and E_e to denote the respective self-energies of hole and electron on conduction levels (cf. Fig.2.3).

Apart from the energy diagram of electronic ionized states of a molecular crystal (Fig.2.3) which clearly shows the physical meaning of the formation of conduction states E_e and E_h in the modified model, it is frequently useful in practice to use a diagram of the type shown in Fig.2.4. Here the conduction state E_h of holes is fixed at the same level as the highest occupied state S_0 of the neutral molecule. Polarization energy P_h due to the hole will then be represented as the difference between ionization thresholds of the molecule and the crystal, similarly as in the diagram Fig.2.2.

The energy diagram, as in Fig.2.4, is more convenient than that in Fig. 2.3 for comparing the energies of neutral (b) and ionized (d) states of the crystal, particularly in studies of charge carrier photogeneration. It may, however, easily lead to the fallacy of considering level E_h as an analogue of valency level E_v of covalent crystals.

Figure 2.4 shows also CT states which we shall discuss in greater detail in terms of the modified model in Sect.2.8.

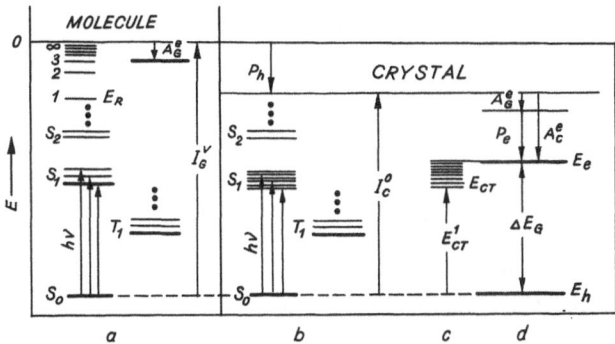

<u>Fig. 2.4.</u> Energy level diagram of an isolated molecule (a), of neutral
states (b), CT states (c), and ionized states (d), in an ideal molecular
crystal, according to the modified Lyons model [1.63,69,2.39]

2.3 Electronic Polarization of a Molecular Crystal by a Charge Carrier

2.3.1 Some General Considerations

As already demonstrated in Sect.2.2, electronic polarization of the surrounding lattice by a localized charge carrier is the basic factor determining its self-energy and, correspondingly, the position of conductivity
levels in the energy diagram of the crystal. From this aspect there is every reason to consider electronic polarization as the cornerstone of Lyons'
phenomenological model.

This raises the question of criteria of necessity to take into account
electronic polarization in considering electronic processes in a particular
crystal.

It may be shown that the Heisenberg uncertainty relation could furnish
such general criteria.

Thus, for example, a relaxation time τ_e necessary for polarization of
electron orbitals of the atoms and molecules of the surrounding lattice by
a charge carrier may be evaluated by means of the expression

$$\tau_e \approx \frac{\hbar}{\Delta\varepsilon_{ex}} \quad , \tag{2.6}$$

where $\Delta\varepsilon_{ex}$ is the excitation energy of an electronic exciton [2.36]. In
traditional semiconductors $\Delta\varepsilon_{ex}$ is determined by the width ΔE_G of the

forbidden energy gap: $\Delta E_G \approx \Delta\varepsilon_{ex}$. In molecular cystals $\Delta\varepsilon_{ex}$ corresponds, in first approximation, to the excitation energy ΔE_{S1} of a singlet S_1 exciton, i.e., $\Delta\varepsilon_{ex} \approx \Delta E_{S1}$.

For typical molecular crystals of the anthracene type we have ΔE_{S1} of the order of 2-4 eV. This means that τ_e in this case lies within the range of $(0.16-0.33) \times 10^{-15}$ s, thus being of an order of $10^{-16}-10^{-15}$ s (cf. Fig. 1.22).

If one uses the band model, the transfer time $\delta\tau$ of a charge carrier between two neighboring lattice sites may be described by the expression

$$\delta\tau \approx \frac{\hbar}{\delta E} \; , \tag{2.7}$$

where δE is the width of the respective conduction band.

For wide-band semiconductors, such as germanium or silicon δE is of the order of an eV, and thus we have $\delta E > \Delta E_G$. Accordingly, in these crystals $\delta\tau < \tau_e$ and, consequently, there may not be enough time for forming an electronic polarization "cloud" around a fast-moving charge. In such a case electronic polarization effects may be neglected. (This is in accordance with the concept of delocalization of free charge carriers in Ge and Si crystals moving as coherent Bloch waves).

The situation changes drastically in the case of narrow-band semiconductors with $\delta E \ll \Delta E_G$. We then have, according to (2.6,7), $\delta\tau \gg \tau_e$, and electronic polarization has to be taken into account.

Thus for naphthalene and anthracene crystals calculated values of bandwidth δE in one-electron approximation are of the order of 0.01-0.1 eV [1.54 -58] (cf. Sect.1.5.1). From (2.7) we get, accordingly, $\delta\tau$ values of the order of $10^{-4}-10^{-13}$ s, i.e., in this case actually $\delta\tau \gg \tau_e$. This evidences pronounced localization of charge carriers in anthracene-type crystals, necessity to take electronic polarization into account, and, consequently, the inapplicability of one-electron approximation.

If a hopping model is used, expression (2.7) should be replaced by (1.17) which gives a mean hopping time of $\tau_h \approx 3 \times 10^{-14}$ s (cf. Table 1.9).

As may be seen, this representation also gives $\tau_h \gg \tau_e$ (for anthracene we have $\tau_e = 2 \times 10^{-16}$ s and $\tau_h = 3 \times 10^{-14}$ s) (cf. Fig.1.22).

It follows thus that electronic polarization is indeed the main factor determining the self-energy of a charge carrier in anthracene-type molecular crystals.

2.3.2 Dynamic and Microelectrostatic Approaches to Electronic Polarization in Molecular Crystals

Electronic polarization of a molecular crystal may be treated in terms of quantum dynamical and classical microelectrostatic approaches.

Dynamical approximation is regarded as more strict and consistent from the theoretical point of view. It presents, however, considerable difficulties in obtaining an adequate quantitative description of the phenomenon, since it boils down to quantum electrodynamics of many-electron systems.

In its most general form the dynamic theory of electronic polarization in insulating crystals was proposed by TOYOZAWA [1.152]. He admits, in Bloch wave representation, delocalization of a charge carrier "dressed" into a polarization "cloud". Such a dynamic formation of a carrier which moves together with its electronic polarization cloud is called an *electronic polaron* [1.152] (cf. Fig.1.21).

In this way TOYOZAWA [1.152] has shown that electronic polarization in insulating crystals may be well described dynamically in terms of electronic interaction with the quantum field of Frenkel excitons. In ideology and mathematical formalism, as well as in physical content the electronic polaron theory [1.152] is analogous to lattice polaron theory (cf., e.g. [1.132,2,28,40]). The only essential difference is that under lattice polarization the atoms and molecules of the lattice are displaced from their equilibrium positions, i.e., the charge interacts with the phonon field of the crystal. Contrary to this, under electronic polarization only the electron orbitals of the molecules in the crystal get polarized (cf. Fig.1.20), i.e., the charge interacts with the electronic exciton field. There is also a corresponding difference in relaxation time scales: it is of the order of $\tau_e \approx 10^{-15}$-10^{-16} s for electronic polarization, and of the order of $\tau_\ell \approx 10^{-11}$-$10^{-12}$ s for lattice polarization in anthracene-type molecular crystals (cf. Fig.1.22).

The dynamic electronic polarization of a crystal can thus be reduced to the well-known type of problems in quantum electrodynamics — that of interaction of particles with a quantum field.

The electronic polaron concept of Toyozawa was later extended by HAKEN and SCHOTTKY [2.41] to the problem of the Wannier exciton in an insulator in order to study the electronic polarization effects on electron and hole interaction. After the work of HAKEN and SCHOTTKY a number of authors have extended the Toyozawa electronic polaron concept to problems of localized electrons in insulators, such as alkali and silver halides and molecular crystals of noble gases. A detailed discussion of these studies is given in the review article by WANG et al. [2.42].

BENDERSKY [2.43] was first to use Toyozawa's theory for estimating the energy of electronic polarization in anthracene-type molecular crystals. The author used tight-binding approximation according to which the excess electron is considered as localized on a particular molecule in the crystal, similarly to valency electrons. Owing to formal difficulties a number of rather rough approximations had to be introduced (for instance, dipole-dipole interaction between induced dipoles was neglected). The results obtained in [2.43] were consequently only of semi-quantitative nature and yielded only the right order of magnitude for electronic polarization energy P, not providing a possibility of obtaining more accurate values.

More accurate calculated data on electronic polarization energy in molecular crystals may be obtained using an alternative approach, namely, improved methods of microelectrostatics. Before we proceed along these lines, however, the validity and the applicability limits of microelectrostatic approximations ought to be discussed.

The microelectrostatic approach of electronic polarization energy calculations in ionic crystals was first proposed by MOTT and LITTLETON [2.44] as early as at the end of the thirties (see also [2.45]). The excess charge carrier in such an approach was assumed to be localized on a particular ion and its interaction with the induced dipole moments on each of the surrounding polarizable ions was calculated in a self-consistent way.

Microelectrostatic methods were later successfully used by RITTNER et al. [2.46] for polarization energy calculations in ionic crystals, and by DRUGER and KNOX [2.47] for rare-gas solids. The first serious attempts to calculate electronic polarization in anthracene-type molecular crystals within the framework of microelectrostatic approach were made by LYONS and MACKIE in the early sixties [2.31,32]. These first calculations formed a theoretical basis for the Lyons phenomenological model (cf. Sect.2.2) and for further extended studies of microelectrostatic approach (cf. Sect.2.4).

The problem of the feasibility of static approximation in electronic polarization calculations was theoretically discussed in great detail by FOWLER [2.48].

FOWLER used TOYOZAWA's [1.152] electronic polaron theory and HAKEN-SCHOTTKY's [2.41] theory of r-dependent dielectric function for calculating electronic polarization in some high-resistance solids, such as alkali and silver halides and molecular crystals of noble gases. It was convincingly shown by FOWLER [2.48] that both dynamic theories actually may be reduced to the static MOTT-LITTLETON approximation [2.44]. This means that the classical microelectrostatic approach is perfectly adequate for the treatment of electronic polarization in solids possessing pronounced tendency towards charge carrier localization. In any case the "classical" value of electronic polarization energy should serve as upper limit of the value of P, since, allowing the excess carriers to move would, presumably, only diminish the magnitude of their polarization energies. In the limiting case, when the excess carrier moves so fast that the electronic orbitals of the neighboring molecules do not manage to get polarized, the energy of electronic polarization may approach zero value [2.48]. Thus a convenient criterion for the applicability of microelectrostatic approximation may be found from the uncertainty relations (2.6), (2.7) and (1.17). If the value of $\delta\tau$, or correspondingly, τ_h considerably exceeds the value of τ_e, ($\tau_h \gg \tau_e$), one is justified to consider electronic polarization as practically "inertialess" and apply static approximation for calculations of P.

In anthracene-type molecular crystals this condition appears to hold in practically all cases for thermalized charge carriers (cf. Fig.1.22). Only for "hot" charge carriers the value of τ_h or, correspondingly, of $\delta\tau$ may be of the same order of magnitude as τ_e. In this case the classical microelectrostatic approximation should be replaced by a more complex quantum dynamical approach. In the other extreme case when $\delta\tau < \tau_e$, the many-electron approach is reduced to one-electron approximation.

We see thus that the most complicated case, from the theoretical aspect, is just the intermediate one, when we have "hot" (yet not too "hot") charge carriers that form a wide spectrum of short-living dynamic states of electronic polarons. This wide dynamic polaron "band" of ca. 1-2 eV width is obviously positioned between the energy levels of "bare" and completely "dressed" charge carriers (see Sect.2.7.4).

It is interesting to note that in the static limiting case the results of calculation according to the dynamic theories of TOYOZAWA and HAKEN coincide with zero order values of Mott-Littleton's static approach [2.48]. Higher order static approximation yields more accurate results than dynamic theories at this limit. This is easily understood since, as said before, it is difficult to account for subtler effects, such as dipole-dipole interaction, using the dynamic approach.

This becomes particularly obvious if one applies Toyozawa's theory to electronic polarization in molecular crystals [2.43]. A treatment based on the tight-binding approximation of an excess carrier as localized on a particular molecule in the crystal automatically reduces the problem to a quasi-static approach. On the other hand, the value of this method is considerably impaired by the impossibility of accounting for dipole-dipole interaction, and preference must be given to more accurate calculations of improved microelectrostatic approximation (cf. Sect.2.4).

ČÁPEK [2.49] has recently discussed the problem of correspondence between the dynamic and microelectrostatic approaches to polarization in molecular solids. He has shown that quantum dynamical treatment of electronic polarization in molecular crystals (including both charge-induced dipole and dipole-dipole interaction terms) leads to expressions completely identical with the corresponding expressions of microelectrostatic approximation (cf. Sect.2.4.2). This demonstrates once more the equivalence of the dynamic and the microelectrostatic approach to electronic polarization energy calculations in molecular crystals.

2.4 Electrostatic Methods of Electronic Polarization Energy Calculation in Molecular Crystals

2.4.1 Microelectrostatic Methods of Zero-Order Approximation

Microelectrostatic methods of evaluating electronic polarization energies are semiempirical, since they employ for calculations experimental data, such as molecular polarizability, molecular dipole moments, crystallographic lattice parameters and optical dielectric constant of the crystal concerned.

Microelectrostatic approximation offers the advantage of presenting a whole choice of methods of calculation, providing various degrees of accuracy. Depending on the desired accuracy, the investigator may satisfy himself with a rough estimate of P, or he may use more accurate methods of zero-order approximation or, finally, he can resort to refined methods of self-consistent polarization field calculations (cf. Sect.2.4.2). In addition, there practically always exists a possibility of assessing the error involved using one or the other of approximations.

The present section gives a brief summary of calculation methods developed in the course of the last 8-10 years. Some earlier work, of rather historical interest at present, is discussed in detail by GUTMANN and LYONS [1.38].

In microelectrostatic approximation the electronic polarization energy P of a crystal may be represented as the sum of all kinds of electrostatic interaction between an excess charge carrier localized on a particular molecule (or a molecular ion), and the surrounding neutral molecules of the lattice. In zero-order approximation this sum may be presented in the following way

$$P = W_{i-d} + W_{d-d} + W_{\mu_0} + W_{Q_0} + W_S + W_{i-Q} + W_M \quad , \tag{2.8}$$

where W_{i-d} is the interaction energy between a localized excess charge carrier (ion) with electrical dipoles induced on surrounding molecules; W_{d-d} is the energy of dipole-dipole interaction between induced dipoles; W_{μ_0} is the interaction energy between the charge carrier (ion) and permanent dipoles of surrounding molecules; W_{Q_0} is the interaction energy between the charge carrier (ion) and permanent quadrupoles of surrounding molecules; W_{i-Q} is the interaction energy between the charge carrier (ion) and quadrupoles induced on surrounding molecules; W_S is the interaction energy due to superpolarization effects, i.e., deviation of molecular polarizability from linear in strong electrical fields; W_M is the energy contribution from higher order multipoles.

A detailed analysis of all the terms of (2.8) contributing towards electronic polarization P of a molecular crystal has been carried out by LYONS et al. [1.38,59] and HUG and BERRY [1.60,2.50].

The basic contribution towards electronic polarization energy P in nonpolar molecular crystals of the anthracene type is due to the first term of sum (2.8), viz. the term of charge-induced dipole interaction W_{i-d}.

The term W_{i-d} may be determined as follows [1.38]. Let the main axes of the kth molecule L, M and N (L being the longer, M the shorter, and N—the axis perpendicular to the plane of the molecule) be inclined to the x, y, z axes of the crystal (x corresponding to the a, y to the b, and z to the c' axis of the crystal), the corresponding cosines of the inclination angles being respectively (ℓ_1,ℓ_2,ℓ_3), (m_1,m_2,m_3) and (n_1,n_2,n_3). Then the polarization energy of the kth molecule $W_{i-d}(k)$ by a charge carrier localized on molecule (0,0,0) equals

$$W_{i-d}^{(k)} = \frac{e^2}{2r^6} [b_1 (\ell_1 r_x + \ell_2 r_y + \ell_3 r_z)^2 + b_2 (m_1 r_x + m_2 r_y + m_3 r_z)^2$$
$$+ b_3 (n_1 r_x + n_2 r_y + n_3 r_z)^2] \quad , \tag{2.9}$$

where b_i (i=1,2,3) are the main components of the tensor of molecular polarizability along the main axes L, M, and N of the molecules, respectively; r
is the distance from the ion (0,0,0) to the kth molecule.

The total energy of the term W_{i-d} is obtained as the sum of $W_{i-d}^{(k)}$ over
all N-1 molecules of the crystal:

$$W_{i-d} = \sum_{k=1}^{N-1} W_{i-d}^{(k)} = \sum_{k=1}^{N-1} \frac{e^2}{2r_k^6} \sum_{i,j} b_i (k_{ij} r_{jk})^2 \quad (i,j = 1,2,3) \quad , \tag{2.10}$$

where k_{ij} are the respective cosines of the angles of inclination.

For a rough estimate of the term W_{i-d} the tensor of polarizability b_i is
sometimes replaced by the mean isotropic polarizability α of the molecule:

$$\alpha = \frac{1}{3} \sum_{i=1}^{3} b_i \quad . \tag{2.11}$$

In this case the expression (2.10) becomes considerably simpler:

$$W_{i-d} = -\sum_{k=1}^{N-1} \frac{e^2 \alpha}{2r_k^4} \quad . \tag{2.12}$$

The term of interaction between induced dipoles W_{i-d} is as follows [1.38]:

$$W_{d-d} = \sum_{k,\ell} \frac{e^2 \alpha^2}{r_k^3 r_\ell^3} [\underline{r}_k \underline{r}_1 - 3r_{k\ell}^{-2} (\underline{r}_k \underline{r}_{k\ell}) (\underline{r}_\ell \underline{r}_{k\ell})] \quad , \tag{2.13}$$

where $\underline{r}_k$, $\underline{r}_\ell$ is the mean distance between the ion and the neutral molecules
k and ℓ, respectively; $\underline{r}_{k\ell}$ is the distance between molecules k and ℓ.

Dipole-dipole interaction partly screens off the field of the localized charge, hence the sign of the term W_{d-d} is opposite to that of the term W_{i-d}, i.e., the term W_{d-d} reduces the total energy of polarization P.

The contribution of the term W_{d-d} is rather considerable and can be as high as 30-40% in the case of aromatic hydrocarbon crystals [1.59]. One cannot therefore neglect the effect of dipole-dipole interaction in electronic polarization calculations, even in rough estimates.

Let us consider the possible influence of the other terms of the sum (2.8) which are usually disregarded in calculations.

Nonpolar molecules of the anthracene type do not possess a permanent dipole moment, and for them the term W_{μ_0} is zero. The situation is less simple with respect to multipoles of higher order. Centrally symmetrical molecules have no permanent dipole moment and no moments of odd order, but they possess a permanent quadrupole moment and other moments of even order [1.38].

Interaction energy between a localized charge carrier (ion) and the permanent quadrupole of the kth molecules changes proportionally to $1/r_k^3$ [1.38]:

$$W_{Q_0} = -\frac{eQ}{r_k^3} \quad . \qquad\qquad (2.14)$$

The authors of [1.59] attempted to estimate the total energy of the term W_{Q_0} for naphthalene and anthracene. If we assume that a naphthalene molecule has a quadrupole moment of cylindrical symmetry along the L axis and of magnitude 1×10^{-39} cm^2 (3×10^{-26} in cgse units), the summation over a spherical shell yields, at convergence limit, a value of $W_{Q_0} \approx 0.075$ eV [1.59]. Summation according to the assumption that the quadrupole moment has main axes parallel to the M and N axes of the naphthalene molecule yielded values of 0.026 and 0.049 eV, respectively.

For anthracene an estimate of the term W_{Q_0} gave a convergence value of 0.088 eV in the spherical shell. The authors of [1.59] concluded that neglect of the term W_{Q_0} introduces an error not exceeding 0.1 eV, i.e., ca. 6-7%.

Since exact values of the permanent quadrupole moments are not known at present for molecules of aromatic hydrocarbons, we are compelled to reconcile ourselves with this error for the time being in calculating the total energy of electronic polarization P.

Molecular superpolarizabilities β and γ are of importance only at high values of electric field intensity acting on the molecule. Therefore only molecules in the vicinity of the charged center contribute towards the energy of the term W_s. In the case of centrally symmetrical molecules superpolarizability $\beta = 0$, and W_s depends only on the second superpolarizability γ. GUTMANN and LYONS [1.38], using a γ value of the order of 10^{-35} cgse units for naphthalene, showed that the contribution of each molecule neighboring with the ion does not exceed 0.0003 eV. HUG [2.50], using γ values calculated by SCHWEIG [2.51] also showed that the contribution of term W_s towards the total energy is negligible.

Possible contributions from higher induced multipoles should also be considered, e.g., that of W_{i-Q}, i.e., the term due to an induced quadrupole. HUG [2.50] estimated the magnitude of quadrupole polarization for a molecule neighboring with a charged center in naphthalene and found that $W_{i-Q}^{(k)}$ in this case is of the order of 5×10^{-3} eV, which means that term W_{i-Q} can also be neglected.

Term W_M corresponding to even higher order multipoles can obviously be neglected as well.

The above analysis of specific contributions by separate terms of the sum (2.8) shows that only the first two terms of the sum, viz. W_{i-d} and W_{d-d}, must be considered in the calculation of the total electronic polarization energy of the crystal by a localized charge carrier.

In zero-order approximation such an approach is usually used. At first the term W_{i-d} is calculated from (2.10), then the term W_{d-d} is found from (2.13), and the total energy of electronic polarization P_0 in zero-order approximation is found as the sum of these two terms

$$P_0 = W_{i-d} + W_{d-d} \quad . \tag{2.15}$$

An acute problem in the calculation of the term W_{i-d} is that of convergence of sums (2.10) or (2.12) depending on N. On the one hand, the interaction energy between an ion and an induced dipole diminishes in proportion to r^4. On the other hand, the number of molecules inside a sphere of radius r increases rapidly with r.

It was already shown in earlier work (cf. [1.38]) that summation has to be performed over not fewer than 60 molecules, i.e., $N \gtrsim 60$ for correct estimation of W_{i-d}.

In more accurate calculations, e.g., in [1.60], summation was performed over $N = 7000$. The authors estimate the contribution on the part of all the remaining molecules ($N = 7000 \to \infty$) not to exceed 0.1 eV. This means that the polarization cloud spreads over a distance of some 12 unit cell parameters around the ion.

The effect of the chosen limits of summation (2.10) on the value of the term W_{i-d} has also been studied by BATLEY et al. [1.59]. The paper contains a refined calculation of electronic polarization energy P_0 for a number of linear acenes (from naphthalene to pentacene) and of other aromatic hydrocarbons. The term W_{i-d} was calculated according to (2.10) for a limited range of summation, namely for 3,6 and 10 unit lengths of an elementary cell. The sum W_{i-d} for $N \to \infty$ was estimated by linear extrapolation of term values W_{i-d} obtained for limited ranges of summation.

The results of this work show that if summation is performed within a region bounded off by a distance of 10 unit parameters of a unit cell from the ion, i.e., within the range of a parallelepiped of size $170 \times 120 \times 220$ $Å^3$, then the molecules outside this volume and belonging to an infinite crystal contribute, in fact, only some 0.1 eV towards the energy total.

The term of dipole-dipole interaction W_{d-d}, as shown in [1.59], rapidly converges with increasing distance from the ion and upon summation within a distance up to two unit cell parameters around the ion. This provides an accuracy up to $\pm$ 0.01 eV in the calculation of the toal energy of the term W_{d-d}.

In the above calculation of P_0 value in zero-order approximation after (2.10) and (2.13), the charge on the molecular ion is considered as a point charge, and the induced dipole moments on neutral molecules — as point-charge moments.

In the same work [1.59] an attempt was made to estimate the possible effect of actual charge distribution in the molecular ion on the value of the term W_{i-d}. The point charge in the center of the ion was replaced by charges distributed over the atoms of the given ion. These charges were calculated by LCAO MO method in the Hückel approximation (cf. Table 2.5). In this case the total energy of electronic polarization can be represented as follows:

$$P_1 = W_{i-d} + \Delta W_{i-d} + W_{d-d} \quad , \tag{2.16}$$

where ΔW_{i-d} is the calculated correction due to accounting for charge distribution in the ion.

The results obtained in [1.59] show that the calculated values of P_0 and P_1 differ considerably, and this difference increases with increase in size of the molecule, as evidenced by data on linear acenes from naphthalene to pentacene. A comparison of the calculated P_0 and P_1 values with average experimental electronic polarization energy P_{exp} yields an increased energy value for the point approximation of zero order P_0, and a considerably diminished value of electronic polarization energy when allowance is made for charge distribution in the ion.

The overall results of this work show that both methods of calculating P seem to overlook some essential point. For this reason the P_0 and P_1 values obtained deviate for anthracene and its higher analogues, that of P_0 being higher, and that of P_1 being lower than the experimental one, i.e., $P_1 < P_{exp} < P_0$ (cf. Fig.2.5).

The inconsistency of zero-order point approximation is evident. Such an approach assumes that the charge-induced dipoles remain unchanged in the field of other dipoles, and these initial zero-order dipole moments are used for calculating the term of dipole-dipole interaction W_{d-d}. This is, in fact, a very rough approximation. The initial charge-induced zero-order dipole moments $\mu^{(0)}$ change in the field of other neighboring dipoles, and the problem of finding the real value of μ has to be approached by means of self-consistent polarization fields of interacting dipoles.

2.4.2 Method of Self-Consistent Polarization Field

The idea of self-consistency in calculations of electronic polarization energy in microelectrostatic approximation was first proposed in the classical work by MOTT and LITTLETON [2.44] (see also [2.45]). This approach was later used by RITTNER et al. [2.46] for polarization energy calculation in ionic crystals and by DRUGER and KNOX [2.47] in rare-gas (argon, krypton and xenon) solids.

JURGIS and SILINSH [1.61,64] extended the method of self-consistent polarization field (SCPF) to electronic polarization energy calculations in anisotropic anthracene-type molecular crystals.

The essence of the SCPF method consists in considering every induced dipole as being in the field of the localized charge carrier (ion) and in the self-consistent field of other induced dipoles. In this case the polarization energy P of the localized charge carrier can be calculated in isotropic approximation according to DRUGER and KNOX [2.47] by means of the following expression:

$$P = -\sum_{i} \left[\underline{E}_0(r_i)\mu_i + \frac{\mu_i^2}{2\alpha} - \sum_{j>i} \mu_i \lambda_{ij} \mu_j \right] = -\frac{1}{2} \sum_{i} \underline{E}_0(r_i)\mu_i \quad , \qquad (2.17)$$

where $\underline{E}_0(r_i)$ is the charge carrier (ion)-created electric field intensity at the center of molecule i; μ_i, μ_j are the values of induced dipole moments on molecules i and j, respectively; α is the average isotropic polarizability of the molecule [cf. (2.11)]; λ_{ij} is the operator of dipole-dipole interaction:

$$(\lambda_{ij})^{\mu\nu} \equiv \frac{\left[r_{ij}^2 \delta_{\mu\nu} - 3(r_{ij})^{\nu} (r_{ij})^{\mu} \right]}{r_{ij}^5} \quad , \qquad (2.18)$$

r_{ij} being the distance between the dipoles.

The first term of (2.17) describes the interaction energy between the localized charge carrier (ion) and the induced dipole μ_i; the second one corresponds to the self-energy of the induced dipole μ_i and the third to the interaction energy between induced dipoles μ_i and μ_j.

According to the SCPF method the value μ_i of each induced dipole is calculated in a self-consistent field of all other dipoles by means of an iteration procedure.

In the extended SCPF calculations by JURGIS and SILINSH [1.61,64] the following approximations were used:

a) the isotropic SCPF approximation (2.17) was replaced by a modified approach in which anisotropic molecular polarizability of aromatic hydrocarbons was taken into account. The following main components b_i of the polarizability tensor were used: for naphthalene $b_1 = 21.5$ $Å^3$; $b_2 = 17.6$ $Å^3$; $b_3 = 10.1$ $Å^3$ [2.52],
for anthracene $b_1 = 33.9$ $Å^3$; $b_2 = 29.2$ $Å^3$; $b_3 = 12.9$ $Å^3$ [2.53] (cf. Table 2.2);

b) the point-charge and point-dipole approximations were used within the framework of a microelectrostatic model;

c) the total energy P of electronic polarization was calculated in two stages. At first, using the SCPF method, the polarization energy P' was calculated for a sphere of radius $r = 17.4$ $Å$ (for naphthalene) and of $r = 18.9$ $Å$ (for anthracene); this sphere contains 122 molecules surrounding the ion. After that the contribution ΔP of the molecules outside the sphere towards the total polarization energy P was estimated in macroscopic approximation using the expression

$$\Delta P = - \frac{e^2}{2r} (1 - \frac{1}{\varepsilon}) \quad , \tag{2.19}$$

where ε is the optical dielectric permeability of the crystal.

The total energy P of electronic polarization is then

$$P = P' + \Delta P \quad . \tag{2.20}$$

The following values of optical dielectric permeability were used for estimating ΔP: $\varepsilon = 3.00$ for naphthalene, and $\varepsilon = 3.55$ for anthracene (cf. Table 2.4).

The calculated values of the total electronic polarization energy P and of its components P' and ΔP, as obtained by the SCPF method in [1.61] for naphthalene and anthracene crystals and refined later in [1.64], are presented in Table 2.1.

Table 2.1. Calculated values of total electronic polarization energy P (and its components P' and ΔP) for naphthalene and anthracene crystals, as obtained by the SCPF method [1.61,64]

Crystal	P' [eV]	ΔP [eV]	P [eV]
Naphthalene	-0.99	-0.28	-1.27
Anthracene	-1.26	-0.27	-1.53

In the calculations of P for naphthalene and anthracene by the SCPF method the convergence in the iteration procedure was more than satisfactory. The estimated convergence error in this case did not exceed 0.02 eV [1.61].

In order to check the adequacy of the results obtained by the SCPF method the calculated P values (cf. Table 2.1) should be compared with reliable experimental P_{exp} values.

It is traditionally assumed that the most reliable method of checking calculated P values consists in using (2.1), viz. the basic postulate of the Lyons model (cf. Sect.2.2.1):

$$P_h = I_G - I_C \quad , \tag{2.21}$$

where I_G and I_C are the ionization energies of the molecule and crystal, respectively (cf. Fig.2.2).

I_G can be determined experimentally with high accuracy — to ca. 0.01-0.03 eV (cf. Table 2.11). This does not, however, apply to I_C which is found from crystal photoelectron emission measurements. As may be seen, e.g., from Table 2.13, the reported I_C values for anthracene differ considerably. Only if special precautions are taken, the experimental error of I_C determination can be diminished below 0.1 eV (see Sect.2.9 and also [1.144]). For this reason expression (2.21) may thus be used only for a rough assessment of the adequacy of calculated P value.

A different independent method for P or, rather P_e determination was proposed in [1.62], based on intrinsic photoconductivity threshold measurements.

The method employs the principle of equivalence of calculated values of electronic polarization energy for an electron (P_e) and a hole (P_h) in case of point-charge approximation in a homomolecular crystal of the anthracene type [cf. (2.10,12,13)]

$$P_e = P_h = P \quad . \tag{2.22}$$

(It may be shown that the equivalence (2.22) holds in anthracene-type crystals also in the case when P is calculated taking into account charge distribution on the ion. Quantum chemical calculations, as well as experimental NMR data [2.54] show that in such a case the charge distribution is symmetrical on the cation and anion).

For empirical determination of P according to (2.22), the following expression can be used [1.62]

$$P_e = P = \frac{I_G - \Delta E_G - A_G}{2} \quad , \tag{2.23}$$

where ΔE_G is the width of the energy gap, determined from the intrinsic photoconductivity threshold of the crystal; A_G is the electron affinity of the molecule.

Experimental data on ΔE_G have a smaller spread than those on I_C. For anthracene ΔE_G has been determined with an accuracy of ±0.1 eV (cf. Table 2.14). There are also reliable averaged data on A_G values for aromatic hydrocarbon molecules (cf. [1.95] and Table 2.12). It appears therefore that (2.23) gives a more reliable estimate of P than the widely used expression

(2.21). The main point is, however, that in the case of homomolecular crystals of the anthracene type, for which (2.22) holds true, expressions (2.21, 23) provide independent methods of determining P and a possibility of comparing the results.

Figure 2.5 shows a comparatively wide spread of experimental results, particularly of those obtained by the photoelectron emission method. Only the region of overlap of data obtained by both methods appears to be sufficiently reliable. This covers a range not exceeding ±0.1 eV. (For references on reported data of I_G, I_C, ΔE_G and A_G values used, see [1.62]; data on anthracene according to Tables 2.11-14).

As may be seen from Fig.2.5, the P values calculated by SCPF method for naphthalene and anthracene (marked by arrows) are in excellent agreement with experimental values of P determined independently according to (2.21, 23). This is good evidence in favor of the adequacy of SCPF calculations of electronic polarization energies of naphthalene and anthracene crystals. These crystals may consequently be used as model systems for reliable polarization energy calculations (cf. Sects.2.6 and 5.1).

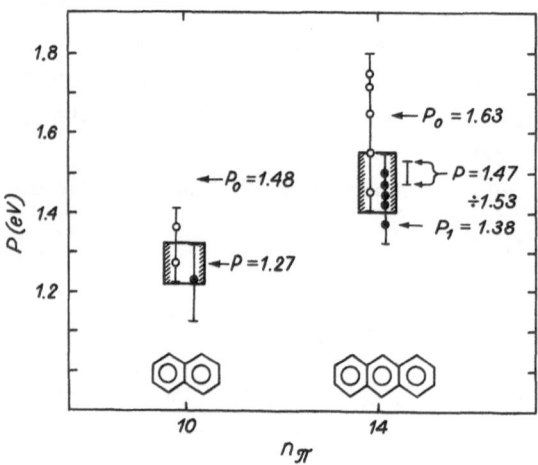

Fig. 2.5. Energy of electronic polarization P of a naphthalene crystal (number of π electrons n_π = 10), and of anthracene crystal (n_π = 14): o - experimental P values, obtained from photo-ionization data; • - those obtained from intrinsic photoconductivity data. P_0 - calculated values of zero-order approximation according to [1.59]. P_1 - calculated values of zero-order approximation with allowance for charge distribution over the ion according to [1.59]; P - calculated values of SCPF approximation according to [1.61,64]

Less satisfactory are the calculation results for higher linear acenes, viz. tetracene and pentacene. In this case the convergence is poor and the calculated P values are apparently overestimated. (Thus, e.g., we have for tetracene the calculated value P = 2.0 ± 0.2 eV, the experimental one being P_{exp} = 1.64 eV). Apparently it is the increased size of molecule and closest packing for triclinic space group (cf. Sect.1.3.3) which makes the point-charge point-dipole approximation nonvalid for higher linear acenes.

Figure 2.5 shows also that calculations using zero approximation [1.59] yield overestimated values of P = P_0. This is another argument in favor of applying the self-consistent polarization field method.

On the other hand, calculations using a model in which charge distribution is accounted for in the ion, but dipoles are treated in point approximation, yield obviously underestimated values of P = P_1 for anthracene and its higher analogues [1.59] (cf. Fig.2.5). It appears that such a "hybrid" model using a quantum-mechanical description of a distributed charge for the ion, and electrostatic point approximation for the dipoles is, physically speaking, too inconsistent to be feasible.

It stands to reason that an improvement of electronic polarization energy calculations, especially in the case of higher linear acenes and other multi-atomic molecules, may be found in the future by employing a three-stage procedure based on three mutually subordinated models of different degree of "subordination" in accuracy. At the first stage, the energy of interaction between the ion and the first coordination sphere of the nearest neighboring molecules should be calculated, preferably by quantum mechanical approach, yielding charge distribution on the ion, as well as on the polarized neighboring molecules. This should be followed by a calculation of interaction energy between the systems of distributed charges of the ion and the neighboring polarized molecules. The task is not a simple one, but well within the possibilities of modern quantum chemistry techniques.

The second stage of the procedure should consist in calculating the interaction between the ion and molecules outside the first coordination sphere, up to distances of r = 50-100 Å. Here, apparently, point approximation within the framework of SCPF method might be applicable.

Finally, in the third stage, the contribution of molecules outside the sphere of radius r should be evaluated in macroscopic approximation, using (2.19).

Such a refined "hybrid" method of calculating electronic polarization energy of molecular crystals is, most likely, going to be effectuated in the nearest future [1.64].

   In this connection it is interesting to mention that ČÁPEK [2.55] has recently evaluated the possible quantum effects in electronic polarization energy due to dipole-dipole correlations (and their distortion by the excess
charge carrier) between the ion and adjacent neutral molecules. It has been
found that the quantum corrections ΔP_{quant} due to dipole fluctuations are
positive ($\Delta P_{quant} > 0$) and of the order ~ 0.1 eV in anthracene and thus diminish the magnitude of $P(P < 0)$ calculated by microelectrostatic approximation. This partly explains the excellent agreement between the calculated
P and experimental P_{exp} values of electronic polarization energy in naphthalene and anthracene crystals (cf. Fig.2.5). The quantum corrections ΔP_{quant}
and possible contribution of neglected terms of higher multipoles [cf. (2.8)]
are of opposite sign and apparently compensate each other.

   On the other hand, the value of ΔP_{quant} should be increased in case of
higher linear acenes. This means that, taking into account the possible
quantum corrections, the overestimated values of P for higher polyacenes,
calculated in microelectrostatic approximation, might be improved.

   Recently BOUNDS and MUNN [2.56] have developed a novel method for polarization energy calculations which gives an explicit self-consistent algebra-
ic expression for the polarization energy of a point charge localized in a
perfect molecular crystal in terms of Fourier transformed lattice multipole
sums. An exact macroscopic expression is also derived for a long-range contribution. The results of polarization energy calculations by this approach
confirm those obtained by the SCPF method [1.61,64] for anthracene and naphthalene (see Fig.2.5) and are in good agreement with experimental data. The
method may be extended to calculate the forces and torques in molecules near
an ion and electron-phonon coupling due to polarization fluctuations [2.56].

2.5 Determination of Molecular Polarizability Tensor

In calculations of electronic polarization energy P in a molecular crystal
(cf. Sect.2.4) the main components of the molecular polarizability tensor
b_i ($i = 1,2,3$) along the principal axes L, M and N of the molecule are used
as parameters [cf. (2.9)].

   Reported values of b_i have been obtained by different experimental methods, as well as by semiempirical calculations of various degrees of perfection and accuracy. Therefore they differ considerably from each other (cf.

Table 2.2). This naturally raises the question of reliability criteria for these data.

Another problem arising in this connection is that of the possible effect of changes in the parameters of anisotropic polarizability b_i of a molecule on the calculated value of electronic polarizability energy P of a crystal.

It was also stated by CUMMINS et al. [2.57] that the principal axes of effective polarizability of a molecule in an anthracene crystal can deviate from the principal polarizability axes of an isolated molecule. This raises doubts as to the feasibility of using the components of the polarizability tensor of an isolated molecule for calculating P of a crystal.

We shall attempt to find an answer to these questions in the following sections.

2.5.1 Experimental Methods

The mean value of isotropic molecular polarizability α can be measured comparatively simply by determining the molecular refraction of a diluted solution of the given substance

$$\alpha = \frac{3}{4\pi N_A} R \quad , \tag{2.24}$$

where N_A is the Avogadro number; $\alpha = 0.3964R$ (in 10^{-24} cm^3 units).

The determination of the main components of the polarizability tensor b_i presents, however, a much more difficult task. Since the mean polarizability α of the molecule can be divided into three main components of anisotropic polarizability [cf. (2.11)], three independent methods of measurement must be applied for finding b_i in order to obtain three independent equations with respect to b_i.

LE FEVRE et al. used the independent methods of measuring molecular refraction, Kerr effect and molecular optical anisotropy in order to find the three polarizability components b_i for a number of polar molecules in diluted solutions of nonpolar solvents (cf. review articles [2.52,58], as well as [2.53,59,60]).

The values of b_i (i = 1,2,3) are found, in this case, from the following three independent equations:

$$P_E = R = \frac{4\pi N_A}{9} (b_1 + b_2 + b_3) \quad ; \qquad (2.25)$$

$$K_M = \frac{4\pi N_A}{405kT} \left\{ \frac{P_D}{P_E} \left[b_1 - b_2 \right)^2 + (b_2 - b_3)^2 + (b_3 - b_1)^2 \right]$$

$$+ \frac{1}{kT} \left[(\mu_1^2 - \mu_2^2)(b_1 - b_2) + (\mu_2^2 - \mu_3^2)(b_2 - b_3) \right.$$

$$\left. + (\mu_3^2 - \mu_1^2)(b_3 - b_1) \right] \right\} \quad ; \qquad (2.26)$$

$$\gamma^2 = \frac{1}{2} \left[(b_1 - b_2)^2 + (b_2 - b_3)^2 + (b_3 - b_1)^2 \right] \quad , \qquad (2.27)$$

where $P_E = R$ is the molecular electronic polarization [molecular refraction, cf. (2.11,24)]; K_M is the molecular Kerr constant; P_D is the deformational polarization of the molecules; μ_1, μ_2, μ_3 are the main components of molecular dipole moments along the principal axes of the molecule; γ^2 is the molecular optical anisotropy.

This method cannot, unfortunately, be applied directly for the determination of b_i values of nonpolar molecules for which $\mu_i = 0$. In such a case (2.26,27) become interdependent. These difficulties were ingeniously circumvented by LE FEVRE and his co-workers. They used (2.25,26) for direct determination of components b_1 and b_2 of nonpolar aromatic hydrocarbons. As regards component b_3, they found it by means of calculation using the dependence of b_3 on the number [2.53] or on length of C-C bonds [2.59] in the molecule. Another rather ingenious method proposed by LE FEVRE consists in the following. At first all three components b_i of a polar derivative of the given aromatic hydrocarbon are determined, using (2.25-27). (Such derivatives could be, e.g., bromo- or iododerivatives of anthracene). Then calculations are applied for determining the corresponding b_i values of the nonpolar molecule (e.g., anthracene [2.60].

CHENG et al. [2.61] suggested an alternative independent experimental method for determining b_i of nonpolar molecules, viz. measurements of magnetic anisotropy (Cotton-Mouton effect). In this case the following equation must be used instead of (2.26):

Table 2.2. Calculated and experimental values of the main components of the polarizability tensor b_i (i = 1,2,3), and of mean molecular polarizability α of naphthalene, anthracene, tetracene and pyrene molecules in the ground state (in 10^{-24} cm^3 units)

b_1	b_2	b_3	α	Method of determination	Ref.
	Naphthalene				
22.7	16.2	2.9	13.9	Calculated by CNDO method (cf. Table 2.5)	[2.65]
16.6	10.3	-	9.0	Calculated by PPP-SCF method;(π-electron approximation, (cf. Table 2.5)	[2.65]
21.5	17.6	10.1	16.4	Measurement of Kerr effect and molecular refraction in solutions	[2.52]
21.8	16.6	11.3	16.6		[2.60]
20.2	18.8	10.7	16.6	Measurement of Cotton-Mouton effect, molecular refraction and optical anisotropy in solutions	[2.61]
24.4	18.2	9.6	17.4	Double refringence method in crystal	[2.62]
	Anthracene				
40.3	26.7	4.10	23.7	Calculated by CNDO/S method	[2.65]
30.6	15.4	-	15.3	Calculated by PPP-SCF method; (π-electron approximation)	[2.65]
33.9	29.2	12.9	25.3	Kerr effect and molecular refraction measurements in solutions	[2.53]
35.9	24.5	15.9	25.4		[2.60]
35.2	25.6	15.2	25.3	Measurement of Cotton-Mouton effect, molecular refraction and optical anisotropy in solutions	[2.61]
35.8	25.5	13.9	25.1	Method not stated	[1.60]
40.2	25.7	11.9	25.9	Double refringence method in crystal	[2.62]
	Tetracene				
48.2	34.7	15.6	32.8	Measurement of Kerr effect and molecular refraction in solutions	[2.53]
	Pyrene				
34.6	29.0	17.0	26.8	Measurement of Cotton-Mouton effect, molecular refraction and optical anisotropy in solutions	[2.59]
34.2	34.2	16.3	28.2		[2.61]

Note: $b_1 = \alpha_L$, $b_2 = \alpha_M$, $b_3 = \alpha_N$, where L is long, M the short and N the axis normal to the plane of the molecule.

$$C_M = \frac{2\pi N_A}{45kT} [(k_1 - k_2)(b_1 - b_2) + (k_2 - k_3)(b_2 - b_3)$$

$$+ (k_3 - k_1)(b_3 - b_1)] \quad , \tag{2.28}$$

where C_M is the Cotton-Mouton constant; k_1, k_2, k_3 are the main components of the tensor of molecular magnetic susceptibility.

Table 2.2 shows that b_i values obtained by CHENG et al. [2.61] are in good agreement with the data of LE FEVRE et al. [2.52,53,60]. This indicates the equivalence of both methods for determining b_i in aromatic hydrocarbons.

The methods discussed above yield anisotropic polarizability b_i for isolated molecules. However, these b_i values might undergo changes with formation of a crystal owing to intermolecular interactions.

An interesting approach of determining anisotropic polarizability of molecules in a crystal was proposed by VUKS [2.62]. The author used double refringence, based on the following assumptions.

For anisotropic crystals a modified Lorentz-Lorenz formula is usually applied. It has no strict theoretical backing but has found solid experimental verification

$$\frac{n_g^2 - 1}{n^2 + 2} + \frac{n_m^2 - 1}{n^2 + 2} + \frac{n_p^2 - 1}{n^2 + 2} = \frac{4}{3} \pi N_A (\alpha_g + \alpha_m + \alpha_p) \quad , \tag{2.29}$$

where n_g^2, n_m^2, n_p^2 are the main components of the optical dielectric permeability tensor along the g, m, p axes of the optical indicatrix; α_g, α_m, α_p are the main components of the molecular polarizability tensor along the g, m, p axes; n^2 is the mean optical dielectric permeability

$$n^2 = \frac{1}{3} (n_g^2 + n_m^2 + n_p^2) \quad . \tag{2.30}$$

VUKS [2.62] proposed a modified approximated formula which is in complete agreement with (2.29) and directly correlates separate components of the two tensors viz. n_i^2 and α_i. The formula allows one to determine the anisotropic molecular polarizabilities α_g, α_m and α_p from refraction measurements of the crystal

$$\frac{n_i^2 - 1}{n^2 + 2} = \frac{4}{3} \pi N_A \alpha_i \quad (i = g,m,p) \quad . \tag{2.31}$$

After the polarizabilities α_g, α_m, α_p along the axes of the dielectric permeability tensor g, m, p have been found, the main molecular polarizabilities α_L, α_M, α_N along the L, M and N axes of the molecules can be determined from the following transformations:

$$\alpha_g = \alpha_L(L_g)^2 + \alpha_M(M_g)^2 + \alpha_N(N_g)^2 \quad ;$$

$$\alpha_m = \alpha_L(L_m)^2 + \alpha_M(M_m)^2 + \alpha_N(N_m)^2 \quad ; \tag{2.32}$$

$$\alpha_p = \alpha_L(L_p)^2 + \alpha_M(M_p)^2 + \alpha_N(N_p)^2 \quad ,$$

where $(L_g)^2 = \cos^2 (L_g)$, and so on.

In [2.62] experimental values of anisotropic refraction coefficents n_i were used for evaluating, according to (2.31,32), the main components of molecular polarizability α_g, α_m, α_p along the axes of the dielectric permeability tensor, as well as the polarizability components α_L, α_M, α_N along the molecular axes in naphthalene and anthracene crystals.

This calculation requires, at first, finding the directions of optical indicatrix axes g, m, p and of angles formed by these axes with the L, M, N axes of the molecules. This is done using crystallographic data on the direction of the L, M, N axes of the molecule with respect to the crystal axes a, b, c (cf. Sect.1.3.1). Table 2.3 contains, as an example, the values of angles between the axes g, m, p and L, M, N for an anthracene crystal.

For an anthracene crystal of C_{2h}^5 symmetry the b axis is the axis of symmetry and therefore coincides with one of the axes of the optical indicatrix, in this case with the m axis. The g axis runs along the bisectrix of the angle between the long axes L and L' of two molecules in the unit cell.

Table 2.3. Values of angles (in degrees) between the principal axes g, m, p of the dielectric permeability tensor, and the L, M, N axes of the molecule in an anthracene crystal [2.62]

Axis	L	M	N
g	7	83.7	93.1
m	97	26.6	115.5
p	90	64.2	25.8

The calculated values of α_g, α_m, α_p and of α_L, α_M, α_N, as well as mean values of α for naphthalene and anthracene crystals are presented according to [2.62] in Table 2.4. In his calculations the author of [2.62] used reported data on experimental anisotropic refraction coefficients of naphthalene [1.90] and anthracene [2.63,64] crystals, measured at λ = 546 nm (cf. Table 2.4).

Table 2.4. Values of refractive indices n_i and of molecular polarizability along the optical indicatrix axes α_i (i = g,m,p) and along the molecular axes α_j (j = L,M,N) for naphthalene and anthracene crystals [2.62].

Crystal	n_i (i=g,m,p) at λ=546 nm			$n^2 = \varepsilon$	α_i (i=g,m,p) [10^{-24} cm^3]			α_j (j=L,M,N) [10^{-24} cm^3]			$\bar{\alpha}$ [10^{-24} cm^3]
	n_g	n_m	n_p		g	m	p	L	M	N	
Naptha-lene	1.945 [1.90]	1.722 [1.90]	1.525 [1.90]	3.02	23.9	16.9	11.4	24.4	18.2	9.6	17.4
Anthra-cene	2.22 [2.63,64]	1.816 [2.63,64]	1.557 [2.63,64]	3.55	40.0	23.4	14.5	40.2	25.7	11.9	25.9

Note: $n^2 = \varepsilon$ is the mean optical dielectric permeability of the crystal; $\bar{\alpha}$ is the mean isotropic polarizability of the molecule. $\alpha_L \equiv b_1$; $\alpha_M \equiv b_2$; $\alpha_N \equiv b_3$ (cf. Table 2.2). NAKADA [2.66] later obtained the following values of refractive indices n_i for anthracene crystal at λ = 589.3 nm: n_g = 2.04 ± 0.08; n_m = 1.78 ± 0.01; n_p = 1.55 ± 0.01

As can be seen from Table 2.4, the values of anisotropic molecular polarizability along the optical indicatrix axes g, m, p and along the corresponding L, M, N axes of the molecule differ only within one percent. But there is another much more important result. A comparison between the data in Tables 2.2 and 2.4 clearly shows that the values of anisotropic molecular polarizabilities α_L, α_M, α_N are pretty close, both for isolated molecules and for molecules in a crystal. The values of α_M and of the mean polarizability α coincide pretty closely in both cases, whilst the α_L component is only slightly larger in the crystal (by ca. 10-15 %), and the α_N component only slightly smaller than in case of an isolated molecule.

This shows that the anisotropic polarizabilities of a molecule do not undergo any substantial changes on formation of a crystal. One is therefore entitled to use α_L, α_M, α_N values for isolated molecules in calculations of electronic polarizability of a crystal. This is not altogether surprising, considering the weak intermolecular interaction in anthracene-type crystals, which cannot affect to any considerable extent the parameters of the electronic structure of the molecule (cf. Sect.1.4).

2.5.2 Theoretical Methods

Semiempirical polarizability calculations of conjugated molecules are based on the assumption that it is possible to separate the polarizability of σ and π electrons .

σ electrons are tightly localized, and their polarizability α_σ retains additivity. Hence, α_σ can be evaluated empirically by summing the polarizability of separate σ bonds.

π electrons, on the other hand, are delocalized, and their polarizability α_π is not subject to additivity. It is therefore necessary to calculate α_π for a particular molecule.

Let us discuss briefly the basic principles for calculating π-electronic polarizability α_π. Such a calculation considers only change in π-orbital energy.

If the frequency of the electromagnetic field does not correspond to the electronic transition band of the molecule, then the field effect on orbital energy is inconsiderable and can be treated by means of perturbation theory. In this treatment the first-order perturbation term disappears, and electronic polarization can be expressed as second-order perturbation [2.67-69]

$$(\alpha_\pi)_x = 4e^2 \sum_{p=1}^{m} \sum_{q=m+1}^{n} \frac{<\psi_p|x|\psi_q>}{E_q - E_p} \quad , \tag{2.33}$$

where $(\alpha_\pi)_x$ is the π-electron polarizability along the x coordinate; E_q is the energy of the unoccupied q orbital; E_p is the energy of the occupied p orbital; $<\psi_p|x|\psi_q>$ is the matrix element in the x direction between wave functions ψ_p and ψ_q, representing, respectively, ground state E_p and excited state E_q of the π electrons.

The molecular orbitals of the π electrons can be represented, according to LCAO MO theory, as linear combinations of atomic wave functions

$$\psi_p = \sum_j c_{pj}\varphi_j \quad (p = 1,2,\ldots,n) \quad , \tag{2.34}$$

where c_{pj} are orthogonalized coefficients of the pth molecular orbital on the jth atom.

Using (2.34), the matrix elements in (2.33) can be expressed in the following form:

$$<\psi_p|x|\psi_q> = \sum_{j=1}^{n} c_{pj}c_{qj}x_j \quad , \tag{2.35}$$

where x_j is the x coordinate of the jth atom.

From (2.35) we obtain the following expression for polarizability $(\alpha_\pi)_x$:

$$(\alpha_\pi)_x = 4e^2 \sum_{p=1}^{m} \sum_{q=m+1}^{n} \sum_{j=1}^{n} \sum_{k=1}^{n} \frac{c_{pj}c_{qj}c_{pk}c_{qk}}{E_q - E_p} x_j x_k \quad , \tag{2.36}$$

where c_{pk} and c_{qk} are the coefficients of the pth and q^{th} molecular orbital on the kth atom, and x_k is the x coordinate of the latter.

Taking into account atom-atom polarizability, the expression (2.36) can be transformed, according to Coulson and Longuett-Higgins, in the following way [2.67-69]:

$$x_{jk} = -4 \sum_{p=1}^{m} \sum_{q=m+1}^{n} \frac{c_{pj}c_{qj}c_{pk}c_{qk}}{E_q - E_p} \quad . \tag{2.37}$$

Using (2.37) the expression (2.36) can be simplified:

$$(\alpha_\pi)_x = -e^2 \sum_j \sum_k x_{jk} x_j x_k \quad . \tag{2.38}$$

Similar expressions can be obtained for the y and z coordinates:

$$(\alpha_\pi)_y = -e^2 \sum_j \sum_k x_{jk} y_j y_k \quad ; \tag{2.39}$$

$$(\alpha_\pi)_z = -e^2 \sum_j \sum_k x_{jk} z_j z_k \quad . \tag{2.40}$$

We see thus that the calculation of polarizability of π electrons in conjugated molecules is reduced to calculation of molecular orbitals of the ground and the excited states and to determining the corresponding energies.

Any standard LCAO MO method can be used for calculating α_π in π approximation (cf. Table 2.5). The simplest one is the Hückel method. More accurate results are obtained by the LCAO MO self-consistent field (SCF) method such as the PPP-SCF method in π approximation [2.65] (cf. Tables 2.2,5).

To the calculated values of π electron polarizability of the molecule those of σ electrons must be added. The value of α_σ can be found empirically as the sum of σ-bond polarizabilities, if their polarizability components along the x, y, z axes are known.

σ-bonds have cylindrical symmetry, hence their x and y components are equal. Therefore one usually considers the polarizability parallel ($\parallel$) and at right angles ($\perp$) to the direction of the σ bond.

Table 2.5. Brief characteristics of the principal methods of quantum chemical calculations in molecular orbital theory approximation (LCAO MO) [cf. (2.34)]

Method	Basic characteristics	References
Hückel method	Semiquantitative method; considers π electrons only	Described in detail in [2.71, 72]
Pople-Pariser-Parr SCF method (PPP-SCF)	Semiempirical method using self-consistent-field approximation; considers π electrons only	Detailed description in [2.73-77]
Extended Hückel method (EHM)	Empirical method of calculation; considers σ and π electrons	[2.78,79]
Complete neglect of differential overlap (CNDO)	Semiempirical method using self-consistent-field approximation; considers all σ and π valency electrons of the molecule. The various modifications of the method (e.g., CNDO-1, CNDO-2, CNDO-S, etc.) differ in parametrization of calculations	[2.80]. Detailed description in [2.81]
Intermediate neglect of differential overlap (INDO)	A more accurate semiempirical method, as compared to CNDO. INDO method has a number of advantages in the calculation of electron spin distribution	Cf. [2.81]
Ab initio calculation	Complete calculation without any empirical parameters, using self-consistent field approximation; considers all electrons of the molecule	Cf., e.g. [2.82]

Table 2.6 presents experimental values of α_σ for some typical σ bonds. These values can be used for an additive estimate of the contribution of σ bonding towards the total polarizability of conjugated molecules. Thus, for instance, using the data from Table 2.6, we obtain the mean total polarizability of σ electrons in an anthracene molecule $\alpha_\sigma = 13.4 \times 10^{-24}$ cm^3. If this value is added to the calculated value of π electron polarizability obtained by the PPP-SCF method after MATHIES and ALBRECHT [2.65], i.e., $\alpha_\pi = 15.3 \times 10^{-24}$ cm^3 (cf. Table 2.2), then we have for the mean total polarizability of an anthracene molecule a value of $\alpha = \alpha_\sigma + \alpha_\pi = 28.7 \times 10^{-24}$ cm^3. A comparison with Table 2.2 shows that this method yields results not inferior to those obtained by CNDO calculations, and the α value thus obtained is close to experimental.

Table 2.6. Mean polarizability α_σ and its longitudinal $\alpha(||)_\sigma$ and perpendicular $\alpha(\perp)_\sigma$ components for some typical σ bonds [2.69]

Type of σ bond	$\alpha_\sigma \times [10^{-25} \text{ cm}^3]$		
	α_σ	$\alpha(||)_\sigma$	$\alpha(\perp)_\sigma$
C-C	5.13	7.8	3.8
C-H	6.64	8.72	5.6
N-H	7.53	5.8	8.4
C-N	6.10	-	-
C=O	7.28	6.84	7.5

Table 2.6 shows that σ polarizability is almost isotropic. The main contribution in anisotropic polarizability is provided by π electrons (cf. Table 2.2).

For more accurate quantum chemical calculations CNDO and INDO methods are used (cf. Table 2.5), in which polarizability of all valency electrons, both σ and π, are accounted for [2.65] (cf. Table 2.2).

According to Table 2.2, CNDO calculations of b_i after MATHIES and ALBRECHT [2.65] yield satisfactory agreement with experiment.

A comparison of b_i values calculated by CNDO and PPP-SCF methods (cf. Table 2.2) shows that the basic contribution toward the polarizability value of such molecules as naphthalene and anthracene is due to π electrons, especially along the L and M axes, amounting to ca. 70% of the total polarizability value.

An alternative semiempirical method for calculating effective molecular polarizabilities in anisotropic molecular crystals, the so-called local electric field method [2.70], was used by CUMMINS et al. [2.57]. The authors found that the principal axes P,Q,R of effective polarizability of molecules in anthracene crystals are situated at an angle of ca. $10°$ with respect to the principal axes L, M, N of the molecule. The calculated values of molecular polarizability components in anthracene are, according to [2.57] $\alpha_L = 66.7 \times 10^{-24} \text{ cm}^3$; $\alpha_M = 21.4 \times 10^{-24} \text{ cm}^3$; $\alpha_N = 12.3 \times 10^{-24} \text{ cm}^3$, the mean polarizability $\bar{\alpha}$ being equal to $33.5 \times 10^{-24} \text{ cm}^3$. This shows that the calculations in [2.57] yield a considerably overestimated value of the component $\alpha_L = b_1$. It exceeds more than twice the b_1 value of an isolated molecule [2.53,60,61], as well as the value b_1 obtained by means of double refringence of the crystal [2.62] (cf. Table 2.2). The calculated mean polarizability of anthracene $\bar{\alpha} = 33.5 \times 10^{-24} \text{ cm}^3$, as obtained in [2.57] is also considerably higher than the experimental value of $\bar{\alpha}$ presented in Table 2.2 where it varies only within a rather narrow range, viz. $\alpha = (24-26) \times 10^{-24}$ cm^3.

Such a pronounced discrepancy in results obtained by the local field method in [2.57], as compared to experimental data and quantum chemical calculations, is, most likely, due to the kind of approximations employed in this method.

Therefore it is difficult to agree with the claims of the authors of [2.57] that effective polarizability of a molecule in a crystal differs considerably from that of an isolated molecule. The data in Table 2.2 provide ample evidence to the contrary. We therefore suggest that one can safely use experimental data on molecular polarizability according to Table 2.2 and consider them as sufficiently reliable for calculating electronic polarization energy P in anthracene-type crystals.

2.6 Selection of Molecular Polarizability Components b_1 for Electronic Polarization Energy Calculations

Table 2.2 shows that the reported data on polarizability tensor components b_i, as obtained by various methods, present a rather wide range of different values for both model components —naphthalene and anthracene. The question arises, naturally, to what extent does a change in b_i values affect the calculated value of electronic polarization energy P, and which set of b_i may be regarded as the most reliable one.

A detailed analysis of the effect of using different sets of reported polarizability tensor components b_i on calculated values of P in anthracene has been carried out by SILINSH and JURGIS [1.64].

The SCPF method [1.61] was used in these studies (cf. Sect.2.4.2), the region of self-consistency being limited to a sphere of radius R = 21.4 Å. The total polarization energy P was determined as the sum of P' and ΔP according to (2.20). For estimating ΔP [cf. (2.19)] the value of ε = 3.55 was used (cf. Table 2.4).

Different reported sets of b_i values for anthracene (cf. Table 2.2) used for calculations, as well as calculated values of total electronic polarization P and its components P' and ΔP are presented in Table 2.7.

90

Table 2.7. Effect of different sets of polarizability tensor components b_i on calculated values of total electronic polarization energy P (and its components P' and ΔP) for an anthracene crystal [1.64]

No. of b_i set	b_i [10^{-24} cm^3]			Refer-ence	P' [eV]	ΔP [eV]	P [eV]
	$b_1=\alpha_L$	$b_2=\alpha_M$	$b_3=\alpha_N$				
1	33.9	29.2	12.9	[2.53]	1.29	0.24	1.53
2	35.9	24.5	15.9	[2.60]	1.26	0.24	1.50
3	35.2	25.6	15.2	[2.61]	1.26	0.24	1.50
4	40.2	25.7	11.9	[2.62]	1.23	0.24	1.47
5	40.3	26.7	4.10	[2.65]	1.12	0.24	1.36

Table 2.7 demonstrates that the use of experimental b_i sets for an isolated molecule (sets No.1-3) and the b_i set for a crystal (set No.4) in the calculation of electronic polarization energy P of an anthracene crystal yields P values coinciding within ± 0.03 eV, viz. P = 1.50 ± 0.03 eV. This range of calculated P values obtained using different experimental b_i sets, according to Table 2.7, is depicted in Fig.2.5. It shows that the calculated P values lie within the limits of reliably established experimental values of P. Consequently, the experimental sets of b_i value of an isolated molecule can be safely used in calculations of electronic polarization energy in a crystal, the error not exceeding 5%.

The sets of b_i values obtained theoretically by the CNDO method (cf. Table 2.7, set No.5) yield a considerably underestimated value of P. As follows from Table 2.7, the calculated value of P is most strongly dependent on the polarizability components b_2 and b_3 which determine the polarization interaction with the nearest neighbors in the ab plane of the crystal. Since it is just the b_3 component which has a lower value in CNDO calculations (cf. Table 2.7, set No.5), a corresponding lowering of calculated P value must necessarily result. On the other hand, the b_1 component, which determines mainly polarization interaction between ab planes, only slightly affects the total energy P of electronic polarization of the crystal.

The following conclusions, essential for calculations of electronic polarization energy in molecular crystals, can be drawn from an analysis of the data presented in Tables 2.1, 7, and Fig.2.5.

1) The SCPF method in point approximation yields sufficiently reliable values of electronic polarization energy P for anthracene and naphthalene crystals. An anthracene crystal can therefore be used as a model system for SCPF electronic polarization energy calculations in approaching various problems of charge carrier self-energy in an ideal and a defective molecular crystal.

2) Experimental values of the main components of the polarizability tensor b_i of an isolated molecule can safely be used in calculations of electronic polarization energy P of a molecular crystal.

2.7 Extended Polarization Model of Ionized States in Molecular Crystals

2.7.1 Intrinsic Electronic Polarization of a Molecule by a Localized Charge Carrier

In the modified Lyons model (cf. Sect.2.2.2) empirical molecular parameters I_G^V and A_G^e and electronic polarization energies of the crystal by electron P_e and hole P_h are used for the construction of an energy diagram of ionized states of an ideal molecular crystal (cf. Fig.2.3).

As has been shown in Sect.2.4, P values can be obtained through calculation, as well as from experiment (cf. Fig.2.5). It is less simple in the case of molecular parameters I_G^V and A_G^e. These empirical values are obtained as the energy difference between particular electronic states of the molecule, and they do not bear any direct information on the origin of these states. An additional analysis of the physical nature of these states is therefore necessary.

The vertical ionization energy I_G^V of a molecule can be calculated theoretically by the self-consistent field (SCF) method in Hartree-Fock one-electron approximation (cf. Table 2.5). Such a calculation is based on the well-known KOOPMANS theorem [2.83]. According to this theorem the orbital energy ε_i of the ith ground state, as calculated in Hartree-Fock approximation, is equal in magnitude and opposite in sign to the ionization energy I_G^V of the given state

$$\varepsilon_i = -(I_G^V)_K \quad , \tag{2.41}$$

where index V denotes the vertical ionization energy of the molecule, at which equilibrium configuration of nuclei in the neutral molecule is conserved, and index K shows that the given ionization energy has been determined according to the Koopmans theorem.

However, (2.41) holds only if in the process of ionization the intrinsic polarization of the electron orbitals of the created positive ion does not occur. HOYLAND and GOODMAN [2.84] showed theoretically that the Koopmans theorem definitely does not hold in the case of aromatic hydrocarbons. The σ and π orbitals of the positive ion get deformed, and the corresponding energy W_P^h of intrinsic electronic polarization of the molecule lowers the ionization energy of the molecule with respect to the theoretical value of ionization energy $(I_G^V)_K$ according to Koopmans' approximation:

$$I_G^V = (I_G^V)_K - W_P^h \quad , \tag{2.42}$$

where I_G^V is the experimentally observed value of vertical ionization potential of the molecule (see Fig.2.6a).

According to theoretical estimates in [2.84], the energy of intrinsic electronic polarization energy W_P^h of an anthracene molecule by a localized positive charge equals 1.1 eV, i.e., it is approximately of the same order as the total electronic polarization energy value of surrounding molecules in the anthracene lattice: P_h = 1.5 eV (cf. Table 2.1 and Fig.2.5).

The diagram of Fig.2.6a shows, however, clearly that the intrinsic electronic polarization of the molecule by a localized positive charge is accounted for automatically if the experimental value I_G^V is used for the construction of the energy diagram of ionized states.

The conclusions obtained in [2.84] on the inadequacy of Koopmans' theorem in the case of aromatic hydrocarbons were confirmed at a later stage by CNDO [2.85], as well as by ab initio [2.82] calculations (cf. Table 2.5) which provided more accurate data for comparison between calculated $(I_G^V)_K$

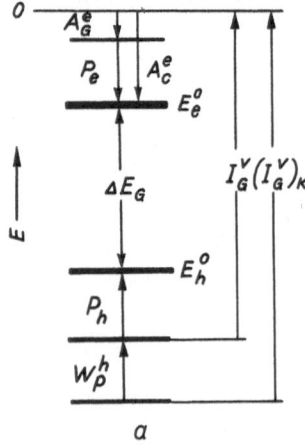

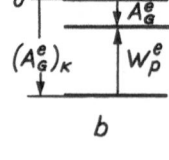

Fig. 2.6. Schematic diagram of electronic ionized states in a molecular crystal, taking into account self-polarization effects of the molecule by a localized hole (a) or electron (b)

and experimental I_G^V values. Thus, ab initio calculations according to
BUENKER and PEYERIMHOFF [2.82] yield for a naphthalene molecule a value of
$(I_G^V)_K$ = 9.30 eV, whilst the experimental value is only 8.12 eV [2.86].
Hence, the energy gain due to intrinsic electronic polarization of the naphthalene molecule by a localized positive charge is $W_P^h \approx 1.2$ eV.

A similar situation exists concerning another important energy parameter
— the electron affinity A_G^e of the molecule. According to Koopmans' theorem
the orbital energy of the unoccupied $(i + 1)$st level ε_{i+1} is equal to the
electron affinity A_G^e of the molecule

$$\varepsilon_{i+1} = (A_G^e)_K \quad , \tag{2.43}$$

where A_G^e is the electronic component of electron affinity of the molecule
[cf. (2.4)].

As shown by the authors of [2.84], expression (2.43) also does not hold
for aromatic hydrocarbons. The energy of intrinsic polarization W_P^e of the σ
and π orbitals of the molecule by a localized electron lowers the calculated
value of $(A_G^e)_K$, and the experimentally observed value of the molecule's
electron affinity is expressed as

$$A_G^e = (A_G^e)_K - W_P^e \quad . \tag{2.44}$$

This is illustrated by the diagram in Fig.2.6b.

For an anthracene molecule the estimated W_P^e value, according to [2.84],
equals ca. 1.8 eV.

If, as can be seen from Fig.2.6, the experimental value of electron affinity A_G^e of the molecule is used for constructing the ionized state energy
diagram of a crystal, the effect of intrinsic polarization of the molecule
by a localized electron is automatically accounted for.

The inadequacy of Koopmans' theorem is an additional proof of the inapplicability of one-electron approximation of the band theory to aromatic hydrocarbon crystals. If one-electron approximation is inapplicable for molecules, it must be all the more so for crystals consisting of such molecules
(cf. [2.87]).

2.7.2 Vibronic Relaxation and Ionic State Formation

According to the modified Lyons model (cf. Sect.2.2.2 and Fig.2.3) the ionized states in an ideal molecular crystal are primarily considered as purely
electronic conduction states, characterized by full conservation of equilibrium configuration of the neutral molecule nuclei coordinates during the
time τ_h of charge carrier localization. (Such purely electronic conductivity
states will be further denoted as E_h^0 for holes and E_e^0 for electrons, see
Figs.2.6, 7). Such separation is convenient and is based on different
charge carrier interaction time scales with electrons (electronic polarization) and intramolecular vibrations (vibronic coupling) (see Sect.1.5.5 and
Fig.1.22). If, however, the charge carrier localization time τ_h is sufficiently large, as compared to vibronic relaxation time τ_v, viz. $\tau_h \gtrsim \tau_v$,
strong vibronic coupling can take place resulting in formation of a genuine
ionic state.

As shown by DEWAR et al. [2.88,89], vibronic relaxation of a molecule
leads to a redistribution of the localized charge which, in turn, causes
changes in bond length and in corresponding vibration frequencies. As a result the molecule passes from the equilibrium configuration of atomic nuclei
of the neutral state into an equilibrium configuration of the ionic state.
The energy gain due to vibronic relaxation (similarly as in the case of
electronic polarization) lowers the self-energy value of the localized
charge carrier and determines the position of the ionic conductivity states
in the energy diagram of the crystal.

An extended polarization model of ionized states in molecular crystals,
including vibronic relaxation effects and corresponding ionic conductivity
states, has been proposed by SILINSH [1.69] (see Fig.2.7).

According to [1.69] the extreme modulation of the self-energy value of
the localized charge carrier due to vibronic relaxation can be evaluated in
static approximation from the bonding energy E_b of an electron by a neutral
molecule, gained in the process of ionic state formation

$$(M_0 e)^- \rightarrow M^- + E_b \quad . \tag{2.45}$$

This process is schematically depicted in Fig.2.7b for a single vibrational mode.

The equilibrium configuration coordinates q_i of the ion and q_0 of the
neutral molecule are connected through the corresponding intramolecular

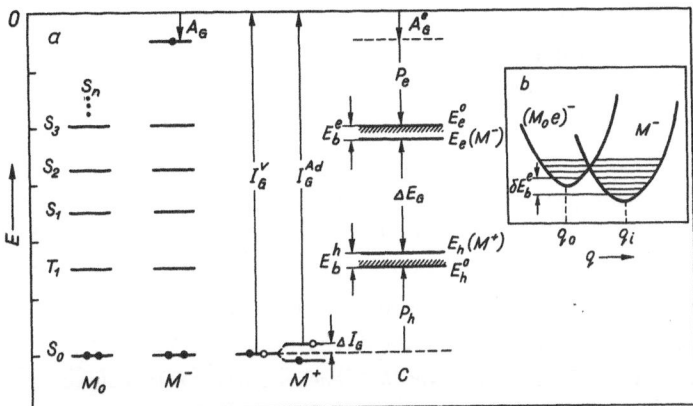

Fig. 2.7. Energy level diagram of an ideal molecular crystal with allowance for vibronic relaxation and formation of ionic states, according to an extended polarization model (a). The inset (b) gives schematic representation of electron bonding upon formation of the ionic state (figure shows potential curves for one separate vibrational mode only) [1.69,2.39]

vibration frequences ω_i of the ion and ω_0 of the neutral molecule by the well-known empirical Mecke relation (cf. [2.90])

$$q_i^2 \omega_i = q_0^2 \omega_0 = \text{const} \quad . \tag{2.46}$$

Thus, e.g., the formation of an ionic state in aromatic hydrocarbons leads to lengthening of C-C and C-H bonds ($q_i > q_0$, cf. Fig.2.7b). This causes lowering of their characteristic frequencies, as compared to the neutral molecule ($\omega_i < \omega_0$). It is, accordingly, possible to estimate the value of energy E_b for zero vibrational energy ($v = 0$, v being the vibrational quantum number) by considering the total frequency change of all normal vibration modes of the molecule passing from the neutral state into the ionic one

$$E_b = \sum_{k=1}^{3N-6} (\delta E_b)_k = -\frac{\hbar}{2} \sum_{k=1}^{3N-6} [(\omega_0)_k - (\omega_i)_k] \quad , \tag{2.47}$$

where δE_b is the change in energy for the kth vibrational mode (cf. Fig. 2.7b); N is the number of atoms in the molecule. The summation must be performed over all (3N-6) normal vibrational modes of the molecule.

Experimentally the characteristic frequencies of the molecule $(\omega_0)_k$ and of its ion $(\omega_i)_k$ can be found from vibrational infrared and Raman spectra.

For linear polyacenes infrared and Raman spectra have so far been studied mainly with respect to neutral molecules. Reliable, if insufficient data on ion spectra have been reported only in recent years [2.93,95-97].

In the series of linear polyacenes vibrational modes have been studied in great detail only for naphthalene [2.91,92,94]. The naphthalene molecule has D_{2h} symmetry (cf. Sect.1.3.2) and 48 normal vibrational modes. Of the latter 20 are active in ir absorption (vibrations of symmetry type B_{1u}, B_{2u} and B_{3u}), 24 are active in Raman scattering (vibrations of symmetry type A_g, B_{1g}, B_{2g} and B_{3g}). Four modes (of symmetry type A_u) are not optically active [2.91,92]. Almost all optically active vibrational modes of naphthalene have been reliably identified. As regards the naphthalene ion, only 14 vibrational modes, active in the infrared spectrum, and 6 modes, active in Raman scattering, have so far been identified for the anion (see Table 2.8).

Table 2.8 contains reported data on averaged experimental values of vibrational modes of the naphthalene molecule and its anion. The table presents also estimated, according to (2.47), values of $(\delta E_b)_k$ and total value of E_b for 20 reported vibrational modes.

As can be seen from Table 2.8, a frequency change of 20 vibrational modes yields, according to (2.47), a total vibronic relaxation energy of 0.05 eV. Of this value 2/3 are due to 14 vibrations of odd symmetry (u) and 1/3 to 6 vibrations of even symmetry (g). It stands to reason that the contribution of the remaining 28 vibrational modes also amounts to ca. 0.05 eV, thus yielding a total of electron bonding energy of the naphthalene anion $E_b^e \approx$ 0.10 eV [1.69].

Quantum-mechanical calculations of bonding energy E_b^h of a positive charge upon formation of equilibrium configuration of a naphthalene cation give, according to DEWAR et al. [2.88], a value of $E_b^h = 0.15$ eV. (It ought to be borne in mind that in the case of linear polyacenes $E_b^e = E_b^h$, owing to symmetric charge distribution on the negative and the positive ions, (cf. Sect. 2.4.2 and [2.54]).

The anthracene molecule has 66 normal vibrational modes. The 16 vibrational modes identified for the anthracene anion give a contribution of ca. 0.03 eV towards the total vibronic relaxation energy E_b^e (see Table 2.9). One can therefore assume that the E_b^e value of the anthracene anion should be ca. $\lesssim 0.1$ eV [1.69]. Quantum-mechanical calculations according to HOYLAND and GOODMAN [2.84] give a value of $E_b^h \approx 0.09$ eV and, according to

Table 2.8. Averaged experimental values of reported vibrational modes of a naphthalene molecule $(\tilde{\omega}_0)$ and of its anion $(\tilde{\omega}_i)$ and estimates of increase in bonding energy of an electron by the anion (δE_b) due to frequency change of kth vibrational mode [1.69]

Type of vibration symmetry	Corresponding components of dipole moment P_λ and of polarizability tensor $\alpha_{\mu\nu}$	Vibrational mode ω_k	$(\tilde{\omega}_0)_k$ [cm^{-1}] [2.91-93]	$(\tilde{\omega}_i)_k$ [cm^{-1}] [2.93]	$\Delta\tilde{\omega}_k =$ $= (\tilde{\omega}_0)_k$ $-(\tilde{\omega}_i)_k$ [cm^{-1}]	$(\delta E_b)_k$ [eV]
A_g	α_{xx}, α_{yy}, α_{zz}	ω_4	1471	1470	~ 0	-
		ω_7	1028	1026	2	0.0003
B_{1u}	P_x	ω_{17}	3060	3050	10	0.0012
		ω_{18}	3029	3023	6	0.0007
		ω_{20}	1389	1320	69	0.0086
		ω_{21}	1267	1195	72	0.0089
		ω_{22}	1128	1093	35	0.0043
		ω_{23}	877	840	37	0.0046
B_{2u}	P_y	ω_{31}	1510	1485	25	0.0031
		ω_{32}	1369	1291	78	0.0097
		ω_{33}	1210	1174	36	0.0045
		ω_{35}	1011	1001	10	0.0012
		ω_{36}	618	595	23	0.0028
B_{3g}	α_{xy}	ω_{37}	3051	2962	89	0.0110
		ω_{38}	3006	2876	130	0.0161
		ω_{40}	1445	1446	~ 0	-
		ω_{41}	1242	1235	7	0.0009
B_{3u}	P_z	ω_{45}	968	928	40	0.0050
		ω_{46}	789	715	74	0.0092
		ω_{47}	481	461	20	0.0025
					763	0.047≈0.05

Note: B_{1u} are vibrational modes, polarized parallel to the short (M), B_{2u} — to the long (L) axis of the molecule, but B_{3u} — parallel to the axis (N) normal to the plane of the molecule [2.91,92,94]; on inversion the components of the electric dipole moment change sign and belong to odd (u) types of symmetry, whilst the components of the polarizability tensor do not change sign and thus blong to even (g) types of symmetry. Vibrational modes B_{1u}, B_{2u} and B_{3u} have been found from infrared absorption spectra, and modes A_g and B_{3g} from Raman scattering.

DEWAR et al. [2.88], $E_b^h = 0.11$ eV for the anthracene cation. These calculated data can thus be seen to agree satisfactorily with empirically estimated E_b values for naphthalene and anthracene [1.69].

Part of the total energy E_b^h of positive charge bonding in the cation can be found experimentally from the difference between the vertical I_G^V and adiabatic I_G^{Ad} ionization potentials of the molecule $\Delta I_G = I_G^V - I_G^{Ad}$ (cf. Fig. 2.7a). However, in the general case $\Delta I_G < E_b^h$, since mainly only backbone

Table 2.9. Averaged experimental values of reported vibrational modes of an anthracene molecule $(\tilde{\omega}_0)$ and of its anion $(\tilde{\omega}_i)$, and estimated increase in bonding energy by an anion (δE_b), due to frequency change of kth vibrational mode [1.69]

Supposed type of vibration symmetry	Vibrational mode	$(\tilde{\omega}_0)_k$ [cm^{-1}]	$(\tilde{\omega}_i)_k$ [cm^{-1}]	$\Delta\tilde{\omega}_k =$ $=(\tilde{\omega}_0)_k$ $-(\tilde{\omega}_i)_k$ [cm^{-1}]	$(\delta E_b)_k$ [eV]	$\left[\dfrac{\Delta\tilde{\omega}_k}{(\tilde{\omega}_0)_k}\right]^2$ [in 10^{-4} units]	$(\delta W_b)_k$ [meV]
B_{3u}	ω_{39}	956.6	850.2	106.4	0.0132	124	0.74
	ω_{40}	885.1	779.6	105.5	0.0131	··142	0.78
	ω_{41}	735.9	699.5	36.4	0.0045	24	0.11
	ω_{42}	475.6	466.4	9.2	0.0011	4	0.01
B_{2u}, B_{1u}	ω_{46}	3055.3	3040.2	15.1	0.0019	2	-
	ω_{48}	1625.3	1568.3	57.0	0.0071	12	0.12
	ω_{49}	1455.4	1451.8	3.6	0.0004	-	-
	ω_{50}	1315.4	1288.4	27.0	0.0033	4	0.04
	ω_{58}	1540.2	1518.6	21.6	0.0027	2	0.02
	ω_{59}	1462.0	1451.8	10.2	0.0013	-	-
	ω_{60}	1400.1	1387.1	13.0	0.0016	-	-
	ω_{61}	1340.0	1319.8	20.2	0.0025	2	0.02
Backbone molecular vibrations		1562	1546	16	0.0020	1	0.01
		1404	1363	41	0.0051	9	0.08
		1261	1235	26	0.0032	4	0.03
		1187	1160	27	0.0033	5	0.04
				535.2	0.033 ≈ ≈ 0.03		~ 2.00

Note: vibrational modes B_{1u}, B_{2u} and B_{3u} have been found from infrared absorption spectra [2.95], and backbone vibrations from Raman scattering [2.96,97]; $(\delta W_b)_k$ is zero energy of interaction between a free electron and an intramolecular vibration estimated according to (2.48) for T = 0 K.

vibrational modes of the molecule take part in making up the value of ΔI_G. It should be mentioned that for anthracene the most probable ΔI_G value is ca. 0.06 eV (see Table 2.11).

The dynamic aspects of interaction of an electron with intramolecular modes (vibrational coupling) have been treated by MUNN and SIEBRAND [1.153-156]. They showed that, if a localized charge carrier moves by a hopping mechanism in a crystal of the aromatic hydrocarbon type, it interacts mainly with out-of-plane bending modes of the C-H bonds of the molecule. As a result of such dominating quadratic coupling within the vibrational coordinates the corresponding frequencies undergo changes (cf. also [2.17]). This brings about a lowering of self-energy value of the carrier, and an enhancement of its localization effect [1.154,156]. This latter effect, as stressed

in [1.156], is considerably stronger than in case of linear interaction
with lattice vibrations. The bonding energy δW_b for quadratic coupling between the charge carrier and intramolecular vibrations can be evaluated, according to [1.154], by means of the expression

$$\delta W_b = -(v + \frac{1}{2})\, \hbar\omega_0 (\Delta\omega/\omega_0)^2 \quad , \tag{2.48}$$

where v is the vibrational quantum number; $\Delta\omega$ is the frequency shift of the
given vibrational mode of the molecule due to interaction with a localized
charge carrier.

Table 2.9 presents the values of zero interaction energy of a free electron with intramolecular vibrational modes of an anthracene molecule, according to (2.48), for $v = 0$ at $T = 0$ K.

According to Table 2.9, the total interaction energy W_b of an electron
with 16 reported vibrational modes of an anthracene molecule is only ca. 2
meV, 80% of which are due to out-of-plane vibrations of B_{3u} symmetry. The
value of W_b must rise with temperature owing to increase in the population
of higher vibrational levels with $v > 0$. Thus, in realistic situations, the
energy of such a dynamic "vibronic polaron" may be of the order of 10^{-3}-
10^{-2} eV (cf. Fig.1.22).

We see thus that, depending on localization time τ_h of the charge carrier
which determines its interaction time scale with intramolecular vibrations,
the decrease of the self-energy value of the carrier due to vibronic polarization can lie between fractions of kT at room temperature (in dynamic
short-time interaction) up to E_b values of 2-3 kT in the case of formation
of a quasi-stationary ionic state of the molecule (cf. Fig.1.22).

2.7.3 Extended Polarization Model Including Ionic States of Electronic Conductivity

It follows from the previous section that the modified Lyons model (cf.
Sect.2.2.2) must be extended still further, including in the ionized state
energy diagram also ionic states of electronic conductivity. Such an extended polarization model has been introduced by SILINSH [1.69], according
to which two limiting levels of stationary conductivity are considered both
for holes and electrons. Thus, for a positive charge carrier (hole) we have
the level E_h^0 of purely electronic conductivity. The position of this conductivity level in the energy diagram of the crystal is determined by the

self-energy of a localized but unbonded hole the transfer of which occurs by hopping (at room temperature) from one neutral molecule to another without vibronic relaxation taking place (cf. Fig.2.7a). The state of the localized charge carrier in this case can be symbolically described by the expression

$$M_0 + h \rightarrow (M_0h)^+ \quad , \tag{2.49}$$

which denotes that the neutral molecule M_0 retains its equilibrium configuration of nuclei during the charge carrier localization time τ_h. The interaction time-scale condition for the formation of such a state is $\tau_h < \tau_v$ (cf. Fig.1.22).

The conductivity level $E_h(M^+)$ of a bonded hole (a positive ion) is shifted by the positive charge bonding energy E_b^h above the conductivity level E_h^0 (see Fig.2.7a). The transfer of such a bonded hole, presumably, also occurs by hopping from molecule to molecule, only with the difference that during the localization time τ_h complete vibronic relaxation takes place, and a positive ion M^+ is formed. The state of the bonded positive charge carrier during localization can be described by expression

$$M_0 + h \rightarrow M^+ \quad , \tag{2.50}$$

which denotes that the neutral molecule M_0 is relaxed to ionic equilibrium state M^+ during the charge carrier localization time τ_h, having a conditional value of $\tau_h > \tau_v$ (cf. Fig.1.22).

For a negative charge carrier we, similarly, have the conductivity level E_e^0 of an unbonded electron, corresponding to the state

$$M_0 + e \rightarrow (M_0e)^- \quad , \tag{2.51}$$

and the conductivity level $E_e(M^-)$ of a bonded electron (negative ion) at E_b below the E_e^0 level (see Fig.2.7a) according to

$$M_0 + e \rightarrow M^- \quad . \tag{2.52}$$

Hence, the conductivity levels E_e^0 and E_h^0 are purely electronic ionized states of the crystal, and levels $E_e(M^-)$ and $E_h(M^+)$ — genuine ionic states of electronic conductivity in the crystal.

The extended polarization model [1.69] still further removes the intrinsic inconsistencies of the initial Lyons model (cf. Sect.2.2.1) and symmetrizes it by introducing two conductivity levels, both for electrons and for holes, separated by the energy E_b of ionic state formation (cf. Fig.2.7a). This model has been borne out at present in a number of indirect experimental results (see, e.g. [2.35]) and agrees with general theoretical considerations on the formation of ionized electronic states in aromatic hydrocarbon crystals [1.42,62,63,69].

As already stated, the modulation of the self-energy value of a charge carrier upon vibronic relaxation must not necessarily reach maximum depth of the genuine ionic state, i.e., levels $E_h(M^+)$ and $E_e(M^-)$, respectively (cf. Fig.2.7a). The interval between levels $E_e^0 - E_e(M^-)$ and $E_h^0 - E_h(M^+)$ rather ought to be considered as the effective width of the initial electronic levels E_e^0 and E_h^0, caused by dynamic "vibronic polarization" of the molecules by the charge carrier.

If, in the construction of energy diagrams of ionized states in molecular crystals, one uses experimental adiabatic value of ionization potential I_G^{Ad}, as well as empirical data on electron affinity A_G of the molecule, according to (2.4), it follows that the energy terms E_b^e and E_b^h of ionic state formation are partially automatically accounted for and the conductivity levels thus obtained are the conductivity levels $E_e(M^-)$ and $E_h(M^+)$ of ionic states. The forbidden energy gap ΔE_G, as determined from intrinsic photoconductivity threshold value is, in first approximation, also the energy interval between conductivity levels $E_e(M^-)$ and $E_h(M^+)$ (cf. Sect.2.9).

2.7.4 Dynamic Electronic Polaron States in a Molecular Crystal

The stationary conductivity levels E_e and E_h in the modified Lyons model (cf. Fig.2.3) correspond to the self-energy value of a quasi-free electron and hole in thermal equilibrium state, i.e., at the bottom of the potential well of electronic polarization. The effective width of the given conductivity levels E_e and E_h is determined, first, by vibronic coupling of the charge carrier with intramolecular vibrations, its value being of the order of kT (cf. Sects.2.7.2,3); secondly, by dynamic fluctuations of electronic polarization energy ΔP due to thermal lattice vibrations, also of the order of ca. 0.03 eV, according to estimates by GOSAR and CHOI [1.158] and VILFAN [1.157] (cf. Sect.1.5.5 and Fig.1.22).

On the other hand, "hot" charge carriers created by field injection or photoexcitation appear to have a wide spectrum of energy values above E_e . level ("hot" electrons) and below E_h level ("hot" holes) (cf. Fig.2.3). The kinetic energy (velocity) of the charge carrier determines, in this case, the effective value of dynamic electronic polarization energy P_{dyn} and, accordingly, the self-energy value of a hot charge carrier forming a wide "band" of short-lived states. The conceptual basis for such a wide spectrum of dynamic electronic polaron states has already been discussed by FOWLER [2.48] (cf. Sect.2.3.2). In a dynamic approach to electronic polarization one easily distinguishes two limiting cases. If the excess charge carrier is allowed to move rapidly, its effective polarization energy P_{dyn} would be, obviously, decreased in absolute value. In the limiting case, when the excess carrier moves so fast that the electronic orbitals of the neighboring molecules do not manage to get polarized, the effective electronic polarization energy should approach zero value, i.e., $P_{dyn} \rightarrow 0$ [2.48]. The other limiting case is set by microelectrostatic approximation for thermalized charge carriers for which the dynamic polarization energy P_{dyn} is equal to the "classical" value of P, i.e., $P_{dyn} = P_{stat}$ (cf. Sect.2.3.2). Thus the effective width of the spectrum of dynamic electron polaron states can be given by

$$0 \leq P_{dyn} \leq P_{stat} \quad . \tag{2.53}$$

The existence of such a wide spectrum of dynamic electronic polarization states is a logical consequence of the Lyons phenomenological model (cf. Fig.2.3) and is consistent with the very essence of the model.

There is a certain amount of indirect experimental evidence in favor of the existence of such a wide band of short-lived electronic polaron states in molecular crystals which emerge in hot charge carrier relaxation processes. Thus BELKIND and BOK [2.98] interpreted the wide band of photoemitted electron energy distribution, being of the order of $\Delta = I_G - I_C \approx 1.5$ eV in aromatic molecular crystals, as due to electronic relaxation of a "hot" localized hole, created in the photoionization process (cf. also [1.144]). The energy gained in the random electronic relaxation process of the "hot" hole may be, apparently, transfered to the emitted electron via Coulomb interaction field (the mean electron escape depth L in aromatic crystals being smaller than the critical Coulomb interaction radius r_C [cf. (2.59)]), thus causing the wide energy spectrum of emitted electrons observed. The wide hot hole polaron band concept is used also by KARL and FEEDERLE [2.99] for the interpretation of the possible pathways of energy dissipation in the process of photoionization of the tetracene monopositive ion in the anthracene lattice.

SCHLOTTER and BAESSLER [2.100] have observed laser-induced high-mobility photoconduction in anthracene single crystals. Their data may also be regarded as experimental evidence of excited charge carriers in a broad conduction state band above the stationary conductivity level.

On the other hand, it is also difficult to describe hot electron thermalization processes in molecular crystals (cf. Sect.2.8.3) without applying the concept of a wide polaron spectrum of electronic relaxation situated above the stationary conductivity level E_e (cf. Figs.2.3 and 9).

There does not exist at present a comprehensive dynamic electronic polaron theory applicable for molecular crystals (cf. Sect.2.3.2). Such theory, however, will be required in the near future to provide self-consistency to existing phenomenological approaches. Possible avenues for the creation of such a theory have been recently indicated by ČÁPEK [1.126] who considers the effects of dynamic self-consistency of polarization in molecular solids and introduces an energy-dependent electronic polarization energy $P_q(\underline{k})$ which in the most general way may be presented as

$$P_q(\underline{k}) = \frac{1}{2} \sum_{m(\neq 0)} \underline{E}_m \mu_m(\underline{k}) \tag{2.54}$$

where μ_m is the value of induced dipole moment on molecule m, dependent on wave vector $\underline{k}$ of the electron. This approach, however, describes only the dynamic states of already relaxed charge carriers at the vicinity of stationary conductivity level above the static limit (cf. Sect.2.4.2).

2.8 Charge Transfer (CT) States in Molecular Crystals

2.8.1 General on CT-States

A characteristic feature of molecular crystals is the existence of so-called charge transfer states (CT states) playing an important role in general electronic state energy structure and electronic microprocesses of the crystal.

The concept of CT state was introduced by LYONS [1.65] in his initial ionized state model (cf. Sect.2.2.1) and retains its utmost importance also in
the modified one (cf. Sect.2.2.2). The energy levels of CT states are schematically shown in Figs.2.2 and 4. As already discussed in Sect.2.2.1, the
CT state is a nonconducting polar state formed by an electron-hole pair
bound by Coulomb interaction forces (cf. Fig.1.21). The energy E_{CT} of an
equilibrium CT state is therefore lower than the width of the energy gap,
viz. $E_{CT} < \Delta E_G$ (cf. Fig.2.4).

The localization effect of free charge carriers, so characteristic of
molecular crystals (cf. Sect.1.4), holds also for electron-hole pairs of a
CT state. Both the electron and the hole of a bonded CT state are localized
on particular molecules of the crystal.

If the localization time τ_h is sufficiently large as compared to vibronic
relaxation time τ_v ($\tau_h \gtrsim \tau_v$), CT states can exist also as genuine ionic
states (cf. Sect.2.7.2) forming an ion pair, bound by Coulomb interaction
field. The CT state in molecular crystals can therefore not be regarded as
an analogue of a large-radius Wannier exciton (cf. [2.36]). The traditional
exciton concept cannot, on principle, be applied to the CT state, because
the Coulomb field-bound electron and hole (or, correspondingly, the positive
and negative ion) migrate in the crystal noncoherently, by stochastic hopping, without conservation of translational symmetry.

Therefore the CT state is usually treated in terms of the Onsager model
(cf. Sect.2.8.3).

The probability of CT state formation by direct photo-induced charge
transfer is negligible in aromatic molecular crystals. As shown by LYONS
[1.65], the oscillator strength f of a direct optical charge transfer to
the nearest neighboring molecule in an anthracene crystal is of the order
of $f = 10^{-5}$-10^{-6}.

In anthracene-type crystals CT states are usually formed as intermediate
states via autoionization of a molecular exciton followed by subsequent
thermalization of the hot electron (cf. Sect.2.8.3).

Therefore one can introduce also a concept of non-equilibrium "hot" CT
states, consisting of a hot electron and a hole (or vice versa) which ther-
malize within the critical Coulomb interaction radius r_c to equilibrium CT
state (cf. Sect.2.8.3).

CT states also arise as intermediate states in recombination processes
of free charge carriers (cf. Sect.3.8.4).

It has been shown that in molecular crystals with a higher value of intermolecular interaction (such as phthalocyanines, polar derivatives of anthracene, perylene, etc.), apart from dominating optical excitation of neutral exciton states, direct optical electron transfer between neighboring molecules can take place, resulting in formation of dipole CT states. Such direct optical transitions into polar states are, as a rule, weak and concealed against the background of strong absorption by neutral (Frenkel) excitons. They can therefore be detected only by methods of Stark-spectroscopy [2.101]. Thus, using Stark-spectroscopy technique, ABBI and HANSON [2.102] have observed direct optical transitions into CT states in 9,10-dichloroanthracene crystals. The observed CT states have an energy of $E_{CT} = 3.2-3.3$ eV, which approximately corresponds to the E_{CT} value of the lowest CT state in anthracene (cf. Fig.2.15), and a dipole moment of ca. 23 D [2.102]. In thin crystalline perylene layers, on the other hand, direct optical transitions into polar states have been observed by BLINOV and KIRICHENKO [2.103] with energies of $E_{CT} = 3.3-3.4$ eV. These states can, most likely, be attributed to excited CT states, since the ΔE_G value for perylene is 3.10 eV (cf. Fig.2.17). The excited CT state, most likely, consists of an ionic pair, one of the ions being in the ground state and the other in the excited state.

As demonstrated by BAESSLER and KILLESREITER, CT states can also be formed at a molecular crystal-metal interface as a result of Frenkel exciton decay [2.104], or upon dark field injection of charge carriers [2.105] (see also [2.106]). The existence of such CT states has been independently proved by superfine-range weak magnetic field effects on photo- and dark conductivity in thin layer cells of tetrathiotetracene (TTT): Au/TTT/Al by KAULACH and SILINSH [2.107].

Thus we see that molecular crystals may exhibit a variety of different types of CT states.

2.8.2 Evaluation of CT State Energies in Anthracene and Naphthalene Crystals

LYONS [1.65,38] first tried to evaluate the nearest-neighbor ion pair (CT state) electronic polarization and dissociation energies in anthracene and naphthalene crystals, using an electrostatic model of a dipole in a dielectric medium. The location of the first CT state in aromatic crystals was also evaluated in ±0.5 eV approximation by CHOI et al. [2.37].

HUG and BERRY [1.60] have attempted to calculate electron and hole interaction energies in an anthracene crystal for a large set of CT states corresponding to an electron and hole, each located on a particular molecule.
The calculations were performed in the framework of a microelectrostatic
model in zero-order approximation (cf. Sect.2.4.1). Data reported in [1.60]
show that the calculated E_{CT} values seem to be overestimated, most likely
due to approximations used. Thus, the calculated first CT-state energy E_{CT}^1
for anthracene equals 3.84 eV, i.e., lies 0.39 eV above the reported experimental value [1.42] and practically coincides with the estimated energy gap
value ΔE_G (see Fig.2.15).

Later the CT-state energy levels E_{CT} in anthracene and naphthalene crystals were calculated by JURGIS and SILINSH [1.61]. The authors used a microelectrostatic approach in terms of SCPF approximation (cf. Sect.2.4.2).

According to this approximation the electronic polarization energy $P_{e-h}(r)$
of an electron and a hole localized on particular lattice points can be
found, as a function of their mutual distance r, by [1.61]

$$P_{e-h}(r) = -\frac{1}{2} \sum_i \underline{E}_0(r_i)\mu_i \quad , \tag{2.55}$$

where $\underline{E}_0(r_i)$ is the net electric field on the ith molecule, created by a
hole-electron pair; μ_i is the dipole moment induced on the ith molecule, calculated by SCPF procedure from a set of linear equations determining the polarization field.

The self-energy value $E_{CT}(r)$ of a CT state depends on the mutual distance
r of an electron-hole pair in the crystal and can be determined by [1.61]

$$E_{CT}(r) = I_G - A_G + W_c(r) + P_{e-h}(r) \quad , \tag{2.56}$$

where $W_c(r)$ is the Coulomb interaction energy between localized hole and
electron.

Expression (2.56) clearly shows that CT states are essentially many-electron interaction states and are consistent with the modified Lyons model (cf.
Fig.2.4).

The dissociation energy E_d of the CT state equals

$$E_d = \Delta E_G - E_{CT} \quad . \tag{2.57}$$

Owing to partial mutual compensation of polarization fields by the hole and the electron we have

$$P_{e-h} < P_e + P_h \quad . \tag{2.58}$$

The mutual distance $r_c(T)$ at which the Coulomb energy of the electron-hole pair of the CT state is equal to kT (often called critical Coulomb interaction radius) is given by

$$r_c(T) = \frac{e^2}{\epsilon kT} \quad , \tag{2.59}$$

where ϵ is the dielectric permeability of the crystal.

Thus, e.g., for anthracene, at T = 300 K and ϵ = 3.55, r_c equals 150.2 Å.

$P_{e-h}(r)$, $E_{CT}(r)$ and E_d values, as calculated in [1.61] and later refined by the authors of [1.64] for a number of CT states in naphthalene and anthracene crystals, according to (2.55-57) are presented in Table 2.10.

$P_{e-h}(r)$ values were calculated for a limited crystal region containing from 122 to 180 molecules. The term $W_c(r)$ of Coulomb interaction of the electron-hole pair was obtained from the well-known expression in point approximation $W_c(r) = -e^2/r$. (Only for the first CT state the $W_c(r)$ value was also calculated using the model of charge distribution over the corresponding ions after CHOI et al. [2.37]).

As can be seen from Table 2.10, the calculated energy values of the first CT state E_{CT}^1 are in fair agreement with reported experimental E_{CT}^1 values, especially in the case of $W_c(r)$ estimate in the approximation of charge distribution over the ion.

The calculated functional dependences $P_{e-h}(r)$, $W_c(r)$ and $E_{CT}(r)$ for an anthracene crystal are graphically demonstrated in Fig.2.8.

Calculating $P_{e-h}(r)$ dependence in microelectrostatic approximation, with allowance for the anisotropy of molecular polarizability, the potential hypersurface of the CT-state energies $E_{CT}(r)$ proves to be more complex than that of purely Coulomb nature.

Due to this anisotropy local minima appear on the $E_{CT}(r)$ curve even in the ab plane of an anthracene crystal, as shown by an arrow in Fig.2.8 at r = 10 Å.

Similar singularities of the $E_{CT}(r)$ hypersurface have been obtained also in the calculations of HUG and BERRY [1.60]. This shows that the isotropic

Onsager model can serve only as zero-order approximation for describing the CT states in anthracene-type crystals.

Table 2.10. Calculated values of CT-state parameters in anthracene and naphthalene crystals, as dependent on mutual distance r between hole and electron [1.64]

Position of electron (hole situated in 0,0,0 position of the lattice)	r [Å]	$P_{e-h}(r)$ [eV]	$W_c(r)$ [eV]	$E_{CT}(r)$ [eV]	$E_d(r)$ [eV]
		Anthracene			
(1/2,1/2,0)	5.24	-1.06	-2.75*	3.04*	0.75*
			-2.50*	3.29*	0.50*
(0,1,0)	6.04	-1.24	-2.38	3.23	0.56
(1,0,0)	8.56	-1.77	-1.68	3.40	0.39
(1/2,3/2,0)	10.02	-2.06	-1.44	3.35	0.44
(1,1,0)	10.48	-2.10	-1.37	3.38	0.41
r > r_C	r > 160	$P_{e-h} \to 2P = -3.06$	$W_C \to 0$	$E_{CT} \to \Delta E_G = 3.79$	$E_d \to 0$
		Naphthalene			
(1/2,1/2,0)	5.095	-0.84	-2.83	4.31	1.12
			-2.65*	4.48*	0.95*
(0,1,0)	6.003	-1.02	-2.39	4.56	0.87
(1,0,0)	8.235	-1.40	-1.75	4.82	0.61
(1,1,0)	10.19	-1.67	-1.41	4.89	0.54
r > r_C	r > 185	$P_{e-h} \to 2P = -2.54$	$W_C \to 0$	$E_{CT} \to \Delta E_G = 5.43$	$E_d \to 0$

Note: $P_{e-h}(r)$ is the energy of electronic polarization of the crystal by an electron-hole pair; $W_c(r)$ is the energy of Coulomb interaction; $E_{CT}(r)$ is the self-energy value of CT states; $E_d(r)$ is the dissociation energy of CT states. Number with as asterisk (*) means that the values of W_C, E_{CT} and E_d are calculated using a model of charge distributed over the molecular ion.
In the calculations the following parameters were used: for anthracene: I_G = 7.40 eV [1.96,2.108]; A_G = 0.55 eV [1.95]; calculated values P = 1.53 eV (cf. Table 2.1) and ΔE_G = 3.79 eV (cf. Sect.2.10); for naphthalene: I_G = 8.12 eV [2.86]; A_G = 0.15 eV [1.95]; calculated values P = 1.27 eV (cf. Table 2.1) and ΔE_G = 5.43 eV [1.92]. Experimental values: E_{CT}^1 = 3.45 eV for anthracene and E_{CT} = 4.38 eV for naphthalene, according to POPE and BURGOS [2.38] (see also [1.42]).

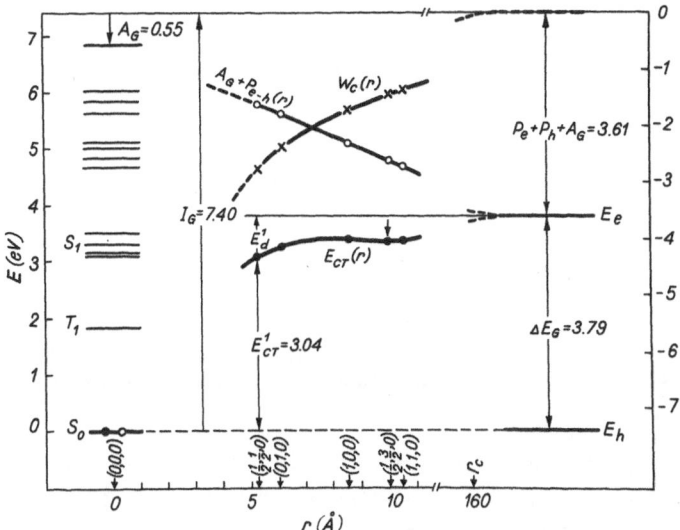

Fig. 2.8. Electronic polarization energy $P_{e-h}(r)$, Coulomb interaction $W_C(r)$ and CT-state energy $E_{CT}(r)$ dependence on distance r of an electron-hole pair in an anthracene crystal. The hole is localized in (0,0,0) position, and the electron in the corresponding lattice point (x,y,0) in the ab plane [1.61, 64])

2.8.3 CT States in Photogeneration Processes

It has been shown that CT states play an important role in photogeneration processes in molecular crystal, especially in the near-threshold spectral region of intrinsic photoconductivity.

Photogeneration of free charge carriers in this spectral region proceeds through various intermediate stages, one of them being the CT state (cf. Fig.2.9).

The first stage is photon absorption and formation of a neutral Frenkel exciton state (cf. Fig.1.21).

In the second stage ionization of the neutral excited state with efficiency Φ_0 can occur creating a pair of quasi-free charge carriers — a localized hole and a hot electron. This process is usually treated in terms of an autoionization mechanism proposed and developed for the description of ionization in molecular crystals by POPE et al. [2.109,110].

According to this model, ionization is considered as a configurational interaction between an excited discrete neutral state and a continuum of ionized states in the spirit of FANO's [2.111] theoretical treatment of autoionization processes in gaseous phase.

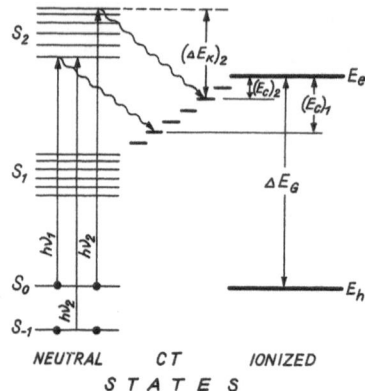

NEUTRAL   CT   IONIZED
       S T A T E S

Fig. 2.9. Schematic diagram of photogeneration stages in a molecular crystal

In the third stage the hot quasi-free electron rapidly loses its initial excess kinetic energy ΔE_k, owing to fast scattering, and becomes thermalized at a mean thermalization length $<r_0>$ inside the critical Coulomb capture radius r_C [cf. (2.59)], thus forming an intermediate CT state (see Fig.2.9).

In the initial phase of autoionization the hot electron may be coherent. Its coherence of motion, however, is very rapidly lost, owing to inelastic scattering on lattice vibrations, presumably on optical lattice phonons, the scattering events taking place practically at every lattice site (cf. Sect. 1.5). Thus the electron becomes strongly localized, and further energy relaxation occurs in an incoherent random walk process.

Finally, after thermalization, the CT state can either dissociate, forming free charge carriers, or, on the contrary, undergo geminate recombination.

The probability of these processes, at given temperature T and field strength $\mathscr{E}$, can be described in terms of the Onsager model.

ONSAGER [2.112] treated rigorously the Brownian motion problem of a pair of charged particles in a Coulomb field and developed a model describing ionizing radiation processes. More recently, the Onsager theory has been used extensively by radiation chemists (see, e.g. [2.113,114]).

BATT et al. [2.115,116] applied this theory to intrinsic photoconductivity in anthracene single crystals in order to gain an understanding of the observed temperature and applied electric field dependences of the free charge carrier generation quantum yields.

Later the validity of the Onsager model for the interpretation of activation energy of photogeneration and electric field dependences in anthracene single crystals has been convincingly proved by CHANCE and BRAUN [2.117,118], BAESSLER and KILLESREITER [2.34], and LYONS and MILNE [2.119,120].

The Onsager model appears to have general validity in describing photogeneration processes also in other organic photoconducting solids; e.g., such as tetracene and pentacene vacuum deposited layers [2.121,122], polyvinyl carbazol [2.123], polydiacetylene-TS single crystals [2.124], as well as amorphous selenium [2.125].

According to the Onsager model, the activation energy of intrinsic photogeneration E_a^{ph} is uniquely determined by the Coulomb bonding energy E_C of the CT state [2.122] (cf. Fig.2.9)

$$E_a^{ph} = -E_C \frac{e^2}{\varepsilon <r_0>} \quad , \tag{2.60a}$$

where $<r_0>$ is the mean thermalization length of the hot electron. Thus, measuring the spectral dependence of photogeneration activation energy $E_a^{ph}(h\nu)$ in a crystal, it is possible to evaluate the spectral dependence of the mean thermalization length $<r_0> (h\nu)$

$$<r_0> (h\nu) = \frac{e^2}{\varepsilon E_a^{ph}(h\nu)} \quad . \tag{2.60b}$$

Subsequently, using these data one can easily calculate the spectral dependence of dissociation probability $\Phi[r_0(h\nu), \mathscr{E}]$ of an electron-hole pair in CT state at given field strength $\mathscr{E}$, according to Onsager's formula

$$\Phi(r_0,\mathscr{E}) = \frac{\Phi_0}{2\gamma r_0} \exp\left(-\frac{e^2}{\varepsilon kTr_0}\right) \sum_{m=0}^{\infty} \frac{\left(\frac{e^2}{\varepsilon kTr_0}\right)^m}{m!}$$
$$\times \sum_{n=0}^{\infty} \left[1 - \exp(-2\gamma r_0) \sum_{k=0}^{m+n} \frac{(2\gamma r_0)^k}{k!} \right] \quad , \tag{2.61}$$

where Φ_0 is the primary quantum yield of the ionization process, i.e. the number of CT states formed per absorbed photon; $\gamma = e\mathscr{E}/2kT$.

In their initial use of the Onsager theory for the interpretation of photogeneration stages in anthracene BATT et al. [2.115,116] used a delta function in order to describe the distribution $g(r)$ of thermalization length r_0 of the hot electron

$$g(r) = \frac{\delta(r - r_0)}{4\pi r_0^2} \quad . \tag{2.62}$$

For delta function distribution the mean thermalization length $<r_0>$ equals r_0 by definition, viz. $<r_0> = r_0$. Later CHANCE and BRAUN [2.118] introduced Gaussian distribution of $g(r)$ as more realistic. The Gaussian distribution, however, is a two-parameter curve (cf. Sect.3.1). Since experiment yields only one parameter, namely $<r_0>$ [cf. (2.60)], one should assume that the distribution parameter σ is small and delta function approximation is thus valid.

Recently a model of exponential distribution of electron-hole separation distances $g(r)$ has been treated analytically by BORSENBERGER and ATEYA [2.118a]. The authors showed that at low electric fields the photogeneration efficiency of exponential and δ distributions are similar in the framework of the Onsager theory but are strongly dependent on assumed thermalized pair configurations at high field intensities.

The spectral dependence of $E_a^{ph}(h\nu)$ in anthracene crystals has been extensively studied by BATT et al. [2.115,116] and later refined by CHANCE and BRAUN [2.118] in a careful set of experiments. These authors cover a wide spectral range of $E_a^{ph}(h\nu)$ dependence from 4.1 to 6.2 eV. The experimental points of the E_a^{ph} dependence in the most important near-threshold region of anthracene, namely, from 4.0 to 4.5 eV, are however, few and fragmentary (cf. Ref.[2.118], Fig.7).

The first set of experiments by BATT et al. [2.115,116] give in the near-threshold region of anthracene (at $\sim$ 4.5 eV) the mean thermalization $<r_0>$ of ca. 80 Å, which seems to be an overestimated value. Later refined measurements by CHANCE and BRAUN yielded a value of $<r_0>$ from 37 to 45 Å, dependent on photon energy in the region from 4.2 to 4.5 eV. Recent two-photon excitation measurements by RYAN et al. [2.118b] yielded a value of $<r_0>$ equal to 58 ± 16 Å in the 4.2-3 eV region of anthracene.

According to HUGHES [2.114] the mean thermalization length $<r_0>$ of low-energy electrons in anthracene is ca. 40 Å, as obtained from geminate recombination studies under x-ray excitation.

On the other hand, LYONS and MILNE [2.120] estimated from indirect measurements (from the field dependences of photogeneration efficiency) the mean $<r_0>$ value in anthracene to be of the order of 20 to 25 Å in the 4.1-4.5 eV spectral region. These indirectly derived data seem to be underestimated. Thus, the most probable values of the mean thermalization length $<r_0>$ of low-energy electrons in the near-threshold region of anthracene obviously lie in the interval from 25 to 45 Å.

The most important feature of the results obtained in these studies is that the E_a^{ph} value steadily decreases and the corresponding mean thermalization length $<r_0>$ increases with increasing photon energy $h\nu$ in the near-threshold region ($h\nu$ = 4.0-4.5 eV) of an anthracene crystal (cf. [2.34,118, 120]).

Similar correlations have been obtained, even more convincingly, in the threshold spectral regions of pentacene and tetracene crystals by SILINSH et al. [2.122]. The authors studied the $E_a^{ph}(h\nu)$ and corresponding $<r_0>$ ($h\nu$) spectral dependences by pulse photoconductivity method in vacuum-evaporated layers of pentacene and tetracene (cf. Sect.5.15).

The obtained $E_a^{ph}(h\nu)$ and $<r_0>$ ($h\nu$) dependences for pentacene layers are presented in Fig.2.10.

As can be seen from Fig.2.10, the E_a^{ph} value decreases from ca. 0.3 eV at $h\nu$ = 2.2 eV (corresponding to the threshold value for pentacene, viz. E_e = = 2.20 eV) to ca. 0.03 eV at $h\nu$ = 2.8 eV. At the same time the average thermalization length $<r_0>$, in accordance with (2.60b), increases from cy. 10 Å to 120-130 Å with increasing photon energy.

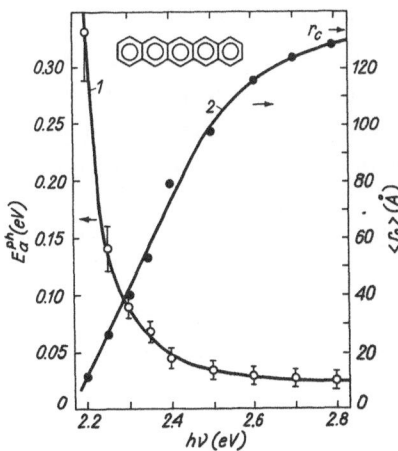

Fig. 2.10. Photogeneration activation energy E_a^{ph} and hot electron mean thermalization length $<r_0>$ dependence on the photon energy $h\nu$ in the near-threshold region of intrinsic photoconductivity in pentacene

These data are in perfect agreement with the schematic diagram of photo-generation stages (cf. Fig.2.9). Increased photon energy $h\nu$ imparts increased excess kinetic energy $\Delta E_k(h\nu)$ to the hot electron, resulting in greater thermalization length $\langle r_0 \rangle(h\nu)$ and, correspondingly, a shallower final CT state E_C.

From the spectral dependence of $\langle r_0 \rangle(h\nu)$ (cf. Fig.2.10) one can easily find spectral dependence $\Phi[r_0(h\nu), \mathscr{E}]$ of CT state dissociation probability at the given field strength $\mathscr{E}$, according to the Onsager formula (2.61). The results are shown in Fig.2.13b.

As can be seen from Fig.2.13, the calculated dependence $\Phi(r_0, \mathscr{E}) = f(h\nu)$ at $\mathscr{E} = 4 \times 10^4$ V/cm is functionally, as well as in absolute magnitude, very close to the spectral dependence of quantum efficiency of intrinsic photo-conductivity $\beta(h\nu)$ at $\mathscr{E} \approx 10^4$ V/cm in the near-threshold region of pentacene. [The best agreement between both spectral curves can be obtained at the value of parameter Φ_0 in the Onsager formula (2.61) equal to 0.5-0.8].

The second important conclusion that follows from the comparison of both spectral dependences is that the $\Phi(h\nu)$ curve can be approximated in the threshold region by the following empirical formula

$$\Phi(h\nu) = (h\nu - E_e)^{5/2} \qquad\qquad (2.63)$$

which is functionally equivalent to the empirical formula (2.69) (cf. Sect. 2.9) for approximation of the $\beta(h\nu)$ curve.

Both approximations yield, as can be seen from Fig.2.13, the same threshold value $E_e = 2.20$ eV for pentacene by linear extrapolation in $(\Phi^{2/5}, h\nu)$ and $(\beta^{2/5}, h\nu)$ coordinates, respectively.

An analysis of $\Phi(h\nu)$ dependence explains also the high value of β for pentacene (cf. Fig.2.13 and Table 2.15). For $h\nu = 2.8$ eV and $\mathscr{E} \approx 10^4$ V/cm $\langle r_0 \rangle$ reaches a value of 120-130 Å, which is close to the critical Onsager distance r_C [cf. (2.59)] ($r_C = 133$ Å) corresponding to complete dissociation of the electron-hole pair (cf. Table 2.15). Hence $\Phi(h\nu)$ and, accordingly $\beta(h\nu)$ are close to unity for pentacene.

Similar coincidence of both $\beta(h\nu)$ and $\Phi(h\nu)$ dependences has been observed also for tetracene layers, the best agreement being obtained at Φ_0 values 0.8-1.0. In this case, however, the corresponding maximum β value is lower, apparently owing to smaller thermalization length $\langle r_0 \rangle$ values for tetracene (cf. Table 2.15).

From the above analysis two important conclusions may be drawn. First, the spectral dependence of quantum efficieny $\beta(h\nu)$ of intrinsic photoconductivity in the near-threshold region of molecular crystals is, apparently, determined by the spectral dependence of dissociation probability $\Phi(h\nu)$ of intermediate CT states, in accordance with the Onsager model.

Secondly, it can be assumed that the functional dependence of the empirical approximation formula for $\beta(h\nu)$ [cf. (2.69)] is also determined by the corresponding functional dependence for $\Phi(h\nu)$, according to (2.63).

An analysis of the $\Phi(h\nu)$ dependence for an anthracene crystal, based on data reported in [2.118], shows that the primary quantum yield of ionization Φ_0 in this case is much lower, viz. ca. $3\text{-}4 \times 10^{-3}$ to fit the experimental photoconductivity curves in the near-threshold region (cf. Table 2.15). This indicates an effective competition between intramolecular energy conversion channels (mainly fluorescence) and the ionization channel. The Φ_0 value obtained is in agreement with earlier estimates by KLEIN and VOLTZ [2.126] ($\Phi_0 \approx 10^{-3}\text{-}10^{-2}$) and CHANCE and BRAUN [2.118] ($\Phi_0 = 5 \times 10^{-3}$). In general, the primary ionization efficiency can be defined as

$$\Phi_0 = \frac{k_{AI}}{k_{AI} + \sum_i k_i} \quad , \qquad\qquad (2.64)$$

where k_{AI} is the rate constant of autoionization and k_i are the rate constants of various intramolecular (radiant and nonradiant) energy dissipation transitions.

Thus the condition for high ionization efficiency is $k_{AI} \gtrsim \sum_i k_i$, which apparently holds for pentacene and tetracene crystals.

Generally speaking, for a wide spectral range the Φ_0 value may be dependent on photon energy either through possible k_{AI} spectral dependence, or, even more likely, through spectral dependences of some of the intramolecular conversion rate constants. In a narrow spectral range, in the near-threshold region (of about 0.5-0.8 eV width) one can, however, regard the Φ_0 as a constant, spectrally independent quantity. This has been evidenced by good agreement between $\Phi(h\nu)$ and $\beta(h\nu)$ curves at $\Phi_0 = $ const, as discussed above.

It is interesting to mention that after initial rise in the near-threshold region, the $\langle r_0 \rangle$ ($h\nu$) dependence reaches a plateau at higher photon energies of about 0.6-1.2 eV above the threshold (cf. [2.34,118,120]). This may be caused, as shown by BAESSLER and KILLESREITER [2.34], by setting in of transitions from lower lying valence states at sufficiently high photon energies, in accordance with the model proposed earlier by GEACINTOV and POPE [2.110]. As can be seen from Fig.2.9, in such a case two sets of thermalized electrons with different excess energies ΔE_k and, consequently, different mean thermalization length $\langle r_0 \rangle$ values, should emerge. The experimentally determined $\langle r_0 \rangle$ value would be, in this case, an effective, averaged one, apparently causing the appearance of the plateau on the $\langle r_0 \rangle (h\nu)$ curve.

A different interpretation of this phenomenon is given by CHANCE and BRAUN [2.118]. These authors suggested that the plateau (or even several ones) can arise as a result of dominating autoionization from particular "metastable" excited states (which might be called autoionization "terraces"). These states are, presumably, reached by rapid vibrational relaxation before ionization occurs (see also [2.118b]).

In real situations, both proposed mechanisms may probably arise, transitions from lower lying valence states, most likely, being the dominant ones.

It should be emphasized that the Onsager model actually describes only the final thermalized CT states, the influence of temperature and electrical field on its fate. As to the very process of hot electron thermalization, and possible field effects on it—these questions remain outside the limits of applicability of the model.

KNIGHTS and DAVIS [2.127] have proposed a model for the description of photogeneration processes in amorphous selenium. According to this model, which is a modified Poole-Frenkel approach, a high-intensity electrical field, acting during the thermalization process, can considerably increase the mean thermalization length $\langle r_0 \rangle$ of a hot charge carrier.

Direct experimental evidence of such high-electric field effect on mean thermalization length $\langle r_0 \rangle$ in pentacene and tetracene layers has been recently produced by SILINSH et al. [2.122]. The results of their studies can be illustrated by $\langle r_0 \rangle$ dependence in electrical field $\mathscr{E}$ in pentacene layers, determined by pulse photoconductivity method in the near-threshold region.

As may be seen from Fig.2.11, the initial electron-hole separation $\langle r_0 \rangle$, having zero-field value of about 40-60 Å at photon energies about 0.2-0.3 eV above the threshold (E_e = 2.20 eV), may be considerably increased by an electrical field at $\mathscr{E} \gtrsim 1 \times 10^5$ V/cm. Thus, at $\mathscr{E} = 2 \times 10^5$ V/cm the field dependent $\langle r_0 \rangle$ value reaches 100-130 Å at photon energies $h\nu$ = 2.3-2.5 eV, i.e., approaches the critical Onsager radius r_c for pentacene at room temperature.

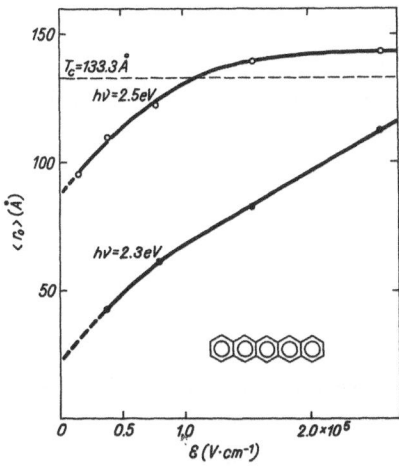

Fig. 2.11. Hot electron mean thermalization length $\langle r_0 \rangle$ dependence on electric field $\mathscr{E}$ in the near-threshold region of intrinsic photoconductivity in pentacene

It may be shown that such field effects are possible under conditions when $er\mathscr{E} \gg kT$ and, additionally, $\mathscr{E}(r)$ is of the same order as the internal Coulomb field, viz. $er\mathscr{E} \sim e^2/r$. This means that the one-dimensional charge carrier drift in an external field dominates over three-dimensional diffusion processes. These conditions, as can be shown, are actually fulfilled at the initial zero-field electron-hole separation distances $\langle r_0 \rangle$ of about 30-50 Å at field strength $\mathscr{E} \gtrsim 10^5$ V/cm. On the other hand, in case of tetracene layers, for which the initial zero-field separation distance is only about 20 Å, the field effect is considerably smaller (cf. Table 2.15).

We have estimated that the observed decrease of E_a^{ph} value with increasing field strength $\mathscr{E}$ in pentacene layers is considerably greater than one could expect from the field dependence of E_a^{ph} at constant thermalization length $\langle r_0 \rangle$ according to the Onsager formula (2.61) and is apparently caused by field-induced increase of effective thermalization length of hot electrons $\langle r_0 \rangle = f(\mathscr{E})$ as shown in Fig.2.11. Such an assumption has been additionally confirmed by our observation of very short photoexcited current pulses in pentacene layers, equal in duration to 20 ns Nd: glass laser light pulses at $\lambda = 530$ nm. In our opinion such short current pulses are of polarization origin. They may be caused by formation of oriented electron-hole pair dipoles during the thermalization process of photoexcited electrons, i.e., $j = dp/dt$, where p is polarization density vector $p = Ne\langle r_0 \rangle$ along the external electric field. It may be suggested that recently reported observations of similar short current peaks in anthracene crystals by SCHEIN et al. [2.127a] which repeat 20 and 2.5 ns Nd: glass laser light pulses at $\lambda = 265$ nm may be also of polarization current origin.

At present there does not exist a comprehensive theory or even a unified model describing both thermalization and post-thermalization stages (cf. Fig.2.9) of charge carrier photogeneration processes in molecular crystals. Thus, a comparative study carried out by LYONS and MILNE [2.120] showed that for the description of the post-thermalization stage of photogeneration in anthracene crystals a modified Poole-Frenkel model, as proposed by KNIGHTS and DAVIS [2.127], does not fit experimental results and the Onsager model is shown to be much more adequate (at least up to fields of 10^5 V/cm). On the other hand, the thermalization stage itself is outside the limits of applicability of the Onsager approach.

HONG and NOOLAND [2.128] have recently proposed a solution of the time-dependent Onsager problem. The most interesting consequence of their analysis is the existence of a critical value of electric field $\mathscr{E}$ above which the distribution fucntion and related quantities no longer exhibit diffusion-like behavior for long times, but rather show pure exponential decay.

It may be concluded that the problem of detailed photogeneration mechanism in molecular crystals is at present an open field for comprehensive studies to both experimental and theoretical physicists.

Following the treatment of KNIGHTS and DAVIS [2.127] it is possible to make a rough estimate of averaged CT state thermalization parameters also in molecular crystals. Thus, if we assume the mean zero-field electron-hole separation distance $<r_0>$ to be ca. 50 Å at $h\nu = 2.4$ eV in the case of pentacene layers (cf. Fig.2.11), and scattering is assumed to occur on optical phonons with typical energy $h\nu_{ph} = 50$ cm^{-1} (cf. Sect.1.5.3), the mean thermalization time τ_{th} can be roughtly estimated according to the uncertainty principle, viz.

$$\tau_{th} = \frac{\Delta E_k}{h\nu_{ph}} \cdot \frac{1}{\nu_{ph}} \quad , \tag{2.65}$$

where ΔE_k is the excess kinetic energy of the hot electron (cf. Fig.2.9): $\Delta E_k(h\nu) = (h\nu - E_e) + E_C$, and ν_{ph} is the frequency of optical phonons.

The ratio $\Delta E_k/h\nu_{ph}$ gives the number n of inelastic scattering events.

The estimated τ_{th} value, according to (2.65) for pentacene layers equals $\tau_{th} \approx 7 \times 10^{-12}$ s.

The averaged number n of inelastic scattering events, during which the excess energy ΔE_k is lost, equals: $n = \Delta E_k/h\nu_{ph} = 0.25/0.012 \approx 20$. If scattering occurs on every molecular site, as it is typical for molecular crystals (cf. Sect.1.5), the averaged hopping time is $\tau_h \approx \tau_{th}/n \approx 3.5 \times 10^{-13}$s. This value is in fair agreement with data presented in Table 1.9.

Finally, the mean electron diffusion coefficient D can be estimated from the expression

$$D = \frac{<r_0>^2}{\tau_{th}}$$

(2.66)

which gives a value of $D = 0.035$ cm^{2}s^{-1} for $<r_0> = 50$ Å in pentacene layers.

Recently SCHEIN et al. [2.127a] have estimated the τ_{th} value in anthracene using the expression $\tau_{th} = <r_0>^2 e/\mu kT$. They found τ_{th} being equal to 24×10^{-12} s taking $<r_0> = 50$ Å, and $\mu = 0.4$ cm^2/Vs at $T = 300$ k.

2.8.4 CT States in Recombination Processes

CT states actually play an important role in the reverse process of charge carrier generation, namely, in recombination of free excess carriers in molecular crystals.

If free charge carriers of both signs are produced in the bulk of the crystal by photogeneration, ionizing radiation [2.113,114] or double injection (see [1.76]), they disappear very rapidly through bimolecular recombination via CT states.

Whenever two charge carriers of opposite sign occasionally approach each other within the distance of Coulomb capture radius r_c [cf. (2.59) and Table 2.15] their bimolecular recombination probability Φ_r increases drastically in accordance with the Onsager formula (2.61); Φ_r being $\Phi_r = 1 - \Phi$.

The approaching charge carriers gradually dissipate their energy rolling down the CT state "staircase" (cf. Fig.2.9), and finally collapse on a single molecule producing excited S_1 and T_1 states in statistical 1:3 ratio (cf. [Ref.[1.45], p.280]). Therefore recombination processes in molecular crystals are, as a rule, accompanied by recombination luminescence, corresponding to S_1-S_0 transition and being of purely molecular nature.

Thus the symmetry of the recombination process is the reverse of the symmetry of photogeneration. If, in the case of the latter, molecular photoexcitation is the first stage, then for recombination molecular luminescence is the final one. In both processes the molecular-solid state dualism, so characteristic of molecular crystals (cf. Sect.1.4), is evident. Therefore, such concepts as recombination through a band-to-band transition or via recombination centers, as accustomed in traditional semiconductor physics are alien to molecular crystals.

The validity of the Onsager recombination model for anthracene-type crystals has been proved beyond doubt quantiatively by measurements of recombination constant values.

The bimolecular recombination constant γ can be defined by [1.45,2.129]

$$- \frac{dn}{dt} = \gamma np \quad , \tag{2.67}$$

where

$$\gamma = \frac{4\pi e (\mu_n + \mu_p)}{<\varepsilon>} \quad , \tag{2.68}$$

$<\varepsilon>$ being the mean value of the dielectric tensor.

The bimolecular recombination constant of charge carriers in anthracene crystals has been determined as $\gamma = 3 \times 10^{-6}$ $cm^3 s^{-1}$ [1.45,2.113,129], which is in a fair agreement with the Onsager model.

The physical basis of the Onsager recombination model in molecular crystals is the small mean path for charge carriers $<\ell> \approx a_0$, as compared to the Coulomb capture radius, r_c, viz. $<\ell> \ll r_c$.

The bimolecular recombination studies are usually performed by drift-current technique [1.45], measurements of bulk recombination limited CV characteristics at double injection [1.76] and recombination luminescence techniques [1.45,76].

Recombination processes have been most thoroughly investigated in naphthalene and anthracene crystals (see, e.g. [1.45,76]).

2.9 Experimental Determination of Energy Structure Parameters in Molecular Crystals

We shall briefly discuss in this section the main experimental methods of ionized and neutral state investigation in molecular crystals illustrated by energy structure determination for pentacene crystals, according to SILINSH et al. [1.99,104,2.39,130].

The basic ionized state parameters of an idealized molecular crystal, namely, ionization energy level I_c (or, correspondingly, the hole conductivity level E_h, cf. Fig.2.3), electron conductivity level E_e (or, correspondingly, the energy gap ΔE_G, cf. Fig.2.4), and electron affinity A_c of the crystal can be primarily determined from intrinsic photoconductivity $\beta(h\nu)$ and photoemission $Y(h\nu)$ quantum efficiency spectra in the threshold region (cf. Fig.2.12).

As shown earlier by GEACINTOV and POPE [2.110] both spectral curves exhibit similar functional dependence on photon energy in the near-threshold region and fairly coincide with each other if the photoconductivity spectral curve $\beta(h\nu)$ is shifted towards higher energies by the value A_C of electron affinity of the crystal. The authors proved this coincidence of $Y(h\nu)$ and $\beta(h\nu)$ spectra to be valid in the case of anthracene and tetracene single crystals. This empirical rule of spectra coincidence is illustrated for vacuum-deposited pentacene layers, according to data reported in [1.99], in Fig.2.12.

As may be seen from Fig.2.12, in pentacene layers a distinct photoconductivity threshold emerges in the vicinity of $h\nu = 2.2$ eV exhibiting a decrease in β value of three orders of magnitude in a narrow threshold region of about 0.3 eV width. A similar distinct threshold emerges also for intrinsic photoemission at 5.1 eV.

The $\beta(h\nu)$ and $Y(h\nu)$ spectra (Fig.2.12) allow one to determine directly the parameter A_C value as well as to evaluate the approximate E_e and I_C

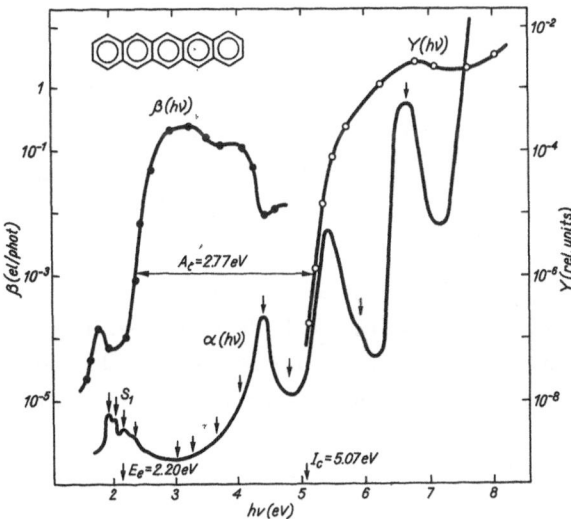

Fig. 2.12 Spectral dependences of quantum efficiency of photoconductivity $\beta(h\nu)$, quantum yield of photoemission $Y(h\nu)$ and optical absorption $\alpha(h\nu)$ for pentacene thin-layer specimens

values. The empirical coincidence rule serves also as one of the main criteria for the identification of genuine threshold regions.

HIROOKA et al. [2.131] and KOCHI et al. [2.132] were first to suggest an
empirical approximation formula for the spectrum of photoemission quantum
yield in the near-threshold region in organic crystals. They showed that
the best approximation can be reached by a cubic formula $Y(h\nu) \sim (h\nu - I_c)^3$.
Later BELKIND and KALENDAREV [2.133] showed, however, that reported intrinsic photocurrent spectra of anthracene and tetracene in the near-threshold
region are more satisfactorily approximated by a 5/2 dependence law, viz.
$j_{ph}(h\nu) \sim (h\nu - E_e)^{5/2}$. The 5/2 law has been later confirmed in our studies
as the best approximation for the near-threshold region of the $\beta(h\nu)$ spectrum in case of a number of other aromatic and heterocyclic crystals [2.39,
134]. Thus, the validity of the 5/2 law has been specially tested for a
large number of pentacene and tetracene layers, and as the most reliable
the following n values were obtained: n = 2.5 ± 0.3 for pentacene and n =
2.4 ± 0.2 for tetracene.

Consequently, for the determination of the intrinsic photoconductivity
threshold, i.e., the electron conductivity level E_e (cf. Fig.2.4), the following approximation formula can be used [1.99,2.39,134]:

$$\beta(h\nu) = (h\nu - E_e)^n \quad , \tag{2.69}$$

with n = 2.5, where $\beta(h\nu)$ is the quantum efficiency of intrinsic photoconductivity at the given field strength $\mathscr{E}$, determined in electrons per absorbed photon:

$$\beta(h\nu, \mathscr{E}) = \frac{j_{ph}(h\nu, \mathscr{E})}{e \cdot k(h\nu) \, I(h\nu) \, G(h\nu)} \quad , \tag{2.70}$$

where $j_{ph}(h\nu, \mathscr{E})$ is the density of photocurrent at given photon energy $h\nu$
and field strength $\mathscr{E}$, $k(h\nu)$ is the absorption coefficient of light in the
semitransparent electrode; $I(h\nu)$ is the incident flux of light in $phot/cm^2 s$;
$G(h\nu)$ is a coefficient determining the portion of light absorbed in the sample

$$G(h\nu) = \int_0^L \exp[-\alpha(h\nu)x]dx = \{1 - \exp[-\alpha(h\nu)L]\} \quad , \tag{2.71}$$

where $\alpha(h\nu)$ is the absorption coefficient, L is the thickness of the sample.

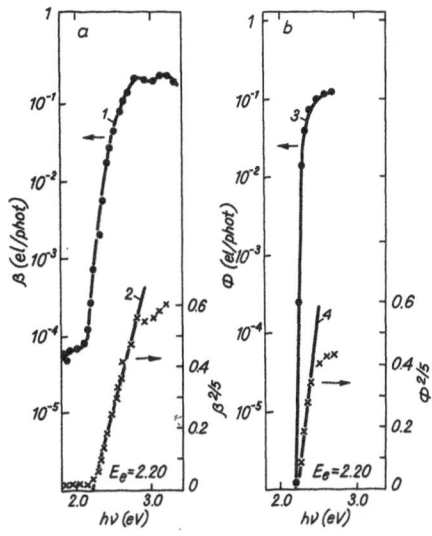

Fig. 2.13. Spectral dependence at near-threshold region of intrinsic photoconductivity for pentacene: 1 - photoconductivity quantum efficiency $\beta(h\nu)$; 2 - curve 1 in $(\beta^{2/5}, h\nu)$ coordinates; 3 - dissociation probability of CT states $\Phi(h\nu)$; 4 - curve 3 in $(\Phi^{2/5}, h\nu)$ coordinates

As can be seen from Fig. 2.13a, a linear extrapolation of (2.69) in $(h\nu, \beta^{2/5})$ coordinates gives the value of E_e (or ΔE_G, respectively, cf. Fig. 2.4) for pentacene equal to $E_e = (2.20 \pm 0.05)$ eV. A theoretical interpretation of formula (2.69) has been discussed in Sect.2.8.3, where it was shown that the empirical dependence (2.69) might be caused by CT state dissociation probability $\Phi(h\nu)$, giving spectral dependence in the near-threshold region similar to (2.69) [cf. (2.63) and Fig.2.13].

Consequently, the $\Phi(h\nu)$ formula (2.63) gives an independent (although cumbersome, cf. Sect.2.8.3) method for the determination of E_e value. As can be seen from Fig.2.13b, this method gives for pentacene an identical value of E_e equal to 2.20 ± 0.05 eV.

The E_e values, determined by formula (2.63) for tetracene and anthracene crystals also coincide within ± 0.05 eV limits with E_e obtained by approximation according to (2.69).

Another independent method of determining E_e or ΔE_G (cf. Fig.2.4) is provided by activation energy E_e measurements of intrinsic dark conductivity, based on the well-known relation $\Delta E_G = 2E_a$. This method yields the following value of ΔE_G for pentacene [1.99]: $\Delta E_G = (2.20 \pm 0.3)$ eV. It should be emphasized, however, that this method may serve as an auxiliary and only in a case when genuine intrinsic dark conductivity is secured. It is, however, a rather difficult job owing to the presence of a considerable number of local trapping states in the forbidden energy gap, usually exhibiting asymmetric distribution (cf. Sect.5.17). In such circumstances genuine intrinsic dark conductivity of the sample can be obtained, as a rule, only under special experimental conditions.

The ionization energy level of the crystal I_C can be determined by two independent methods:

a) from the spectral dependence of intrinsic photoemission quantum yield $Y(h\nu)$, approximated by the following empirical formula [1.99,144,2.132,133, 135,136]

$$Y(h\nu) = (h\nu - I_{C,Y})^n \quad , \tag{2.72}$$

where n = 2.5-3.0.

As can be seen from Fig.2.14, a linear extrapolation in $(h\nu, Y^{2/5})$ coordinates, according to (2.72), gives the value of $I_{C,Y}$ = (5.10 ± 0.05) eV, where the second index Y indicates that the I_C value is determined from photoemission spectral dependence;

b) from energy distribution of emitted photoelectrons, according to Einstein's law [1.99,144]

$$E_{max}(h\nu) = (h\nu - I_{C,E}) \quad , \tag{2.73}$$

where E_{max} is the maximum kinetic energy of emitted photoelectrons. This method yields for pentacene, as can be seen from Fig.2.14a, the value $I_{C,E}$ = (5.00 ± 0.05) eV, where the second index indicates that the I_C value is determined from energy distribution of emitted electrons (cf. Fig.2.14b).

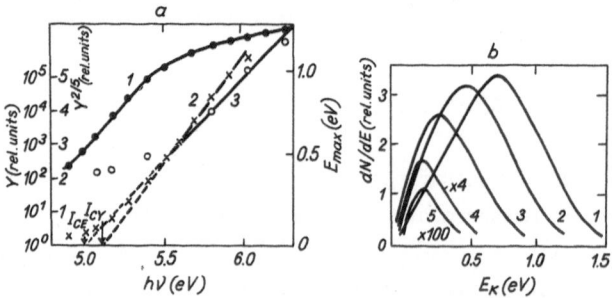

Fig. 2.14. Spectral dependences of photoemission from pentacene layers according to [1.99]: (a): 1 – photoemission quantum yield $Y(h\nu)$; 2 – curve 1 in $(Y^{2/5}, h\nu)$ coordinates; 3 – maximum kinetic energy of emitted electrons $E_{max}(h\nu)$, according to Einstein's law, (b) energy distribution of photoemitted electrons for different photon energy $h\nu$ values: 1– $h\nu$ = 6.26 eV; 2 – 6.02 eV; 3 – 5.80 eV; 4 – 5.39 eV; 5 – 5.07 eV

For the construction of an energy diagram of ionized states for pentacene (cf. Fig.2.16) averaged values of $I_{C,Y}$ and $I_{C,E}$ measurements were used
taking into account also the most reliable reported I_C data (see, e.g.
[2.132]), which gives the mean value of I_C = 5.07 eV. The I_G value of 6.74
eV was used according to BOSCHI et al. [2.137].

The possible theoretical basis of the empirical formula (2.72) has been
discussed by several authors (see, e.g. [1.21,2.132,133,135]. Recently BEL-
KIND et al. [1.149] have proposed a comprehensive theoretical approach to
photoemission from organic solids (see also [Ref. [1.144], p.58]). They
have substantiated the empirical formula (2.72) theoretically, showing that
the value of n = 3 ought to be preferred.

One should emphasize that although the physical nature of the factors
determining the corresponding approximation formulas (2.69,72) for intrinsic photoconductivity and photoemission thresholds are, apparently, quite
different, it is of utmost importance from the methodological standpoint
that their functional dependences are similar (or almost similar), as illustrated by Fig.2.12.

The electron affinity A_C of the crystal can be evaluated by three independent approaches [1.99]:

a) from coincidence measurements of $\beta(h\nu)$ and $Y(h\nu)$ spectra (cf. Fig.
2.12) which give an approximate value of $A_C = Y(h\nu) - \beta(h\nu) = (2.77 \pm
0.10)$ eV;

b) using experimental I_C and E_e values, determined after (2.69,72):
$A_C = I_C - E_e = 5.07 - 2.20 = (2.87 \pm 0.10)$ eV (cf. Fig.2.16);

c) using electron polarization energy: $P_e = P_h = I_G - I_C = (1.67 \pm 0.10)$
eV and electron affinity A_G of the pentacene molecule $A_G = (1.2 \pm 0.1)$
[1.99]: viz. $A_C = A_G + P_e = 1.67 + 1.2 = (2.87 \pm 0.2)$ eV (cf. Fig.2.16).

As can be seen, all three approaches give A_C values for pentacene which
fairly agree within the limits of experimental error.

The vibronic-electronic levels of excited states of the crystal are usually determined by conventional optical methods, and filled valency states
by methods of electron spectroscopy.

In the case of pentacene layers the vibronic structure of the S_1 electronic state and the components of Davydov splitting of the 0-0 transition

at 1.88 and 1.97 eV distinctly appear in the absorption spectrum at 300 K
(cf. Fig.2.12). The vibronic structure of higher excited states could only
be detected from absorption spectra at 77 K (indicated by arrows in Fig.2.12)
[1.99,2.130,138]. The T_1 energy value E_T = 0.81 eV, as reported in [2.139],
is used in the energy diagram of neutral states of pentacene in Fig.2.16.

Such a distinct intrinsic photoconductivity threshold, as observed in
pentacene layers (cf. Fig.2.12), is rather an exception than a general rule.
It is mainly caused by very high photosensitivity of pentacene, the β value
being near unity (cf. Fig.2.12). In a number of cases special preliminary
precautions should be taken in order to single out the genuine threshold.

This is due to the fact that in measurements of β(hν) spectral dependence
in some molecular crystals, particularly in thin-layer cells, other factors
apart from intrinsic photoconductivity can affect the character of β(hν) in
the near-threshold region. These factors have either to be taken into account, or else removed if the correct threshold value dependence has to be
obtained. Thus, the intrinsic photoconductivity threshold can be partly or
totally concealed by some form of nonintrinsic photoconductivity, such as
an exciton mechanism of photo-enhanced conductivity. Therefore it is often
necessary to make a preliminary study of the nature of various photogeneration mechanisms and their particular contribution in the spectral region
in which the threshold is expected. When the extrinsic disturbing factors
have been ascertained, one must find ways and means of allowing for them
or neutralizing them. Thus, for instance, in perylene layers at T = 300 K
an exciton mechanism of photo-stimulated conductivity dominates over the
whole spectral range between 2.0 and 4.0 eV. This completely eclipses the
real threshold of intrinsic photoconductivity (see [2.140]). Since the
exciton mechanism of photogeneration in perylene proceeds with an activation energy of E_a^{ph} = 0.45 eV, it can be quenched by lowering of temperature.
Thus, at T < 220 K an autoionization mechanism becomes dominant within the
spectral region of 3.0-4.0 eV, with intrinsic photoconductivity threshold
at hν = 3.1 eV [2.140] (cf. Fig.2.17).

Another example is thin-layer cells of tetracene Me/Tc/Me in which considerable monopolar injection of holes from the metal electrode takes place
at field intensities of 10^4-10^5 V/cm and where an exciton mechanism of photostimulated conductivity dominates. In this case the real threshold of intrinsic photoconductivity can be singled out only by irradiating the sample
through a blocking semitransparent electrode.

In thin-layer specimens of comparatively low-resistance molecular crystals, such as tetrathiotetracene, contact effects can considerably influence
photogeneration effectivity at room temperatures [2.141,142]. This kind of
influence can be dealt with successfully by lowering the temperature [2.142].
In other cases pulse photoconductivity measurements of β(hν) dependence may
be helpful.

2.10 Energy·Structure of an Anthracene Crystal

Anthracene has been most thoroughly studied as model crystal in many laboratories (cf. Sect.1.3.1). Consequently, there exist a great number of report-

ed data on the energy parameters of neutral and ionized states in anthracene
crystals. We shall discuss in this section the plausibility of various reported data, particularly on ionized state parameters, in order to choose
the most reliable ones for the construction of an energy diagram for an
anthracene crystal.

There are no problems concerning neutral states of an anthracene crystal,
which have been studied in great detail by conventional spectroscopic techniques. Reliable data can be found in recent review articles (see, e.g.
[1.43]) and safely used for the construction of neutral state energy levels
of the crystal (see Fig.2.15). The situation is less favorable regarding
the main parameters I_G, A_G, I_C and ΔE_G of ionized states which are of particular interest in the present work.

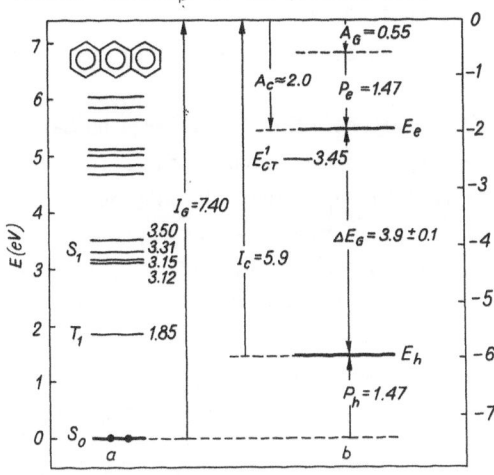

Fig. 2.15. Energy level diagram
of neutral (a) and ionized (b)
states for an anthracene crystal

Table 2.11. Experimental values of ionization energy I_G of an anthracene
molecule

I_G [eV]	Method of determination	Reference
7.47	PS	[2.137]
7.47	PS	[2.143]
7.41	PS	[2.86]
7.38	PI	[1.96]
7.42	PI	[2.108]
7.41	RS	[2.144]
7.40	CT	[2.145,146]

Note: PS - method of photoelectron spectroscopy (I_G found from peak of energy distribution of electrons); PI - photoionization method (I_G found from
threshold value of spectral dependences of photoionization quantum yield);
RS- method of Rydberg spectra (I_G found from convergence limit of Rydberg
series in spectrum); CT - method of charge transfer spectra

The primary object is to choose reliable reported experimental data on the molecular parameters I_G and A_G.

The ionization energy I_G can be determined by a number of independent methods, the most preferable being electron spectroscopy and photoionization, usually providing an accuracy up to ± 0.01-0.03 eV, as well as some special spectroscopic techniques, such as Rydberg spectra.

Table 2.11 presents the most reliable reported data on I_G values for anthracene, obtained by these methods.

It must be borne in mind that photoelectron spectroscopy (PS) yields, as a rule, vertical ionization potential I_G^V, whereas photoionization (PI) and spectroscopic methods yield adiabatic potential I_G^{Ad}. Thus, for anthracene, according to Table 2.11, the most probable value of I_G^V equals 7.45-47 eV whilst I_G^{Ad} is, most likely, equal to 7.40 eV.

In constructing the energy diagram of anthracene (cf. Fig.2.15) the value of $I_G = I_G^{Ad} = 7.40$ eV has been used.

As regards electron affinity of an anthracene molecule, a large number of rather different A_G values has been reported by various authors (cf. e.g. [1.38,92]). Methodologically the most reliable A_G values for anthracene appear to be those determined by electron capture technique, as reported by BECKER and CHEN [1.95] and LYONS et al. [2.147] and presented in Table 2.12.

For experimental determination of polarization energies P_e and P_h (cf. Sect.2.4.2), as well as for direct construction of an energy diagram (cf. Fig.2.15) two more essential energy parameters of the crystal are necessary, viz. the ionization energy I_C and the effective energy gap ΔE_G between the conduction levels of holes E_h and electrons E_e.

Experimental values of I_C for anthracene are given in Table 2.13, and experimental and calculated values of ΔE_G in Table 2.14.

Table 2.13 reveals a considerable spread of reported I_C values. Partially it may be caused by the bend of conductivity levels at the crystal surface due to surface charge which can affect the value of the photoelectron work function (for details cf. [1.144]). Anthracene type crystals, being of p type (cf. Sect.5.17), usually exhibit a downward bend of conduction levels.

Table 2.12. Experimental values of electron affinity A_G of an anthracene molecule

A_G[eV]	Reference
0.55 ± 0.01	[1.95]
0.57 ± 0.02	[2.147]

Table 2.13. Experimental values of ionization energy I_C of an anthracene crystal

I_C[eV]	Method of determination	Reference
5.65	From threshold of spectral dependence of photoelectron emission; single crystal sample	[2.148]
5.68	Extrapolation of spectral dependence of photoemission quantum yield Y(hν) in (Y$^{1/3}$, hν) coordinates; single crystal sample	[2.132]
5.75	Idem; sublimated layers	[2.149]
5.85	From threshold of spectral dependence of photoemission; sublimated layers	[2.150]
5.95	Idem; single crystal obtained from solution	[2.150]
5.98	Analytical evaluation from correlation of I_C for polyacene series	[1.92]

Table 2.14. Experimental and calculated values of energy gap ΔE_G between conductivity levels E_h and E_e in an anthracene crystal

ΔE_G[eV]	Method of determination	Reference
3.85	From threshold of spectral dependence of intrinsic photoconductivity	[2.151]
4.00	Idem	[2.152]
3.91	Extrapolation of spectral dependence of quantum yield of intrinsic photoconductivity β(hν) in coordinates (β$^{1/3}$, hν) after [2.110]	[1.92]
3.86	Idem, after [2.115,116]	[1.92]
4.1 ± 0.1	Refined ΔE_G value, obtained by method of photoinjection and accounting for activation energy of CT-state decay	[2.34]
4.0 ± 0.02	Analytical evaluation based on experimental data of autoionization in anthracene	[2.34]
3.97 ± 0.22	Analytical evaluation based on correlation function of ΔE_G for polyacene series	[1.92]
3.85 ± 0.06	Obtained from calculated P value (cf. Sect.2.4.2)	[1.64]
3.90 ± 0.15	Obtained from experimental P value (cf. Sect.2.4.2)	[1.64]

This leads to a diminished value of the measured I_C. Therefore, the highest I_C value obtained may be regarded as the most reliable one, viz. I_C = 5.85-95 eV. This conclusion is confirmed by an analytical evaluation of I_C value for anthracene by BELKIND and GRECHOV [1.92], based on a correlation function for I_C for the polyacene series (cf. Table 2.13).

The most reliable averaged experimental ΔE_G value for anthracene is, according to Table 2.14, equal to $\Delta E_G = 3.9$ eV, the error not exceeding ± 0.10 eV.

The energy gap ΔE_G in anthracene-type crystals can be estimated independently, as shown by SILINSH and JURGIS [1.64] applying calculated values of polarization energy P and empirical molecular parameters I_G and A_G, by means of the following expression:

$$\Delta E_G = I_G - A_G + P_h + P_e = I_G - A_G - |2P| \quad . \tag{2.74}$$

Using the most reliable reported molecular parameters $I_G^{Ad} = 7.40$ eV and $A_G = 0.55$ eV (cf. Tables 2.11 and 12), and calculated values P = 1.50 ± 0.03 eV (cf. Table 2.7), one obtains, according to (2.74) a value of ΔE_G equal to 3.85 ± 0.08 eV [1.64]. On the other hand, if the value P = 1.475 eV, obtained from averaged photoemission and photoconductivity data, according to Fig.2.5, is substituted in (2.74), we have ΔE_G = 3.90 ± 0.15 eV.

Thus we see that the calculated value of ΔE_G in anthracene is in good agreement with the averaged experimental ones obtained by different experimental procedures (cf. Table 2.14).

The above-mentioned averaged energy parameters I_G, A_G, I_C and ΔE_G have been used for constructing the energy diagram of ionized states of an anthracene crystal, shown in Fig.2.15.

2.11 Energy Structure of Aromatic and Heterocyclic Molecular Crystals

In this section we shall present comparative energy diagrams of neutral and ionized states for a number of aromatic and heterocyclic molecular crystals in order to show the influence of molecular structure and presence of heteroatoms on the energy parameters of the crystal.

Figure 2.16 contains a comparative diagram of energy levels of neutral and ionized states of linear polyacenes, viz. anthracene (Ac), tetracene (Tc) and pentacene (Pc). The energy diagram of Ac is given according to averaged reported data presented in Sect.2.10 (cf. Fig.2.15). For the construction of energy diagrams of Tc and Pc the averaged data of our measurements (cf. [1.67,99,104,2.39,130]), as well as reported ones (cf. [1.92,97, 2.110,138,139]) have been used·(for Pc data see also Sect.2.9).

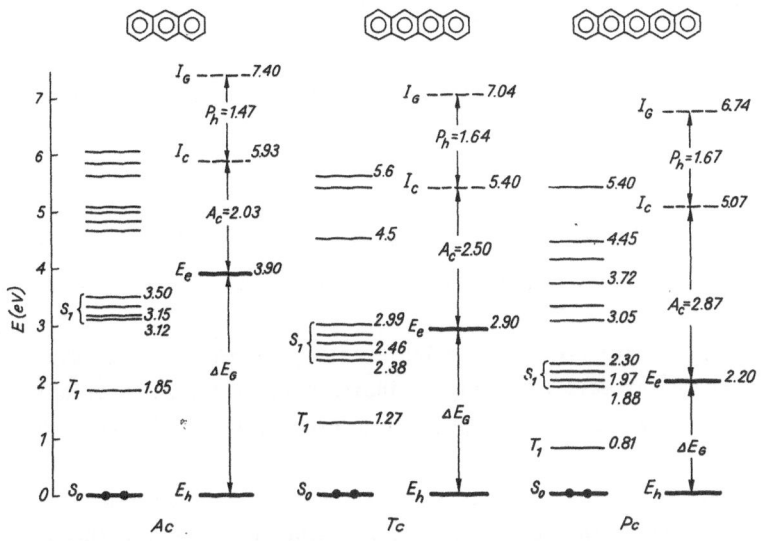

Fig. 2.16. Energy level diagrams of neutral and ionized states for anthracene (Ac), tetracene (Tc), and pentacene (Pc) crystals

A comparison of the energy diagrams of these linear polyacenes (cf. Fig. 2.16) clearly demonstrates that with increase in the number of condensed benzene rings (or, accordingly, with increase in number of π electrons) there is a steady decrease in the value of ionized state energy parameters I_G, I_C and ΔE_G accompanied by a similar steady increase in value of parameter A_C. The decrease is particularly marked in the case of the ΔE_G, its value dropping from the $\Delta E_G = 3.90$ eV for Ac to $\Delta E_G = 2.20$ eV for Pc. This is mainly caused by simultaneous decrease in I_G and increase in A_C and P_h values in the linear polyacene series (cf. also [1.92]).

A number of other empirical correlations can also be observed: with increase in number of condensed aromatic rings the level of the first singlet 0-0 transition E_{S_1} as well as the level of the first CT state E_{CT}^1 [1.62] (cf. Sect.2.8.2) asymptotically tend towards the electron conductivity level E_e. The value of S_1 and T_1 levels steadily decreases whilst the Davydov splitting of the 0-0 transition of the S_1 state increases from $E_D = 0.03$ eV for Ac to $E_D = 0.09$ eV for Pc (cf. Fig.2.16).

According to the polarization model (cf. Sect.2.2.2), the following additivity condition of ionized state parameters of the molecule (I_G and A_G) and crystal (I_C, A_C and ΔE_G) has to be observed:

$$I_G + A_G = I_C + A_C = \Delta E_G + 2A_C = K_C \quad , \qquad (2.75)$$

where K_C is a characteristic constant of the polarization model of ionized states.

In the case of linear polyacenes (Ac, Tc, Pc) the characteristic constant equals $K_C = (7.93 \pm 0.06)$ eV, i.e., condition (2.75) is satisfied within the limits of an error of ± 0.06 eV. This result may be regarded as an important additional proof of the validity of the polarization model for molecular crystals.

As illustrated by data presented in Table 2.15, the molecular structure of linear polyacenes affects (directly or indirectly) also such important crystalline state parameters as the primary quantum yield of autoionization Φ_0 [cf. (2.61)], the mean thermalization length $<r_0>$ of low-energy electrons in photogeneration processes (cf. Sect.2.8.3), Coulomb capture radius r_C [cf. (2.59)], as well as the quantum efficiency $\beta(h\nu)$ of intrinsic photo-conductivity in the near-threshold region. Some aspects of this problem have already been discussed in Sect.2.8.3. Here we should like to emphasize only the obvious correlations between the above-mentioned parameters and molecular structure of the given polyacene series and the impact of these correlations on the photosensitivity of the crystal. Thus, anthracene is a poor photoconductor, owing to low values of both Φ_0 and $<r_0>$. Tetracene is

Table 2.15. Primary quantum yield of autoionization Φ_0, mean thermalization length $<r_0>$ of hot electrons in photogeneration process, Coulomb capture radius r_C at 300 K and quantum efficiency of intrinsic photoconductivity β at near-threshold region of anthracene (Ac), tetracene (Tc) and pentacene (Pc)

Compound	Φ_0 [el/phot]	$<r_0>$ [Å]	r_C [Å]	β [el/phot]
Ac	$(3-5) \times 10^{-3}$ [2.118,126]	25-45 [2.118,120,114]	150.2	$(1-2) \times 10^{-4}$ [1.45,2.118,120]
Tc	0.8-0.9	38-47	140.3	$\sim 4 \times 10^{-2}$
Pc	0.5-0.8	120-130	133.3	$\gtrsim 3 \times 10^{-1}$

Note: The Φ_0, $<r_0>$ and β values are determined at electric field intensities $\mathscr{E}$ ca. 10^4 V/cm at the near-threshold spectral region ca. 0.7-0.8 eV above the intrinsic photoconductivity threshold E_e of the crystal. In the case of Ac, measurements have been performed on single crystal samples, and in the case of Tc and Pc on layers of oriented crystallites.

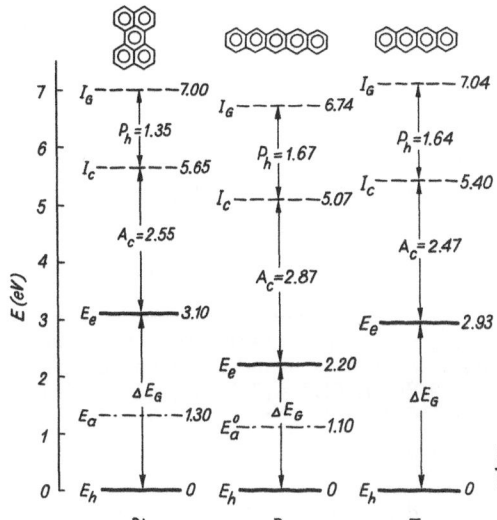

Fig. 2.17. Energy level diagrams of ionized states for perylene (Pl), pentacene (Pc) and tetracene (Tc) crystals [1.119]

a moderate photoconductor: although the Φ_0 value is high, the photogeneration efficiency in near-threshold region is considerably decreased owing to low value of $<r_0>$. Finally, pentacene is an excellent photoconductor, due to high values of both parameters: Φ_0 = 0.5-0.8 and $<r_0> \approx r_C$.

Table 2.15 illustrates the main conditions for obtaining organic molecular crystals with high photosensitivity.

Figure 2.17 shows the energy diagram of ionized states for a perylene (Pl) crystal according to ALEKSANDROV et al. [1.119] in comparison with those for Pc and Tc. As can be seen from Fig.2.17, the energy structure of perylene (Pl) crystals, a pericondensed pentacene analogue, is nearer to tetracene, as regards its energy parameters I_G, I_C, A_C and ΔE_G, than to pentacene containing the same number of benzene rings. It appears that Pl belongs to a different correlational series of pericondensed arenes.

It has been usually taken for granted that ionized state parameters I_C and E_e do not actually depend on the crystalline state of the specimen, and their measurements may be performed with equal success on single crystals as well as on polycrystalline samples. These assumptions have been substantiated by close agreement between I_C and E_e measurement data for single crystal and polycrystalline specimens. (see, e.g., the I_C data for anthracene in Table 2.13 as reported by MARCHETTI and KEARNS [2.150]).

<u>Table 2.16.</u> Ionized state parameters I_C, E_e and A_C in tetracene (Tc) and pentacene (Pc) crystals of different crystalline state [1.104]

Compound	Crystal-line state	E_e [eV]	I_C [eV]	A_C [eV]
Tc	SC	2.88 ± 0.05[a]	5.35 ± 0.05[b]	2.47 ± 0.10
	OL	2.90 ± 0.05	5.35 ± 0.05	2.45 ± 0.10
	QA	2.80 ± 0.05	5.40 ± 0.05	2.60 ± 0.10
Pc	OL	2.20 ± 0.05	5.07 ± 0.05	2.87 ± 0.10
	QA	2.10 ± 0.05	5.05 ± 0.07	2.95 ± 0.10

Notation: SC- single crystal (thickness L = 23μm [2.110]); OL - oriented polycrystalline layers with ab plane of crystallites parallel to substrate surface (average grain dimensions d = 0.2-1μm [1.104,2.153]); QA - quasi-amorphous layers ($\sim$ 50 $\lesssim$ d $\lesssim$ 500 Å [1.104,2.153]) (cf. Sect.5.15).

a) determined by extrapolation of β(hν) dependence in $\beta^{2/5}$, hν coordinates, using data reported by GEACINTOV and POPE [2.110];

b) determined by extrapolation of Y(hν) dependence in $Y^{2/5}$, hν coordinates, using data reported by GEACINTOV and POPE [2.110].

BALODE et al. [1.104] have made a special study of the possible influence of the crystalline state of Tc and Pc samples on the values of ionized state parameters E_e, I_C and A_C (cf. Table 2.16). The determination of E_e and I_C in oriented (OL) and quasi-amorphous (QA) layers of Tc and Pc have been performed by photoconductivity and photoemission methods described in Sect. 2.9. The methods of obtaining OL and QA specimens of Tc and Pc by evaporation in vacuo have been described in Sect.5.15. Brief characteristics of the given layers are given in Table 2.16.

The parameters E_e and I_C obtained for OL and QA layers of Tc in [1.104] are compared in Table 2.16 with those for a Tc single crystal, according to reported data [2.110].

Table 2.16 shows that the ionized state parameters E_e and I_C actually coincide within ± 0.05 eV error for a single crystal sample, polycrystalline layers of oriented crystallites with averaged grain dimensions d = 0.2-1μm, and polydispersed (quasi-amorphous) layers with 50 $\lesssim$ d $\lesssim$ 500 Å.

These results are not unexpected: they are in consistence with the very essence of the polarization model. We should remember that the main contribution towards the electronic polarization energy P, determining the values of parameters E_e and I_C, is due to the nearest environment of the localized charge, with radius of the order of 20 to 30 Å. Thus, as can be seen from Table 2.1, about 120 molecules surrounding the localized charge carrier

within the sphere of radius r = 17-18 Å provide more than 80% of the total
polarization energy P.

Consequently, if a free charge carrier is generated by ionization in a
crystallite with dimensions of the order of 100-200 Å in a polycrystalline
layer, its polarization energy P will be practically the same as in the
case of an infinite single crystal.

These conclusions are of considerable methodological importance. They
show that the basic energy parameters E_e, I_c and A_c of an ideal crystal can
be determined by photoconductivity and photoemission measurements performed
in real crystals, viz. in single crystal samples of definite dimensions, as
well as in polycrystalline layers.

Certain precautions, however, should be observed. In order to get the
most reliable data it is preferable to perform the measurements on a single
crystal sample, or, at least, on layers of oriented crystallites with grain
dimensions of the order of 0.5-1μm. In the case of I_c measurements one
should be certain that the electron escape depth L is of an order of at
least 20-30 Å. In the case of lower L values the bulk polarization energy
of a hole (P_h) may be decreased by near-surface polarization effects (cf.
Sect.5.5).

In the case of heterocyclic compounds the heteroatoms appear to have a
strong effect on the energy structure of the crystal.

This can be demonstrated by the energy diagrams of two isostructural analogues, derivatives of indene, viz. 1-phenylamino-2-phenylindene-1-one-3
(R-O) and 1-phenylamino-2-phenylindene-1-thione-3 (R-S) shown in Fig.2.18
(cf. [2.134] and references therein). These two compounds differ only in
the heteroatoms O and S. However, as can be seen from Fig.2.18, a replacement of heteroatom O by S in the molecule causes considerable changes both
in neutral state spectra, as well as in parameters of ionized states of the
crystal.

Considerably stronger effect of heteroatoms on the energy structure of
the crystal can be illustrated by a comparison of the parameters of two
isostructural tetracene derivatives — tetrathiotetracene (TTT) and tetra-
seleniumtetracene (TSeT) with those of Tc (cf. Fig.2.19). In TTT and TSeT
crystals the values of parameter I_c and, particularly, of ΔE_G are considerably lower than in a Tc crystal. A similar picture is observed in the case
of replacement of the heteroatom S by Se in TSeT: the I_c and ΔE_G values become considerably lower, whilst those of A_c and P_h increase. The comparatively narrow energy gap, viz. ΔE_G = 2.0 eV for TTT and ΔE_G = 1.5 eV for

136

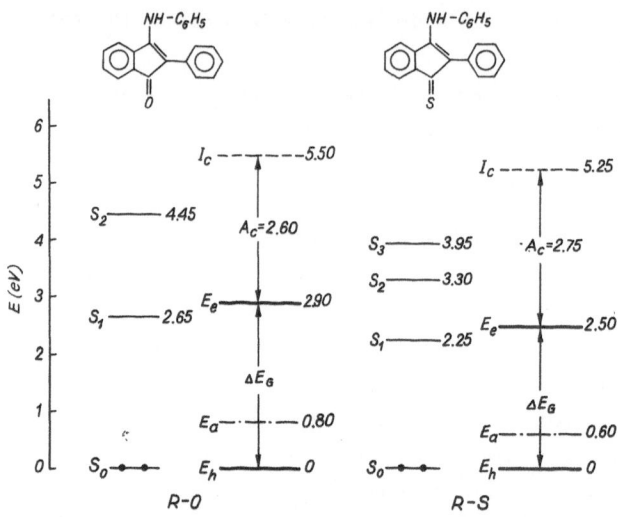

Fig. 2.18. Energy level diagrams of neutral and ionized states for 1-phenyl-amino-2-phenylindene-1-one-3 (R-O) and 1-phenylamino-2-phenylindene-1-thione-3 (R-S) crystals [2.134]

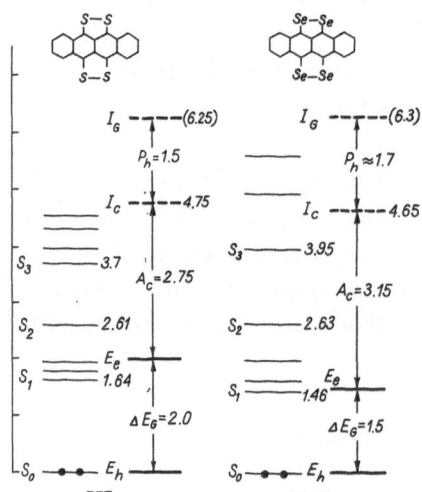

Fig.2.19. Energy level diagrams of neutral and ionized states for tetra-thiotetracene (TTT) and tetraselenium-tetracene (TSeT) crystals

TSeT is largely responsible for the pronounced semiconducting properties of TTT and TSeT crystals and layers [2.141,142].

The transbisbindonilene (TBB) molecule is large, symmetrical and almost planar with a strongly developed conjugated system containing 42 π electrons. The energy diagram of TBB crystals exhibits (cf. Fig.2.20), similarly to

TTT and TSeT, relatively low values of parameters I_C and ΔE_G, as compared to polyacene crystals (cf. Fig.2.16, see [2.134]).

Finally, Fig.2.20 presents energy diagrams of crystals of rather exotic compounds, viz. indandione-1,3-pyridinium betaine (IPB) and its analogue-betaine-2-N-pyridinium-4-aza-indandione-1,3 (N-IPB). Compounds IPB and N-IPB belong to the polar intramolecular salt series, in which cation and anion are covalent bonded in one molecule. These molecules possess a large permanent dipole moment and high molecular polarizability as indicated by the comparatively high values of parameters P_h and A_c, leading to low ΔE_G values of the crystals. A high molecular polarizability of these compounds is, obviously, due to strong conjugation and presence of heteroatoms in the conjugated chain. Under such conditions it is natural to expect different values of electronic polarization energies of an electron and a hole, i.e., $P_h \neq P_e$, owing to absence of symmetry and presence of strongly electronegative heteroatoms. This circumstance considerably complicates calculations of the energy structure of such heterocyclic compound crystals [2.134].

It is noteworthy that, unlike aromatic crystals, in TBB, IPB and N-IPB crystals the ΔE_G value is lower than that of E_{S_1}, viz. $\Delta E_G < E_{S_1}$ (cf. Fig.

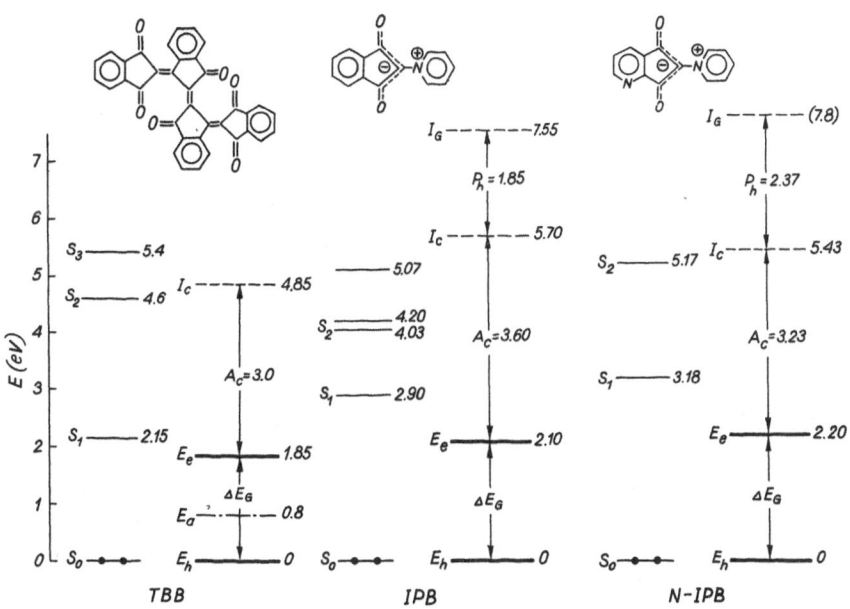

Fig. 2.20. Energy level diagrams of neutral and ionized states of trans-bisbindonilene (TBB), indandione-1,3-pyridinium betaine (IPB) and betaine-2-N-pyridinium-4-aza-indandione-1,3 (N-IPB) [2.134]

2.20). Hence, intrinsic photoconductivity in these crystals can be observed already in the long-wave absorption band [2.134].

This shows that the presence of heteroatoms in a molecule imparts qualitatively new properties on a crystal, thus not permitting any extrapolation of correlations obtained in the study of aromatic crystals to heterocyclic molecular crystals [2.134].

3. Role of Structural Defects in the Formation of Local Electronic States in Molecular Crystals

... And now remains
That we find out the cause of this effect, Or
rather say, the cause of this defect, For
this effect defective comes by cause.

Shakespeare
Hamlet (Act II Sc. II)

It has become obvious in the course of the last 10-15 years of research that the energy structure of molecular crystals cannot be adequately described in terms of discrete levels of ionized states (discrete conduction levels for electrons E_e and holes E_h) alone, without allowing for any local levels inside the energy gap ΔE_G. Such an approach gives an oversimplified picture and is feasible only in the case of an ideal (perfect) crystal (cf. Figs. 2.3b and 3.1b)

The majority of papers dealing with energetics of real molecular crystals, of anthracene in particular (cf. [1.21,38,40,44,45,47,48,62,75,76,3.1])) show that inside the energy gap ΔE there exists a large number of trapping states, both for charge carriers and excitons.

These local states can roughly be divided into two types: 1) traps with discrete energy levels and 2) traps with a quasi-continuous energy spectrum, the latter being often formally described in terms of exponential distribution [1.75,76,3.1].

Information on energy and kinetic parameters of trapping states can be obtained using traditional experimental procedure, such as the methods of space charge limited currents (SCLS) (cf. Sect.5.8), thermo- and photostimulated conductivity (TSC and PSC), photoemission (PE), etc. (cf. Sects. 5.12,13). These experiments do not, however, yield any information on the physical nature and microstructure of the centers responsible for the appearance of local energy levels. There were only some intuitive conjectures, based on general considerations and sometimes on indirect evidence, suggesting that discrete trapping levels are most likely connected with impurity centers, whilst quasi-continuous traps are due to structural defects of the crystal [1.70,3.2,3].

HELFRICH and LIPSETT [1.70] were, to our best knowledge, the first to express the idea that structure-caused trapping centers might be due to local lattice compression. HELFRICH [1.75] suggested that, according to one-electron approximation, the conduction bands must become wider in the compression regions, thus forming local trapping centers. On the other hand, BATLEY and LYONS [2.154] have shown that homogeneous lattice compression causes another effect, namely, an increase in electronic polarization energy P of the crystal by the charge carrier.

In terms of such an approach, BAESSLER et al. [3.4], using the macroscopic Born formula for rough estimation of the electronic polarization energy, have demonstrated that shortening of average intermolecular distance produces an increase in electronic polarization energy of the crystal by a charge carrier localized in the deformed lattice region. This led the authors to the suggestion that regions of local lattice compression can act as traps for electrons and holes.

In 1970 we [1.66] and, independently, SWORAKOWSKI [3.5], proposed a more detailed phenomenological model. According to this model local charge carrier trapping states with quasi-continuous energy spectra are due to local electronic polarization energy variations for charge carriers located in the regions of structural irregularities of the crystal.

However, unlike SWORAKOWSKI [3.5], who used traditional exponential approximation for describing polarization-caused quasi-continuous energy spectra of traps, we emphasized mainly the statistical nature of local state formation [1.66]. This brought us to the idea that the random distribution of structural irregularities in the crystal leads to a statistical dispersion of electronic polarization energy. As the most feasible approximation of energy spectra for such local states of structural origin we proposed the Gaussian distribution model [1.66].

Before embarking on a more detailed treatment of the problem, we consider it essential to outline here our general approach. We should like to emphasize that the hypothesis discussed in this section is for us of utmost importance, since it constitutes the basis of our whole concept on the nature of local states in molecular crystals. We have endeavored to develop this concept in the following chapters, and to prove its validity.

3.1 Statistical Aspects of the Formation of Local States
of Polarization Origin

In an ideal crystal, with perfect periodicity of structure and strictly
equal unit cell parameters throughout the lattice, polarization energies
of electron P_e and a hole P_h are also strictly fixed and have discrete values for a given crystal. This means, that conduction levels E_e and E_h also
have discrete values (Fig.3.1b). However, every real crystal contains various structural irregularities and lattice defects. Accordingly, translational symmetry does not apply throughout the whole bulk of the crystal.
Unit cell parameters also exhibit local deviations, and intermolecular distances r_i in some arbitrarily chosen region in the crystal will have some
kind of statistical distribution.

But every deviation of intermolecular distance Δr_i from its mean value
$\bar{r}_i$ ($\Delta r_i = r_i - \bar{r}_i$, where r_i is the distance between the molecule with localized charge carrier (ion) and the ith molecule) produces a corresponding
change in electronic polarization P of the crystal by the charge carrier

$$\Delta P = P\,(\bar{r}_i + \Delta r_i) - P\,(\bar{r}_i) \quad .$$

The value of ΔP, as obtained in approximation of point charge interaction
with induced dipoles, is given by formula [cf. (2.12)] [1.66]

$$\Delta P = \frac{\partial}{\partial r_i}\left(-\sum_{i=1}^{N-1}\frac{e^2\alpha}{2\bar{r}_i^4}\right)\Delta r_i = 2e^2\alpha\sum_{i=1}^{N-1}\frac{\Delta r_i}{\bar{r}_i^5} \quad , \tag{3.1}$$

where e is the charge of the electron, and α is the isotropic polarizability
of the molecule.

If we admit that deviations in intermolecular distances Δr_i caused by
structural irregularities of the crystal, are of random, statistical nature,
it is natural to assume that the total set of states $N(r_i)$ follows the law
of normal Gaussian distribution [1.66]

$$N(r_i) = \frac{N_0(r_i)}{\sqrt{2\pi}(\sigma_r)_i}\exp\left(-\frac{\Delta r_i^2}{2(\sigma_r)_i^2}\right) , \tag{3.2}$$

142

where $(\sigma_r)_i^2$ is the dispersion of Gaussian distribution of intermolecular distance r_i.

From (3.1,2) it is possible to obtain an expression of the dispersion parameter (root-mean-square standard deviation) σ_p of electronic polarization energy

$$\sigma_p = 2e^2\alpha \sqrt{\sum_{i=1}^{N-1} \frac{(\sigma_r)_i^2}{(r_i)^{10}}} \qquad (3.3)$$

On the basis of (3.1-3) one can postulate a Gaussian distribution of local ionized states in the energy diagram of the crystal

$$N(P) = \frac{N_0(P)}{\sqrt{2\pi}\sigma_p} \exp\left(-\frac{\Delta P^2}{2\sigma_p^2}\right) , \qquad (3.4)$$

where $N_0(P)$ is the total density of states; ΔP is the deviation of electronic polarization energy from the mean value $\bar{P}$.

The Gaussian spectrum of local ionized states in the energy diagram of a molecular crystal is demonstrated in Fig.3.1c.

Distributions (3.2,4) postulate that every single local center of structural imperfection in the crystal causes the appearance of a corresponding

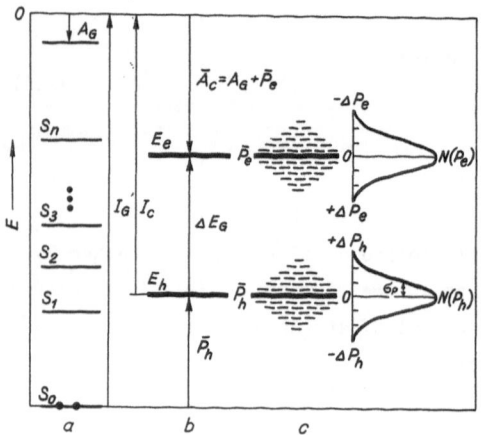

Fig. 3.1. Schematic energy diagram for a real molecular crystal with local states: (a) energy levels of neutral states of crystal; (b) discrete energy levels of ionized states of a perfect crystal; (c) local states of Gaussian distribution due to statistical dispersion of polarization energy, caused by structural imperfections of crystal lattice [1.66]

local state in the energy diagram of the crystal. In other words, the statistical dispersion of the energy eigenvalue of the charge carrier which
is localized on a definite lattice point of the crystal is caused by random
irregularities of the crystal lattice around the localized charge.

From the physical point of view, the model of Gaussian distribution is
more feasible for the description of local states in molecular crystals
than the widely used exponential approximation. Local states with increased
polarization energy ($\Delta P > 0$), which correspond to compressed defect regions
in the crystal, form charge carrier traps inside the energy gap ΔE_G. On the
other hand, local states with diminished polarization engergy ($\Delta P < 0$), which
correspond to dilated defect regions in the crystal, form energetically
unfavorable "antitraps" for charge carriers situated above the electron
conduction level E_e and below the hole conduction level E_h (see Fig.3.1c).

Finally, states having mean polarization energy values ($\bar{P}_h$, $\bar{P}_e$) corresponding to the most regular parts of the crystal form the conduction levels
for holes (E_h) and electrons (E_e), centered at the peak of the Gaussian distribution "bell" (see Fig.3.1c).

It should be borne in mind, however, that in early studies [1.66,3.5] the
possible role of structural imperfections in the formation of local states
was discussed only in a very general fashion, without regard to any definite
kinds of structural defects in molecular crystals.

As a matter of fact, at the beginning of the seventies the data on structural defects in organic molecular crystals were scarce and fragmentary.

At the same time, research on the energy spectra of charge carrier trapping states in a number of molecular crystals, especially in anthracene,
was rather intensive already in the sixties and the early seventies (cf.
[1.38,75,76]). These investigations, however, were carried out by traditional techniques of charge carrier trapping measurements, such as space
charge limited currents, thermo- and photostimulated currents, etc. These
provided abundant data on the energetic and kinetic parameters of trapping
centers, but yielded practically no information on the physical nature of
traps, especially of those with a quasi-continuous energy spectrum. In fact,
at that period research on trapping centers was carried out with complete
disregard for any microstructural studies of the crystal.

Serious studies on topology of microstructural defects in anthracene-
type molecular crystals were actually started only at the beginning of the
seventies, mainly by J.M.THOMAS and his group in Aberystwyth (U.K.) [1.77-80,
3.7-10]. THOMAS and WILLIAMS [1.77] were also first to suggest that edge
dislocations in molecular crystals may act as trapping centers for charge

carriers and excitons. They also attempted to find correlations between dislocation and trapping center densities for charge carriers in an anthracene crystal [1.77]. They did not, however, offer any suggestions as to the possible physical mechanisms of charge carrier trapping in dislocation defects.

The phenomenological model proposed by SILINSH [1.66] and SWORAKOWSKI [3.5] opened new avenues for interconnecting energetic and microstructural aspects in the study of local states. This had a considerably stimulating effect on further search for correlations between specific structural defects in a molecular crystal (dislocations, stacking faults, etc.) and corresponding polarization-caused local states [1.62,63,67,69,21,22,3.6,11-13].

Further development of the Gaussian model of local state distribution [1.66] by NEŠPUREK and SILINSH [1.68] showed the feasibility of introducing two basic kinds of distribution, viz. $G_e(E)$ and $G_g(E)$ (see Fig.3.2).

The shallow, quasi-continuous local trapping states with $\Delta P > 0$ at the conduction level *edge* and their symmetrical counter-part - "antitrapping" states with $\Delta P < 0$ (cf. Fig.3.1) may be approximated by a Gaussian distribution, centered symmetrically around the conductivity levels E_e and E_h with values of $E_t = 0$ and $E_t = \Delta E_G$, respectively (see Fig.3.2).

On the other hand, quasi-discrete, statistically distributed trapping states located in the forbidden energy *gap* may be approximated by another Gaussian distrubution $G_g(E)$ centered at E_t below the conductivity level E_e (see Fig.3.2). The Gaussian model thus makes it possible to consider the integral profile of local state distribution within the forbidden energy gap ΔE_G as a superposition of separate Gaussian distributions.

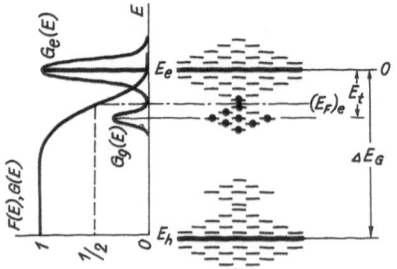

Fig.3.2. Schematic energy diagram for a molecular crystal with Gaussian distribution of local states. $G_e(E)$ - Gaussian trap distribution centered at conductivity level; $G_g(E)$ - Gaussian trap distribution situated in forbidden energy gap [1.68]

Both kinds of Gaussian distribution $G_e(E)$ and $G_g(E)$ may be described by
the formula [1.68]

$$N(E_i) = \frac{(N_t)_i}{\sqrt{2\pi}\sigma_i} \exp\left(-\frac{[E-(E_t)_i]^2}{2\sigma_i^2} \right) , \tag{3.5}$$

where $(N_t)_i$ is the total density of local states of the corresponding distribution, σ_i is the distribution parameter; $(E_t)_i$ is the position of
Gaussian distribution peak inside the energy gap (see Fig.3.2). At $E_t = 0$
or $E_t = \Delta E_G$ the $G_g(E)$ type distribution coincides with that of the $G_e(E)$
type.

Such a generalized Gaussian model includes into the common description
also limiting cases of local state distribution. Thus, at the limit, when
the parameter $\sigma \to 0$, the $G_e(E)$ distribution gives an idealized picture of
discrete conductivity levels E_e and E_h for a perfect crystal. On the other
hand, the $G_g(E)$ type of distribution at the limit, when $\sigma \to 0$, describes a
discrete trapping state at E_t, which may be considered as a δ function.

The Gaussian approach is physically self-consistent in the framework of
a generalized phenomenological ionized state energy model for real crystals,
developed in the present work, according to which the energy of conductivity levels as well as that of the local states of structural origin are determined by electronic polarization of the crystal by charge carriers.

A more detailed discussion of the Gaussian model and of phenomenological
SCLC theory developed for trapping states of Gaussian distribution will be
presented in Chap.5.

3.2. General Considerations on the Role of Specific Structural Defects

A molecular crystal possessing a perfectly regular structure is an idealized
abstraction. Such an abstraction is, nevertheless, of considerable heuristic
value in simplifying to a large extent the analysis of the energy structure
of the electronic states of the crystal (cf. Chap.2).

However, the concept of an ideal crystal, even though easy to define, is
an experimental rarity and invariably we have to talk about "real" crystals,
viz. crystals containing various types of imperfections. These imperfections are usually introduced during the growth or subsequent treatment of

the crystal and have the general property of disturbed regularity or periodicity of the crystalline structure [1.81].

The technological possibilities of crystal engineering in obtaining a
perfect crystal are physically limited by the principles of thermodynamics
since there exist specific concentrations of each type of defect at a particular temperature which have been termed "thermodynamic" defects (cf.
Sect.3.3). From the crystal engineering point of view an ideal crystal can
only be approached asymptotically by means of ultra-high purification and
perfection of conditions of growth and treatment of the crystal.

It is only in living nature, in self-organizing molecular biosystems,
such as biopolymers (DNA, proteins) or electron transfer cascades in
mitochondria and chloroplasts, that one finds a system closely approaching
ideal perfection, since any defect here may lead to irreversible even
lethal changes in the whole molecular biosystem.

In molecular biosystems perfection is achieved through the prinicple of
self-organization, the latter being effected at the expense of decrease in
entropy in the system itself, accompanied by a corresponding greater increase in entropy of the environment. One can expect in the nearest future
attempts to model the principle of self-organization of molecules in biosystems as a technological method of obtaining perfect molecular crystals.

According to the generally accepted classification, structural defects
in lattice regularity can be divided into point, line, planar and bulk
defects (cf. [3.14,15]). The following specific forms of structural imperfections can be distinguished in molecular crystals:

a) *Point Defects*, such as monovacancies, "orientational" point defects
[1.81], interstitial molecules (Frenkel defects), dimers, vacancy pairs,
small clusters [1.77,80,81,3.16,17], "self-trapped" charge carriers (see
Sect.5.16), impurity centers consisting of "guest" molecules or atoms in
lattice sites or interstitialcies, which are often termed "chemical" defects [1.45,3.5] etc.;

b) *Line Defects* — dislocations [1.77,81] and disclinations [3.18,19];

c) *Planar Defects*, such as linearized, small- and medium-angle tilt
boudaries, interphase boundaries, stacking faults, crystal surface [1.77,80,
81];

d) *Bulk Imperfections* which may consist of metastable second-phase inclusions or polymorphic inclusions in hybrid bi- or multiphasic organic crystals [1.80,86,87,88], clusters, impurity aggregations, multivacancies and
various macroscopic bulk defects (pores, fissures, etc.).

Each molecule, being displaced from its equilibrium position within the
defect region, forms a local state with respect to the regular lattice if
it possesses a changed value of electronic polarization energy P of a charge

carrier localized on the given molecule. Thus each specific structural imperfection of the crystal — vacancy, dislocation, tilt boundary, stacking fault, etc. — has its own set of local states equal to the number N of molecules forming the given structural defect. Some of these local states may, naturally, be degenerate in the energy spectrum owing to symmetric arrangement of a number of molecules in the defect region.

This reduces the problem of the full energy spectrum of local states in molecular crystals to the determination of the coordinates of displaced molecules in the defect lattice regions. Once these coordinates are known for all molecules displaced from their regular sites, there are no serious difficulties in evaluating the complete energy spectrum of these local states.

Unfortunately, it is at present possible to evaluate the coordinates of displaced molecules only in the case of the very simplest structural defects, such as planar stacking faults of limited dimensions in an anthracene crystal (cf. Sect.3.9.3). For other types of structural defects, even for a single edge dislocation only a qualitative or semi-quantitative picture can be obtained for the configuration of molecules in the core of the defect.

Such qualitative pictures make it, nevertheless, possible to obtain a rough estimate of the effective energy value of local charge carrier trapping states in the given specific structural defects (cf. Sect.5.2). Useful information can also be obtained from the mean lattice energy values in specific structural defects. The energy of a defect permits an estimate of the probability of its formation. In some cases it is even possible to evaluate the relative change in mean pressure $\delta p/p$ in the defect region of the crystal (cf. Sect.3.8.2). Knowing the value of $\delta p/p$ it becomes, in its turn, possible to assess the effective depth of the given trapping center.

Accordingly, we are going to discuss in the following sections topographic (configurational) and energetic aspects of specific structural defects in anthracene-type molecular crystals. These data will eventually be used in Chap.5 for determining the energy spectrum of local charge carrier trapping states.

It follows from the aforementioned local polarization model (cf. Sect. 3.1) that each definite type of structural defect with fixed topological parameters has its corresponding characteristic spectrum of local states. If by chance any of the defect parameters changes, say, the width or the length of the planar stacking fault (cf. Sect.3.9.1), corresponding changes

parameters can assume random values, the energy spectrum of local states in the defect region must also have a definite statistical distribution (cf. Sect.3.1).

Only simple isolated defects can possess a set of discrete local state levels. It may be shown that statistical superposition of such discrete sets of local states forms a quasi-continuous Gaussian-type spectrum (cf. Sect.5.7).

The statistical nature of local states is more pronounced in the case of dislocation ensembles, in which dislocations interact by means of an elastic stress field of random distribution (cf. Sect.3.7.3). In such cases only a statistical approach can adequately deal with the formation of local states.

As a working hypothesis which takes into account the above-mentioned considerations, one can propose the principle of mutual correspondence between statistical distributions of deformational displacements of molecules in structural defects of a crystal, and the energy spectrum of structure-caused local states.

The feasibility of applying Gaussian distribution for describing the spectrum of local states in molecular crystals has already been proved experimentally, as will be shown in Chap. 5. As regards the possibility of Gaussian distribution of deformational displacements of molecules, we shall attempt to substantiate it by means of a general analysis of different randomizing factors in the formation of various structural defects in real crystals, as well as on the basic of the statistical theory of dislocational ensembles (cf. Sects.3.7.3 and 5.7).

3.3 Point Defects (Vacancies) in Molecular Crystals, Their Crystallographic and Electronic Properties

Irregularities in a crystal which spread in every direction within a range of a few intermolecular distances are called point defects. They may be subdivided into impurity and structural defects (see, e.g. [3.20-22]).

The dominating structural point defect in molecular crystals is a lattice vacancy [1.81,3.17,23], viz. a lattice site with a molecule missing. The interstitials, on the other hand, are not favored energetically, since displacement energy of a molecule into intersititial position exceeds more

than threefold that of vacancy formation [3.16,17]. The probability of
Frenkel-type defect formation is therefore insignificant in molecular crystals.

Methods of observation and characterization of point defects in solids
are essentially indirect. Unlike line or planar defects, none of the conventional structural techniques of direct revelation may be employed in this
case. The detection and investigation of point defects is mainly based on
the effects they have on the physical properties of a crystal [1.81].

In this aspect various crystalline solids differ considerably from each
other. Thus, in ionic crystals point defects are of primary importance in
determining optical and electric properties. A vacancy in an ionic crystal
breaks the periodic potential of the crystalline field and can thus create
an active trapping center for charge carriers. Thus, an anion vacancy in
alkali halide crystals forms an effective local trapping center of an electron, a so-called color or F center [3.20]. A vacancy in a covalent lattice
of, say, a diamond-type crystal means a break of four valency bonds. The
lone electrons can form pairs of hybrid orbitals and thus produce hole trapping levels, so-called vacancy levels in the forbidden energy gap [3.22].

The physical properties of vacancies in a molecular crystal differ fundamentally from those in an ionic or in a covalent one. In the case of
nonvalent Van der Waals-type of interaction between the molecules, and if,
additionally, the molecules forming the crystal possess a positive electron
affinity ($A_G > 0$), a charge carrier cannot, on principle, be trapped in a
vacancy. In other words, vacancies in molecular crystals do not form, as a
rule, local trapping states within the energy gap. Localization of charge
carriers or excitons actually takes place only on individual molecules
(cf. Sect.1.4).

Intermolecular space and lattice vacancies are energetically unfavorable
for charge carrier localization. When a charge carrier is localized on a
molecule adjoining a vacancy, its electronic polarization energy decreases
by a value of ΔP. Accordingly, the molecules in the environment of the lattice vacancy form a set of local states with a lower value of P, thus creating a potential barrier around the vacancy (cf. Sect.5.5). Vacancies in
a molecular crystal therefore act as "antitraps" for charge carriers [1.66,
3.11] and also for excitons (cf. Sect.4.3). The only exception may be crystals consisting of molecules with negative electron affinity ($A_G < 0$), such
as benzene.

Consequently, the role of vacancies in determining optical and electronic properties of a molecular crystal is, most likely, rather insignificant (cf., e.g. [1.79]).

On the other hand, it has been shown that lattice vacancies are of primary importance in processes of self-diffusion of molecules in a crystal [1.79,3.16,17]. Vacancies are actually the dominant mobile point defects in molecular crystals [1.79,3.23].

The migration of a vacancy may be described in terms of the self-diffusion model, by means of diffusion coefficient D [3.16,17]

$$D = D_0 \cdot \exp\left(-\frac{E_d}{RT}\right) = \frac{1}{6} fa^2\omega \exp\left(\frac{S_d}{R}\right) \exp\left(-\frac{E_d}{RT}\right)$$

$$= \frac{1}{6} fa^2\omega \exp\left(\frac{S_V+S_M}{R}\right) \exp\left(-\frac{E_V+E_M}{RT}\right) \quad , \tag{3.6a}$$

where E_d and S_d are, respectively, the activation energy and the entropy of the whole diffusion process; E_V and S_V are the activation energy and the entropy of vacancy formation; E_M and S_M are the activation energy and entropy of migration of the vacancy; f is the correlation factor related to lattice geometry; a is the lattice parameter; ω is the lattric vibration frequency.

According to this model we have

$$E_d = E_V + E_M; \quad S_d = S_V + S_M \quad . \tag{3.6b}$$

A comparison between E_d, E_V, S_d and S_V values for benzene crystals leads, according to SHERWOOD [3.16] and FOX and SHERWOOD [3.24], to the following expressions:

$$E_V \approx \frac{1}{2} E_d; \quad S_V \approx \frac{1}{2} S_d \quad , \tag{3.7}$$

Expressions (3.7) confirm the basic concepts of self-diffusion in molecular crystals as the result of formation and migration of lattice vacancies.

Table 3.1 presents self-diffusion parameters for a number of aromatic molecular crystals, as determined by the radioactive tracer method [3.16]. Reliable measurements are typified, according to CHADWICK and SHERWOOD [3.17], by an activation energy E_d which is approximately double the lattice energy of the crystal [$E_d = (1.9-2.4)L_s)$] (cf. Table 3.1). The variations in D_0 value reflect mainly the corresponding variations in the entropy

Table 3.1. Self-diffusion parameters of some aromatic molecular crystals

Crystal	D_m [m²/s]	D_o [m²/s]	E_d [kJ/mol]	E_d [eV]	L_S [kJ/mol]	L_S [eV]	E_d/L_S	Ref.	S_d [3.16] [J/mol/K]	S_d [3.16] [eV/K]	S_d/S_f
Benzene	10^{-13}	1.5×10^5	96.8	1.0	46	0.48	2.1	[3.24]	209	2.16×10^{-3}	6
Naphthalene	10^{-15}	2×10^{11}	179	1.85	75	0.78	2.4	[3.16,29]	334	3.45×10^{-3}	6
Anthracene	10^{-13}	7×10^6	177	1.83	93	0.96	1.9	[3.16,30]	263	2.72×10^{-3}	4
	10^{-13}	3×10^{-4}	92	0.95	92	0.95	1.0	[3.31]			
	10^{-14}	1×10^{-6}	84	0.87	84	0.87	1.0	[3.32]			
	10^{-16}	1×10^6	202	2.08	88	0.91	2.3	[3.16]			
Phenanthrene	10^{-15}	3×10^{13}	202	2.08	84	0.87	2.4	[3.33]	368	3.79×10^{-3}	7

Note: D_m is the mean diffusion coefficient at melting point; D_o is the preexponential factor [cf.(3.6a)]; E_d is the activation energy of diffusion; L_S is the latent heat of sublimation; S_d is the entropy of diffusion; S_f is the entropy of melting.

term S_d which can be regarded, according to (3.6b), as the sum of the entropy terms associated with the formation and migration of the diffusing defect [3.17]. Therefore it is quite natural that S_d exceeds several times the melting entropy S_f (see Table 3.1.). It is particularly noteworthy that the values of S_d for organic solids are extremely large compared with S_d for simpler solids. Such high values of S_d are in reasonable agreement with the vacancy self-diffusion model [3.17].

Information currently available on the experimental magnitudes of vacancy formation and migration parameters is sparse. Nevertheless overall results of radioactive tracing, as well as jump correlation experiments, confirm that self-diffusion in molecular solids proceeds via vacantly associated processes [3.17]. Judging by presently available data, it seems most reasonable to conclude that the basic point defects in molecular crystals are lattice vacancies.

For a detailed description of the experimental technique used in vacancy point defect investigations we recommend to consult review articles by SHERWOOD et al. [3.16,17].

Rates of self-diffusion in molecular solids vary widely, being low in more rigid lattices and extremely high in more disordered systems [3.17].

Thus, it was established, using radioactive tracer methods, that self-diffusion of molecules proceeds at a considerably higher rate along dislocations and grain boundaries, than within the bulk of a perfect crystal. For instance, the ratio between the mean self-diffusion coefficient along subgrain boundary D_g and the self-diffusion coefficient in a perfect crystal D_ℓ has the following values (near melting point): $D_g/D_\ell = 10^5$ for anthracene and $D_g/D_\ell = 10^6$ for naphthalene crystals [3.16,17,25].

This shows that lattice vacancies in molecular crystals enter into exchange reactions with dislocations. Thus, on condensation of vacancies line defects, (dislocations) may be formed. On the other hand, at points of intersections of dislocations "splinters" in form of vacancy aggregations can arise [1.79] (cf. Sect.3.11 and Fig.3.26b).

It follows that lattice vacancies which in isolated state are not capable of forming trapping centers of polarization origin, are in a position to form such centers on condensation. On the other hand, at points of intersection of dislocations vacancy aggregations may form a more dilated local lattice region in the crystal, which should act as "antitrapping" center for charge carriers (cf. Sect.3.11).

There are indications, based on theoretical estimates by CRAIG et al. [1.87], that isolated monovacancies are the most stable vacancy type.

Condensation with formation of small linear multivacancies appears to be energetically unfavored.

If the presence of a single point defect raises the internal energy of the crystal by E_f, then the change in free energy ΔF of the crystal containing n similar defects at temperature T equals [3.21]

$$\Delta F = nE_f - T (S_c + nS_\ell) \quad , \qquad (3.8a)$$

where E_f is the creation energy of the defect; S_ℓ is the vibrational entropy due to defect-caused displacement of nearest neighbors; S_c is the configurational entropy connected with formation of n defects

$$S_c = k \ln W, \qquad (3.8b)$$

W being the number of various positions of point defects over N lattice sites.

It is assumed that n << N and that there is no interaction between defects.

It can be shown [3.21] that, if the number of point defects is not too large, the term S_c prevails over the terms nE_f and nS_ℓ and the free energy increase on introduction of defects can, in principle, be negative ($\Delta F < 0$). This means that at $T > 0$ thermodynamically stable point defects, such as vacancies, can be formed. Thus we have a situation where the formation of point defects in the crystal is thermodynamically favorable owing to the increase in configurational entropy S_c, caused by the presence of a variety of positions in the crystal which can be occupied by these point defects.

According to KELLY and GROVES [3.20], the vacancy density $n_V(E_{min})$ providing minimum free energy of the crystal $F = F_{min}$ can be estimated, minimizing (3.8a), i.e., evaluating $d(\Delta F)/dn_V = 0$

$$n_V(F_{min}) = N_0 \exp\left(\frac{S_\ell}{k}\right) \exp\left(-\frac{E_V}{kT}\right) \quad , \qquad (3.9)$$

where N_0 is the molecule density in the crystal; E_V is the energy of vacancy formation.

The estimated value of $n_V(F_{min})$ for anthracene, obtained from (3.9), taking into account (3.7) and data from Table 3.1, and putting $S_\ell/k \approx 1$ [3.20], equals 4×10^7 cm^{-3} at room temperature and 6×10^{12} cm^{-3} at melting point of anthracene (T = 216.6°C). This means that an anthracene crystal with a vacancy concentration of the order of 10^7-10^8 cm^{-3} has lower free energy at room temperature than a perfect crystal. It is thus thermodynamically not feasible to obtain perfect crystals, free from point defects, at T > 0. Such types of defects having a specific concentration at a particular temperature are often called "thermodynamical" defects [1.81].

A further increase in vacancy density $n_V > n_V(F_{min})$ entails corresponding increase in free energy value ($\Delta F > 0$). In such a case n_V may be evaluated according to the formula

$$n_V = N_0 \; \exp\left(\frac{S_V}{k}\right) \exp\left(-\frac{E_V}{kT}\right) , \tag{3.10}$$

where S_V is the total entropy of vacancy formation.

The estimated values of lattice vacancy density n_V for anthracene and naphthalene crystals, as obtained from (3.10), are presented in Table 3.2.

Table 3.2. Estimated values of lattice vacancy density n_V in anthracene and naphthalene crystals at room temperature (300 K) and at melting point of the crystal

Crystal	Molecule density in the crystal, [cm-3]	Activation energy of vacancy formation $E_V = 1/2\ E_d$, [eV]	Entropy of vacancy formation, $S_V \approx 1/2\ S_d$, [eV]	Vacancy density n_V at room temperature [cm^{-3}]	Vacancy density n_V at melting point of crystal [cm-3]
Anthra-cene	4.2×10^{21}	0.9	~ 1.4×10^{-3}	1.5×10^{14}	2×10^{19}
Naphtha-lene	5.6×10^{21}	0.9	~ 1.7×10^{-3}	7×10^{15}	3×10^{17}

Note: Data from Table 3.1 and expression (3.10) have been used. Melting points: 216.6°C for anthracene, and 80.3°C for naphthalene.

As may be seen from Table 3.2, a considerable vacancy density can be formed in anthracene and naphthalene crystals at room temperature – up to an order of 10^{14}-10^{16} cm^{-3}. Since each vacancy is surrounded by a set of ten and more local "antitrapping " states, the average local state density $N_t(V)$ with decreased polarization energy value P will exceed the value of n_V by another order of magnitude (cf. Sect.5.5).

Recently CRAIG et al. [1.87] have discussed the possible influence of
the kinetics of crystal growth on the initial concentration of vacancies in
real molecular crystals. Equilibrium concentration of vacancies, evaluated
from (3.10), is realistic only if crystal growth is slow and vacancy mobility sufficient to permit reduction of initial nonequilibrium vacancy concentration by diffusion to the surface. A rough estimate shows [1.87] that
in a crystal growing from pure material at melting temperature at a rate
exceeding $\sim 10^{-7}$m s^{-1}, or from room temperature solution at a rate exceeding $\sim 10^{-9}$m s^{-1}, most of the vacancies at the surface will be incorporated
into the crystal as the nonequilibrium bulk concentration of vacancies.
These initial concentrations may decrease because of migration of vacancies
to preexisting defects, such as dislocations or surfaces. Such a process is,
however, extremely slow. Unless the crystal is annealed for hours near the
melting point, the vacancy concentration will not decrease significantly
towards equilibrium value, even though vacancies are mobile enough to move
a few nm in a period of a few hours [1.87].

It should be mentioned that in organic molecular solids some specific
point defects may be formed, e.g., so-called "orientational point defects"
[1.81]. These defects may arise in crystals consisting of molecules not
possessing a center of symmetry. In some lattice sites an incorrect (e.g.,
reverse) orientation of a molecule with respect to the rest of the perfect
structure may take place. This type of point defect may assume increasing
importance in studies of molecular crystals consisting of asymmetrical,
complex organic molecules.

If the crystal consists of molecules part of which may rotate around a
σ bond, rotational point defects and other types of rotation disorders may
arise (see, e.g. [3.26]).

In recent years increasing interest has been focused on another specific
kind of point defects in anthracene-type crystals, namely, so-called "incipient" dimer or predimer defect states in the lattice which may act as
effective trapping centers for excitons (see, e.g. [3.27,28]) and, probably, also for charge carriers. Since the formation of this type of point
defects is usually connected with other types of extended line or planar
defects it will be discussed in some detail in Sect.3.10.

3.4 Dislocation Defects, Their Role in Local State Formation

In real crystals, dislocations occur as a result of inelastic slip of part
of the crystal along certain, so-called slip planes. Dislocations are essentially one-dimensional (line) defects. A dislocation is, by definition,

the boundary line between that part of a crystal which has undergone slip and that which has not [1.77,81].

All dislocations have a slip vector or Burgers vector $\underline{b}$ associated with them, and this vector uniquely describes the magnitude and direction of the slip that has occurred.

There are two main kinds of dislocations—*edge* and *screw* dislocations. In an edge dislocation, the Burgers vector is perpendicular to the dislocation line, but in screw dislocation the line and the vector are parallel to one another. In general, dislocation lines are bent and irregular, and they possess both edge and screw components: a vector $\underline{b}_1$ arising from a pure edge (at right angles to the dislocation line), and a vector $\underline{b}_2$ from the pure screw (parallel to the line) so that $\underline{b} = \underline{b}_1 + \underline{b}_2$ [1.77].

Figure 3.3 shows schematically, according to THOMAS and WILLIAMS [1.77], the most widerspread dislocation in anthracene, namely, an edge dislocation in the ab plane of the crystal with unit Burgers vector of [010] direction. The line of dislocation lies in this case at right angles to the plane of the drawing and is denoted by the symbol ⊥ of edge dislocation.

We should note from Fig. 3.3 that the edge dislocation may be thought in terms of an extra half plane of molecules inserted above the slip plane, giving rise to regions of dilation (D) below and compression (C) above the slip plane [1.77,81].

This circumstance is of extreme importance for explaining the formation of local electronic states in a defective molecular crystal, according to

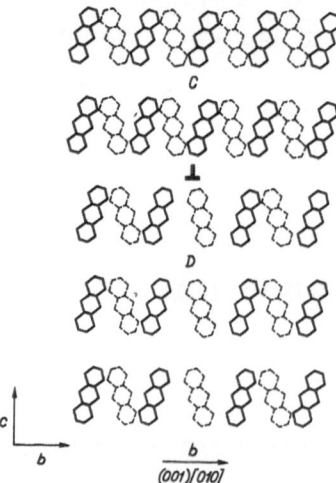

Fig. 3.3. Schematic picture of deformation of an anthracene lattice around an edge dislocation ⊥ of the type (001) [010]. C - region of compression; D - region of dilation [1.77]

our polarization model (cf. Sect.3.1). The regions of compression and dilation at the dislocation core have electronic polarization energies P differing from the mean value of $\bar{P}$ of a perfect crystal. A charge carrier located above the dislocation line in the region C of compressed lattice has increased P value ($\Delta P > 0$). Consequently molecules of the C region form a set of local trapping states for charge carriers. On the other hand, a charge carrier located below the dislocation line in the region D of the dilated lattice has decreased P value ($\Delta P < 0$). This means that molecules of region D form a set of local "antitrapping" states.

Statistical superposition of these local states, arising from different kinds of dislocations throughout the bulk of the crystal may lead to Gaussian distribution of $G_e(E)$ type above and below the discrete conductivity levels for electrons (E_e) and for holes (E_h) of a perfect crystal (cf. Sect. 3.1 and Fig.3.1c).

The qualitative picture of the topography of compression (C) and dilation (D) regions in molecular crystals (cf. Fig.3.3) is based on general quantitative treatment of the distribution of stress fields around the edge dislocation according to continuum elastic theory [3.34]. The diagram in Fig.3.4 illustrates the formation of compressive quasi-hydrostatic stress fields in the quadrant above the dislocation line and tensile stress fields in the quadrant below the dislocation line. The side quadrants show a rather complex picture of stress fields formed by the simultaneous action of combined compressive and tensile stress at right angles to each other. These fields produce corresponding deformational displacements of the molecules in the dislocation core and cause the formation of a set of local states according to the electronic polarization model (cf. Sect.3.1).

In the case of a screw dislocation the stress fields are more uniformly distributed and do not produce regions of considerable compression or dilation of the lattice.

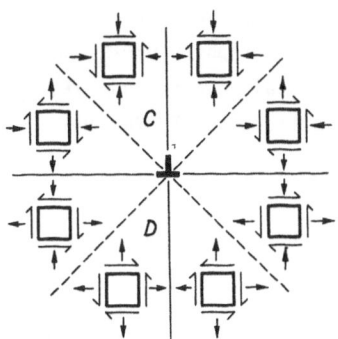

Fig. 3.4. Schematic diagram of stress fields around an edge dislocation: C - quadrant of compression; D - quadrant of dilation [3.34]

These dislocations are therefore of minor interest from the aspect of local state formation.

Considering the electronic properties of the dislocation defects, another important circumstance ought to be borne in mind. In a molecular crystal, due to nonvalent Van der Waals-type of interaction, all molecules situated in the defective region around the dislocation line are identical, including the molecule at the very edge of the extra plane (cf. Fig.3.3). Therefore all molecules in the defective region participate equally in the formation of local states. A quite different situation exists in the case of covalent crystals. Here the atom at the edge of the extra plane has a free valency and thus differs considerably in energy from other atoms of the crystal and forms a discrete trapping state [3.22,35].

We see thus that local states formed in the region of dislocation defects in molecular crystals have an essentially different physical nature from corresponding centers in a covalent crystal (cf. Sect.5.6 and Fig.5.9). It is therefore useless to search for analogies between the electronic properties of dislocation defects in these two different types of crystals, and molecular crystals maintain their singular position, exotic in its own way.

Summing up, we wish to stress once more that the electronics of dislocation defects in molecular crystals has very little in common with electronic properties of dislocation defects in covalent or ionic crystals (cf. Sect. 5.6). However, the purely crystallographic aspects of the theory of dislocations may be applied, if one takes into account such peculiarities of molecular crystals as high anisotropy, low mechanical strength, etc.

For a closer study of dislocation electronics we wish to refer the reader to classical works by its founders [3.35-38] as well as to more recent excellent monographs on the problem [3.20-22].

The basic physical principles of dislocation theory are concisely presented in the excellent review by COTTRELL [3.39]. A modern comprehensive dislocation theory can be found, in the detailed monograph by HIRTH and LOTHE [3.34]. Dislocation defects in molecular crystals and methods of their study are discussed in a number of review articles by THOMAS and WILLIAMS [1.77,79-81].

In recent years a related kind of line defects has come under investigation; they are so-called *disclinations* [3.18,19]. In a disclined solid one part of the structure is displaced by rotation rather than by translation. In a analogy with translationed edge and screw dislocations rotational wedge and twist disclinations may be defined [3.18]. A concise review of

disclinations has been recently given by HARRIS [3.19]. Disclinations are
seldom observed in ordinary three-dimensional crystals of spheric particles.
They occur mainly in arrays of oriented nonspheric complex molecules, e.g.,
in linear organic polymers, in molecular layers of fused benzene rings such
as graphites, in nematic liquid crystals, biomembranes, etc. Disclinations
are important structural elements in many ordered systems other than conventional crystals, such as closed surfaces of biosystems, e.g., protein
coats of viruses [3.19].

Combined translational and rotational displacements in molecular systems
produce a still more complex kind of defects, which are called *dispirations*
[3.18,19]. Like dislocation loops also twist disclination and dispiration
loops, formed by intramolecular rotation of linear molecules around a σ
bond, create local regions of compression in a solid. Such loops may, apparently, serve as local trapping centers for charge carriers.

Disclinations and dispirations may be of increasing importance as structural defects in complex organic molecular solids.

3.5 Energetics of Dislocations in Molecular Crystals

3.5.1 Discrete Configuration of Dislocations

The free energy W of dislocation displacement is not a constant quantity.
It is modulated by the periodic potential of lattice translation. This condition limits the possible Burgers vectors in such a fashion, that only a
discrete set of such vectors is energetically favorable, the one which
transfers the molecules associated with a dislocation from one metastable
state in the lattice into another. The discrete character of dislocational
slip is taken into account in the phenomenological PEIERLS-NABARRO model
[3.40,41] by introducing a variable lattice displacement potential of period $\alpha = x/b$.

In this model the dislocation metastability conditions are described in
terms of Peierls energy W_p which is introduced by a phenomenological equation for free displacement energy $W(\alpha)$ as a function of the displacement
$\alpha = x/b$ [3.34]

$$W(\alpha) = W \left(\frac{x}{b}\right) = \frac{W_p}{2} \left(1 - \cos \frac{2\pi x}{b}\right) = W_p \sin^2 \frac{\pi x}{b} \quad . \tag{3.11}$$

A typical curve of $W(x/b)$ dependence is presented in Fig.3.5a.

Figure 3.5a shows that the free energy of displacement $W(x/b)$ forms minima with a period equal to the Burgers vector b. Thus the Peierls energy
barrier W_p effects the formation of metastable dislocations in translational
lattice sites (see Fig.3.5b).

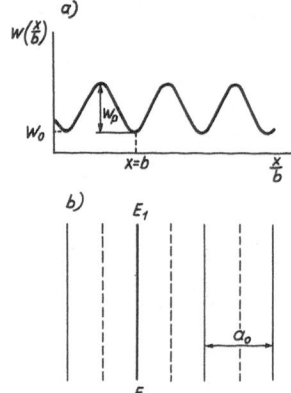

Fig. 3.5. Peierls energy barriers: (a) dependence of free energy of displacement W(x/b) in the dislocation core on the value of displacement x/b; (b) picture of the position of Peierls valleys (———) and of peaks of Peierls barriers (- - -) for a dislocation EE_1 (a_0 - distance between crystallographic planes)

Dislocations with Burgers vectors satisfying the metastability condition (Fig.3.5) are called *characteristic* dislocations of the crystal. It can easily be seen from Fig.3.5 that these are such Burgers vectors $\underline{b}$ which can be expanded in terms of basal vectors $\underline{i}$, $\underline{j}$, $\underline{k}$ of the lattice

$$\underline{b} = \ell\underline{i} + m\underline{j} + n\underline{k} \quad , \qquad\qquad (3.12)$$

where ℓ,m,n = 0,1,2,... .

Such dislocations with $\underline{b}$ values equivalent to integral multiples of the lattice spacing are known as *full* or *perfect* dislocations [1.81]. Of these the energetically most favorable ones are *unit* dislocations with $\underline{b}$ values equal to the translational vector of the lattice, i.e., with value of ℓ,m,n in (3.12) equal to unity or zero.

In anisotropic anthracene-type crystals the most probable direction of slip practically always coincides with the direction of the lattice vector in the plane of closest packing of molecules, i.e., in the ab plane of the crystal (cf. Fig.1.5). Such dislocations are called *basal*. Thus, in crystals of the anthracene space group the dominating basal dislocation is the unit dislocation of (001) [010] type (cf. Fig.3.3 and Sect.3.13).

Multiple dislocations with $\ell,m,n \geq 2$ in (3.12) are energetically unstable, and, as a rule, dissociate into unit dislocations.

In general, any dislocation with a Burgers vector $\underline{b}$ would preferentially, on energetic grounds, dissociate into two dislocations $\underline{b}_1$ and $\underline{b}_2$ provided that b^2 is greater than $b_1^2 + b_2^2$, giving rise to what is known as a dislocation reaction [1.81]. Consequently, to proceed meaningfully in any dislocation study we must consider the specific types of dislocations present in

the particular crystal lattice, since each lattice has its own energetically permissible Burgers vectors and dislocation reactions [1.81]. In specific cases full dislocations may dissociate into *partial* dislocations
having Burgers vectors corresponding to fractions of lattice spacings, provided the condition $b^2 > b_1^2 + b_2^2$ is satisfied. This occurs readily in anthracene crystals where a dominant partial dislocation of (001)1/2 [110]-
type is formed as the result of the dissociation of (001) [100]-type full
dislocations (cf. Sect.3.9.2). An important consequence of these specific
defects is that the region bounded by a pair of such partial dislocations
of opposite sign gives rise to a stacking fault ribbon in the ab plane of
an anthracene crystal (cf. Sect.3.9).

3.5.2 Basic Elastic Properties of Anthracene and Naphthalene Crystals

For calculating dislocation energy and other applications of dislocation
theory it is essential to know the basic elasticity moduli of the given
molecular crystal, in particular the shear modulus μ and the Poisson ratio
ν. This problem is, however, rather complex in the case of anisotropic crystals, such as naphthalene and anthracene. This is due to the fact that the
main expressions, applicable for quantitative studies of dislocations, are,
as a rule, based on isotropic elasticity theory approach [3.34]. Therefore,
dealing with anisotropic crystals, some form of isotropic approximation is
usually applied.

The starting points of isotropic approximation are anisotropic elastic
constants C_{ij} or mechanical compliance moduli S_{ij}.

For crystals with monoclinic symmetry of the anthracene space group
$P2_{1/a}$ we have 13 independent elastic constants [3.42]

$$\begin{vmatrix} C_{11} & C_{12} & C_{13} & 0 & C_{15} & 0 \\ C_{12} & C_{22} & C_{23} & 0 & C_{25} & 0 \\ C_{13} & C_{23} & C_{33} & 0 & C_{35} & 0 \\ 0 & 0 & 0 & C_{44} & 0 & C_{46} \\ C_{15} & C_{25} & C_{35} & 0 & C_{55} & 0 \\ 0 & 0 & 0 & C_{46} & 0 & C_{66} \end{vmatrix} \qquad (3.13)$$

Using the elasticity tensor c_{ij}, the corresponding mechanical compliance
moduli s_{ij} can be calculated

$$s_{ij} = \frac{D_c(\substack{i \\ j})}{D_c} \quad .$$

(3.14)

The reported values of elastic constants c_{ij} and of mechanical compliance moduli s_{ij} of anthracene and naphthalene crystals according to [1.2, 3.42, 43] are presented in Table 3.3.

In the treatment of dislocations in an anisotropic organic crystal like anthracene, it is feasible to use isotropic elastic moduli averaged according to VOIGT [3.34]. In general, averaging can be performed either over elastic constants $c_{ijk\ell}$, or over mechanical compliance moduli $s_{ijk\ell}$.

In most cases, when predislocational local lattice deformations are being considered, averaging over $c_{ijk\ell}$ is preferable [3.34].

Isotropic shear modulus μ_V, averaged after VOIGT, can be found from the following expression:

$$\mu_V = \frac{1}{30} (3C_{ijij} - C_{iijj}) \quad .$$

(3.15)

μ_V values for anthracene and naphthalene crystals estimated according to (3.15), are given in Table 3.4

Table 3.3. Elastic constants c_{ij} and mechanical compliance moduli s_{ij} of naphthalene and anthracene crystals at 300 K and normal pressure

c_{ij} [10^{10}dyn/cm^2]	Anthracene [3.42]	Anthracene [3.43]	Naphthalene [Ref. 1.2, p. 346]	s_{ij} [10^{-12}cm^2/dyn]	Anthracene [3.42]	Naphthalene [Ref. 1.2, p. 346]
c_{11}	8.92	8.47	7.80	s_{11}	16.50	19.0
c_{22}	13.80	11.56	9.90	s_{22}	11.60	56.3
c_{33}	17.00	14.74	11.90	s_{33}	9.05	47.2
c_{44}	2.42	2.63	3.30	s_{44}	49.90	30.9
c_{55}	2.84	2.67	2.10	s_{55}	53.10	378.5
c_{66}	3.16	3.99	4.15	s_{66}	38.20	24.5
c_{12}	4.63	7.07	2.30	s_{12}	-2.75	-4.6
c_{13}	4.49	5.40	3.40	s_{13}	-1.05	-6.7
c_{23}	8.44	4.14	4.45	s_{23}	-4.41	-40.5
c_{15}	-2.58	-1.92	-0.60	s_{15}	11.40	8.8
c_{25}	-2.59	-2.38	-2.70	s_{25}	-3.58	127.2
c_{35}	-2.88	-2.14	2.90	s_{35}	4.19	-119.2
c_{46}	1.14	1.40	-0.50	s_{46}	-18.00	3.7

Table 3.4. Averaged isotropic and basal anisotropic shear moduli and Poisson's ratio of anthracene and naphthalene crystals

Elastic moduli	Anthracene	Naphthalene
μ_V[dyn/cm^2]	3.16×10^{10}	3.21×10^{10}
μ_{11}[dyn/cm^2]	2.60×10^{10}	2.10×10^{10}
ν_{11}	0.17	0.24
μ_{22}[dyn/cm^2]	3.50×10^{10}	—
ν_{22}	0.24	—

Notation: μ_V - isotropic shear modulus averaged according to VOIGT (3.15); μ_{11}, μ_{22} - anisotropic shear moduli in a- and b-directions of the lattice, respectively; ν_{11}, ν_{22} - corresponding Poisson's ratios.

In calculations of μ_V c_{ij} values reported by KITAIGORODSKY [1.2] and by DANNO and INOKUCHI [3.42] were used (cf. Table 3.3).

(For particulars of calculation, especially for application of transition rule from matrix notation $c_{ijk\ell}$ to abridged notation c_{mn} see HIRTH and LOTHE [3.34]).

Considering the energetics of basal unit dislocations of (001) [100] and (001) [010] type in an anthracene crystal, it is preferable to use also corresponding anisotropic shear moduli μ_{11}, μ_{22} and Poisson ratio ν_{11}, ν_{22} values.

These anisotropic moduli may be estimated using corresponding relations between shear modulus μ, Young modulus E, Poission's ratio ν and mechanical compliance moduli s_{ij} [1.82,3.34].

Thus the μ_{11} value, corresponding to the a direction in an anthracene crystal may be evaluated according to expression

$$\mu_{11} = \frac{E_{11}}{2(1 + \nu_{11})} \quad , \tag{3.16}$$

where $E_{11} = 1/s_{11}$ and $\nu_{11} = - s_{12}/s_{11}$.

(A similar expression may be used for evaluation of μ_{22} in the b direction in which s_{11} is replaced by s_{22}).

The evaluated μ_{11} and μ_{22} values, as well as those of ν_{11} and ν_{22}, for anthracene and naphthalene crystals are given in Table 3.4. For these estimates reported s_{ij} values were used (see Table 3.3).

As may be seen from Table 3.4, the averaged isotropic value of shear modulus μ_V for anthracene actually lies between the values of anisotropic moduli for a (μ_{11}) and b (μ_{22}) directions of the crystal.

3.5.3 Energy Estimates for Basal Edge Dislocations in an Anthracene Crystal

The elastic moduli obtained can be applied for estimating the energy of the two main basal edge dislocations of (001) [010] and (001) [100] type in an anthracene crystal. The following dislocation theory expressions were used for these estimates.

The total self-energy of edge dislocation $E_\perp$ may be described as a sum of the following terms [3.22]:

$$E_\perp = E_C + E_{el} + E_D \quad , \qquad (3.17)$$

where E_C is the energy of nonlinear deformation in the dislocation core (so-called core energy); E_{el} is the energy of elastic deformation of the surrounding lattice (elastic strain energy); E_D is the elastic field interaction energy of adjacent dislocations (see scheme presented in Fig.3.6).

Elastic energy of edge dislocation E_{el} may be estimated as follows [3.22,34]:

$$(E_\perp)_{el} = \frac{\mu b^2}{4\mu(1-\nu)} \ln \left(\frac{r_1}{r_C} \right) \quad , \qquad (3.18)$$

where r_1 is the limiting radius of the elastic stress field, r_C is the radius of the dislocation core (usually to be ca. 3$\underline{b}$-5$\underline{b}$).

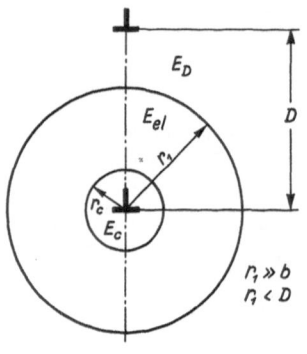

Fig. 3.6. Energy distribution around an edge dislocation: E_C - energy of nonlinear deformation in the dislocation core; E_{el} - elastic deformation energy of surrounding lattice; E_D - energy of elastic interaction field between neighboring dislocations [3.22]

The continuum elastic dislocation theory does not yield the value of core energy E_C directly. However, a comparison with results obtained from the discrete Peierls-Nabarro model shows that the E_C value is approximately equal to the prelogarithmic factor in (3.18) (cf. [3.34]). Consequently the following expression for evaluating the total energy $E_\perp$ of an isolated edge dislocation, in which $r_1 \ll D$ and $E_D = 0$ (cf.Fig.3.6) may be used:

$$E_\perp = (E_\perp)_C + (E_\perp)_{el} = \frac{\mu b^2}{4\pi(1-\nu)} \left(1 + \ln \frac{r_1}{r_C}\right) \quad . \tag{3.19}$$

Tables 3.5 and 6 present results of $(E_\perp)_{el}$, $(E_\perp)_C$ and $E_\perp$ values, estimated according to (3.18,19) for both aforementioned basal unit edge dislocations in anthracene. Corresponding anisotropic shear modulus μ and Piosson's ratio ν values according to Table 3.4 were used in calculations.

Table 3.5. Calculated elastic energy $(E_\perp)_{el}$ values for basal edge dislocations of (001)[010] and (001)[100] type in an anthracene crytal at different selected r_1 and r_C values (cf. Fig.3.6)

Type of dislocation	Burgers vector b [nm]	μ [10^{10} dyn/cm^2]	ν	$(E_\perp)_{el} \cdot [10^{-11}$ J/cm^2] at			
				$r_1 = 10^{-5}$ cm $r_C = 5b$	$r_1 = 10^{-5}$ cm $r_C = 3b$	$r_1 = 10^{-3}$ cm $r_C = 5b$	$r_1 = 10^{-3}$ cm $r_C = 3b$
(001)[010]	0.60	3.5	0.24	0.46	0.53	1.07	1.14
(001)[100]	0.855	2.6	0.17	0.57	0.67	1.41	1.51

Table 3.5 shows that a change in the outer radius r_1 of the elastic deformation region around the dislocation (cf. Fig.3.6) by even as much as two orders of magnitude produces an increase in elastic dislocation energy $(E_\perp)_{el}$ only by a factor of two. The limits chosen for r_1 correspond to typical dimensions of mosaic domains in anthracene single crystals ($r_1 \approx 10^{-5}$ cm) [1.2] or to typical grain dimensions in polycrystalline structures ($r \approx 10^{-4}$-10^{-3}cm) [2.153].

Table 3.6 shows that a considerable part of the total energy of edge dislocations (ca. 20—25%) is concentrated in a small microregion of the dislocation core of size (3-5)$\underline{b}$. It is just this part of the energy which causes nonlinear molecular displacements in the vicinty of the dislocation line and thus leads to the formation of corresponding local states.

Calculation results, as presented in Table 3.6, confirm experimental data showing that basal dislocations of (001) [010] type are energetically

Table 3.6. Calculated core $(E_\perp)_C$, elastic $(E_\perp)_{el}$ and total $E_\perp$ energy values for basal edge dislocations of (001) [010] and (001) [100] type in anthracene crystal

Type of dislocation	Energy unit	$(E_\perp)_C$	$(E_\perp)_{el}$ for $r_1 = 10^{-5}$ cm $r_C = 5\underline{b}$	$E_\perp$	$(E_\perp)_C/E_\perp$ [%]
(001)[010]	J/cm	1.3×10^{-12}	4.6×10^{-12}	5.9×10^{-12}	22
	eV/cm	0.8×10^7	2.9×10^7	3.7×10^7	
	eV/$\underline{b}$	0.49	1.73	2.21	
(001)[100]	J/cm	1.8×10^{-12}	5.7×10^{-12}	7.5×10^{-12}	24
	eV/cm	1.1×10^7	3.6×10^7	4.7×10^7	
	eV/$\underline{b}$	0.97	3.04	4.0	

most favorable in anthracene, whereby this dislocation is dominant among other basal and nonbasal dislocations (cf.Sect.3.13.1). This conclusion is by no means trivial since the macroscopic linear compressibility of anthracene crystal in the [010] direction is some three times lower than in the [100] direction (cf.Table 5.2).

Decisive in this case is the fact that the dislocation energy $E_\perp$ in (3.19) is proportional to the square of the Burgers vector representing the discrete nature of dislocation displacements. It is the [010] direction of slip that has the lowest value of the Burgers vector of unit dislocation ($\underline{b}$ = 0.60 nm) and is thus energetically most favorable.

The dislocation energy is proportional to the length of the dislocation line (cf. Table 3.6), and therefore a dislocation line will always strive to become shorter [1.81].

It can be seen from Table 3.6 that the energy of basal dislocation formation in an anthracene crystal has a value of ca. 2-4 eV/$\underline{b}$. The formation energy of nohbasal (e.g., in the $\underline{c}$ direction where $\underline{b}$ = 1.11 nm) or of multiple dislocations is considerably higer and may reach values of 10-15 eV/$\underline{b}$.

On the other hand, multiple dislocations with Burgers vector values of n$\underline{b}$ are unstable and very effectively dissociate into n dislocations of unit strength $\underline{b}$. This lowers their total energy n^2b^2/nb^2, i.e., n times. Since the Burgers vector $\underline{b}$ is larger in the case of molecular crystals than in the case of atomic ones and possesses values of 0.5-2.0 and 0.2-0.3 nm, respectively, it is to be expected that the formation energy of dislocations must be considerably higher for molecular crystals than for atomic ones. This is, however, almost completely compensated by a lower value of shear moduli μ for molecular crystals. We have, for instance, $\mu_V = 3.2 \times 10^{10}$ dyn cm^{-2} for anthracene (cf. Table 3.4) the corresponding value of germanium being $\mu = 10^{12}$ dyn $\cdot$ cm^{-2} [3.22].

In general the formation energy of dislocations, even of unit basal ones, in anthracene-type crystals exceeds considerably that of point defects, such as of lattice vacancies ($E_V \approx 0.9$ eV) (cf. Tables 3.1. and 2). Therefore the formation of dislocations under conditions of thermodynamic equilibrium is of comparatively low probability, i.e., dislocations, unlike vacancies, may not be regarded as "thermodynamical" defects.

Dislocations are usually formed as a result of various nonequilibrium processes, such as mechanical stress, nonequilibrium thermal effects, and mainly during growth of the crystal. Dislocation density in molecular crystals depends to a large extent on the technology of growing. The highest dislocation density is observed in crystals obtained from melt, the lowest one in those grown from the gaseous phase or by sublimation (cf. Sect.3.13.2 and Table 3.12). In the process of growth from melt local regions of mechanical stress are easily formed possessing sufficient energy for the creation of dislocations.

It is also easy to introduce dislocational deformations into a molecular crystal by the action of external mechanical forces, such as unilateral local pressure, bend, etc. Edge dislocations are rather easily formed in the process of vacancy condensation. Thus, thermodynamically "permitted" point defects are capable of generating high-energy "forbidden" dislocation states through cooperative interaction.

Contrary to general thermodynamic considerations, dislocations, once formed, usually possess unexpectedly high stability and a practically unlimited lifetime. Firstly, multiple dislocations dissociate into energetically more favorable metastable unit dislocations. Secondly, dislocations are capable of forming various stable alignments, such as symmetrical or asymmetrical small-angle boundaries (cf. Sect.3.8), possessing rather high stability [1.77,81].

3.6 Atomic and Molecular Models of the Dislocation Core

3.6.1 Models of Spherical Atoms and Molecules

The traditional dislocation theory of a continuous medium encounters substantial difficulties in describing a dislocation core at $r \leq r_C$ (cf. Fig.3.6).

Stress σ dependence on r, as obtained from the continuum elastic theory, e.g., the expression

$$\sigma = \frac{\mu b}{2\mu r} \quad , \tag{3.20}$$

obviously possesses singularity along the dislocation axis at $r \to 0$. Therefore the dislocation core is usually excluded from consideration in continuum dislocation theory (cf. [3.34,39]). At the same time, it is just the discrete configuration of the dislocation core we are interested in, for that is the region where the displacements of molecules from equilibrium position are largest. Usually the dislocation core at $r < r_c$ is treated separately with the aid of phenomenological models allowing to describe its discrete atomic or molecular microstructure [3.34]. Historically first dislocation core energy evaluations, based on discrete micromodels, were performed for ionic and metal crystals (see, e.g., [3.44,45] and references in [3.34]).

ENGLERT and TOMPA [3.46] were the first who calculated coordinates and energies of atoms in the core of an edge dislocation in an argon crystal — the simplest model of a molecular crystal. Interaction energy e_{ij} between argon atoms i and j at distance r_{ij} is described, according to [3.46], by means of a Lennard-Jones potential (cf. Sect.1.1)

$$e_{ij} = e_A\left[\left(\frac{r_A}{r_{ij}}\right)^{12} - 2\left(\frac{r_A}{r_{ij}}\right)^6\right] \quad , \tag{3.21}$$

where e_A and r_A are parameters characterizing, respectively, depth and minimum position of the potential well.

The reduced total configuration energy of the dislocation core is then determinded by the expression

$$U = \frac{1}{2} \sum \frac{e_{ij}}{e_A} \quad . \tag{3.22}$$

The initial positions of the atoms in the core were estimated from elastic deformation theory for values of $\underline{b} = a_0$ and $\nu = 0.25$.

After that minimization of energy was carried out for each pair of atoms in turn. The calculation results are graphically presented in Fig.3.7.

Fig.3.7 shows that the configuration of reduced equipotential curves ΔU around the dislocation line is rather complicated. ($\Delta U = u - u_0$, where u_0 is

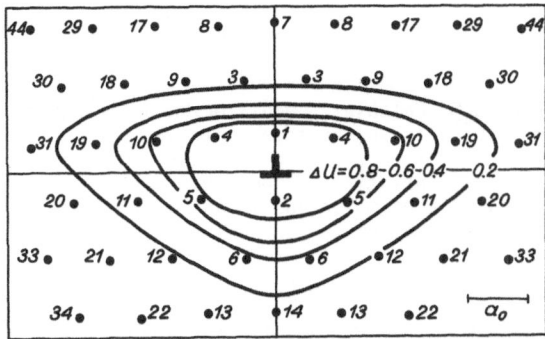

Fig. 3.7. Schematic picture of calculated atomic coordinates and reduced equipotential curves ΔU around an edge dislocation line in an argon crystal [3.46]

the energy of atoms in the xy plane of a perfect argon crystal). The gradient $\delta(\Delta U)/\delta r$ is strongly asymmetrical, especially above the dislocation line in the region of maximum compression of the lattice. The diagram clearly shows that the effective radius of the dislocation core r_c, within which the law of linear elastic deformation does not hold, actually is of the order $r_c \approx 3\underline{b}$. This demonstrates the feasibility of using this value of r_c for dislocation energy calculations in molecular crystals (cf. Sect.3.5.3).

The problem of topography and energetics of molecules inside the dislocation core in anthracene-type molecular crystals can, in principle, be solved in a similar fashion using molecular dislocation models. Such a calculation ought to yield equilibrium displacement coordinates for all molecules in the dislocation core, which, in turn, determine the local changes of electronic polarization energy ΔP for charge carriers localized on molecules in the dislocation core.

This presents, however, a much more complex task than the calculation of model molecular crystals of noble gases. Instead of spherical atoms (such as argon atoms) we have a polyatomic system of complicated configuration (e.g., an anthracene molecule). Instead of a simple expression for the interaction between two spherical atoms, as in (3.21) we should use a considerably more complex potential to describe intermolecular interaction (cf. Sect.1.2). Furthermore, in addition to three variable parameters of molecular coordinates it is necessary to introduce three variable angular parameters of molecular orientation in the lattice. Apparently for this reason rather simplified approximations and models for evaluating the disloca-

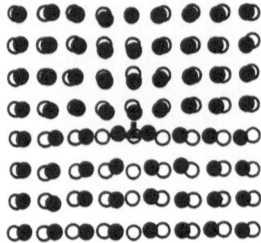

Fig. 3.8. Calculated equilibrium positions of molecules around an edge dislocation line ⊥ in the (001) plane of a hypothetic molecular crystal consisting of spherical molecules. (Positions of molecules in an ideal crystal are marked by empty circles) [3.6]

tion core topology in complex molecular crystals of the anthracene type have been presented up to now. Thus, for instance, in SWORAKOWSKI's work [3.6] the monoclinic anthracene crystal is replaced by a hypothetic model crystal with a primitive regular lattice with lattice constant equal to 0.6 nm, and real anthracene molecules are replaced by hypothetic spherical molecules with an assumed polarizability equal to the mean molecular polarizability of anthracene, viz. $\alpha = 25 \times 10^{-24}$ cm^3. The interaction energy of such spherical molecules is described by means of the Lennard-Jones potential (cf. Sect.1.1).

Equilibrium positions of the molecules in the dislocation core were found by variation method, minimization being performed for 124 molecules around the edge dislocation in the ab plane of the crystal.

Figure 3.8 presents the positions of molecules in the given edge dislocation core. An estimate of local electronic polarization energy for a charge carrier localized on molecules in the dislocation core shows that the edge dislocation forms a wide spectrum of local trapping states for charge carriers (0.02-0.16 eV) in such a hypothetic molecular crystal. The density of these states decreases exponentially with increase in energy. (The energy of electronic polarization in [3.6] was estimated only in terms of charge carrier-induced dipole interaction for 124 molecules surrounding the localized charge (cf. Sect.2.4.1).

The calculations performed in [3.6] qualitatively support the idea of polarization nature of local trapping centers for charge carriers in edge dislocation-type structural defects. They were the first attempts to assess the spectral distribution of local centers in the region of one separate dislocation core. However, the quantitative value of these calculations is rather negligible owing to the oversimplified model of the hypothetic crystal, as compared to a real crystal of the anthracene type.

3.6.2 Polyatomic Molecular Models

For calculating equilibrium configuration and energies of molecules in a
dislocation core of an anthracene-type molecular crystal the method of atom-
atoms potentials seems to be the most promising one (cf. Sect.1.2).

The procedure of calculating equilibrium configuration of molecules by
the atom-atom potential method has reached a state of sufficient perfection
for aromatic hydrocarbon crystals with a regular lattice possessing translational symmetry [1.2].

In this case it is necessary to take into account all possible changes
in the geometry of the lattice and consider the lattice energy U as a function of the periods a,b,c and the angles α,β,γ of the elementary cell as
well as of the Eulerian angles θ,φ,ψ and the coordinates X,Y,Z describing
the position of the reference molecule with respect to the axis of the crystal. (The positions of the atoms inside the molecule are assumed to be
fixed).

As a result we obtain a multidimensional potential surface of lattice
energy U as a function of geometrical parameters of the lattice [1.2]:

$$U = U(a,b,c,\alpha,\beta,\gamma,\theta,\varphi,\psi,X,Y,Z) \quad . \tag{3.23}$$

Since a regular crystal possesses translational symmetry, the variable
parameters must change simultaneously for all equivalent molecules of the
crystal in the minimization process of (3.23). Hence the problem of calculating equilibrium configuration of a regular crystal does not contain more
than 12 variable parameters.

The calculation of equilibrium configuration of the dislocation core in
a defective molecular crystal is undoubtedly a considerably more complicated
problem. In this case translational symmetry is not conserved, and the potential energy minimum in the dislocation core has to be found for each molecule separately. A dislocation core contains, as a rule, several hundred
displaced molecules [at $r_c \approx$ (3-5)$\underline{b}$], and each molecule has, generally,
six variable parameters (X,Y,Z,θ,φ,ψ). This means that the total number of
variable parameters runs into the hundreds.

On the whole, calculation of equilibrium configuration of molecules in
the dislocation core by the atom-atom potential method does not present any
fundamental physical difficulties, but a lot of purely computational ones.

The main features of the essential stages for such computations are at
present sufficiently lucid although difficult to realize in practice.

At first an "idealized" configuration of nonrelaxed molecules at the dislocation core is set up, like the one shown in Fig.3.3 for an edge dislocation in anthracene.

Thus initial data are present for calculating molecular coordinates of relaxed equilibrium configuration by the atom-atom potential method. The energy u_i of the ith molecule in the dislocation core is calculated by minimization procedure as dependent on variable coordinates X_i, Y_i, Z_i and angles $\theta_i, \varphi_i, \psi_i$, leaving the coordinates of the other molecules unchanged

$$u_i = u_i(X_i, Y_i, Z_i, \theta_i, \varphi_i, \psi_i) \quad . \tag{3.24}$$

A similar energy minimization is then performed for energy u_j of the jth molecule, and so on for all other molecules of the dislocation core. The procedure is reiterated until self-consistency.

In case of molecular crystals the potential hypersurface of the lattice energy in the dislocation core can, in principle, possess a number of closely situated energy minima forming a set of metastable states. This circumstance, as well as the large number of variable parameters raises additional problems in the minimization procedure.

It is just these difficulties which have so far been an obstacle for development of a general computational approach to the study of extended defects in anthracene-type crystals, such as edge dislocations.

At present some promising results have been obtained only for simplest planar defects in anthracene crystals, namely, stacking fault ribbons in the ab plane of the crystal. Symmetry elements of configuration of molecules in stacking faults of this kind make it possible to reduce considerably the number of variable parameters and to carry out the calculations within the limits of certain allowances (cf. Sect.3.9).

It may be concluded that the atom-atom potential method is undoubtedly promising for more general computational approach in future studies of relaxation processes and equilibrium configuration of molecules in dislocation cores of anthracene-type crystals, although considerable efforts will be required.

3.7 Dislocation Alignments and Aggregations, Their Configurational and Energetic Properties

> *For do we must what force will have us do.*
> *Shakespeare*
> King Richard II
> (Act III Sc. III)

3.7.1 Interaction Between Dislocations

Edge dislocations interact through their own fields of elastic stress, which are of distant action ($\sim 1/r$). The stress fields are of rather complicated configuration and depend on the direction with respect to the slip plane (cf. Fig.3.4). Hence the resulting interaction between two edge dislocations depends on the angle between their slip planes, on mutual distance and position.

Figure 3.9a shows a schematic diagram of interaction fields between two edge dislocations of equal sign, situated in parallel slip planes. A dislocation situated in the shaded quadrants with coordinates $|y_0| > |x_0|$ is attracted by the dislocation at the origin of coordinates. In a three-dimensional case a cone of attraction is formed for two dislocations of equal sign.

Interacting dislocations can form a very complicated configuration of fields of stress. Depending on the mutual position of dislocations, these stress fields can consist of regions of compression and dilation of the

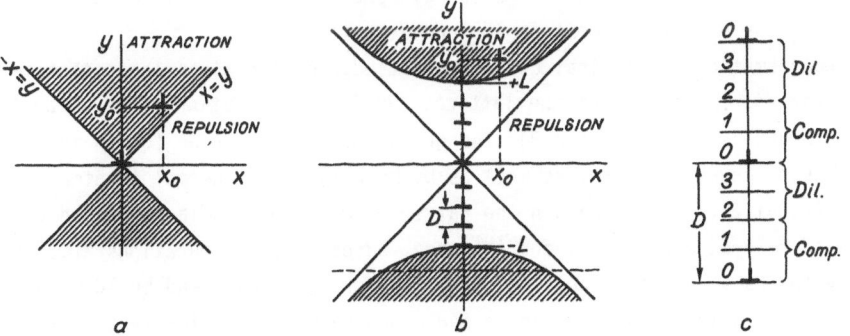

Fig. 3.9a-c. Schematic diagrams of edge dislocation interaction forces: (a) fields of interaction forces between two dislocations of equal sign and situated in parallel slip planes, according to [3.47]; (b) fields of interaction forces between a dislocation and a vertical dislocation alignment of finite dimensions, according to [3.22]; (c) periodic alternation of compression and dilation regions of the crystal along the vertical dislocation alignment

lattice. This means that at such interaction between dislocations, formation of local states with a rather complicated energy spectrum is to be
expected in the interdislocation space. It must be borne in mind that the
mutual position of dislocations can also be changed by means of thermal activation of their motion (cf. [1.77,81,3.34,41]). Thermal treatment of a
crystal can therefore lead to corresponding changes in spectral distribution and density of polarization-caused local states. Such dynamic behavior
of structural charge carrier traps has actually been observed in experiment
(cf. Sect.5.15).

3.7.2 Dislocation Alignments

The existence of a cone of attraction (cf. Fig.3.9a) enhances the formation
of a vertical alignment of edge dislocations of equal sign. The regions of
forces of attraction to the vertical dislocation alignment of length 2L, as
well as repulsion of a unit dislocation from it are shown schematically
in Fig.3.9b. The boundaries dividing the regions of attraction and repulsion form hyperbolas with asymptotes $x = \pm y$. Unit dislocations get attracted
to the vertical alignment if they are inside the shaded region, and repelled
from it, if they remain outside the two hyperboloids. Along the vertical
dislocation alignment there is a periodic alternation of compression and
dilation regions of the crystal, the period of repetition being equal to
the distance between dislocations D (cf. Fig.3.9c).

Depending on the statistical distribution of D along the alignment, we
get a corresponding change in the energy spectrum of local states caused by
microregions of lattice compression and dilation.

The formation of vertical alignments of edge dislocations leads to macroscopic disarrangement of the lattice, causing the creation of small-angle
grain boundaries (cf. Sect.3.8). The grain boundaries can be formed in the
process of crystal growth, as well as by mechanical or thermal treatment of
the crystal [1.74,77,3.48]. In the latter case we speak of a so-called process of *polygonization*. If, for instance, excess edge dislocations arise in
a crystal as a result of bending, they can "creep across" and build up a
vertical alignment, thus forming a new boundary of symmetrically disar-
ranged grains. Polygonization possesses, as a rule, a certain activation
energy. It therefore occurs most effectively in processes of thermal treatment of a crystal. Polygonization usually leads to removal of internal
stresses in the crystal as the result of thermal "training". This process
is of great importance in pre-planned change of the energy spectrum and

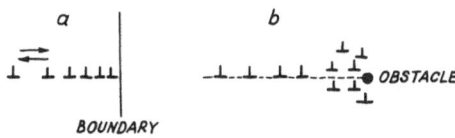

Fig. 3.10a,b. Aggregation of dislocations according to [3.34]: (a) horiz-
zontal dislocation alignment at a
grain boundary; (b) aggregation of
dislocations in front of an obstacle

density of local states of structural origin in molecular crystals (cf.
Sect.5.15).

If edge dislocations form a horizontal alignment [3.34,49] by piling up
in a common slip plane of the crystal (cf. Fig.3.10a) then the regions of
compression and dilation of separate dislocations enhance each other. Such
microregions of strong compression are capable of forming relatively deep
trapping centers for charge carriers.

Asymmetrical horizontal alignments of dislocations may be formed under
effect of some external force, such as an obstacle on the common slip plane
or a grain boundary which does not permit further advance of the head dislocation (see Fig.3.10a). This head dislocation is not only acted upon by
external stress, but also by interaction forces with other dislocations of
the alignment. Thus, a high force density f = F/L arises at the head dislocation, proportional to the applied stress σ and the number N of dislocations in the alignment [3.34]

$$f = \frac{F}{L} = \sigma N b \qquad (3.25)$$

If N is large, it may lead to considerable stresses at the head dislocation, even as high as theoretical shearing strength values. As a result a
region of very strong compression is formed above such a head dislocation,
giving rise to correspondingly deep trapping centers for charge carriers
on the very face of the grain boundary (cf. Fig.3.10a). In front of an obstacle dislocation alignment of complicated configuration can be formed,
producing a correspondingly complicated stress field (cf. Fig.3.10b).

3.7.3 Dislocation Ensembles

Apart from vertical and horizontal dislocation alignments having some definite arrangement or symmetry elements, so-called dislocation ensembles
may frequently be observed in real crystals in which the dislocations are
distributed inhomogeneously, their local density and mutual positions being completely random (cf. Fig.5.12). Such dislocation ensembles and their

stress fields may be treated only by statistical methods [3.50,51] (cf. Sect.5.7).

One of the main statistical characteristics of a dislocation ensemble is the mean density of dislocations $\bar{\rho}$. It is connected with the mean distance between dislocations $\bar{\lambda}$ by the simple expression [3.51]

$$\bar{\lambda} = 1/\sqrt{\bar{\rho}} \qquad (3.26)$$

$\bar{\rho}$ and $\bar{\lambda}$ are, however, only scalar characteristics of the dislocation ensemble. In order to account for spatial orientation of dislocations, an orientational distribution function Φ_b is introduced. This function charac- terizes the full length of dislocations, with Burgers vector $\underline{b}$ crossing an elementary volume inside the small solid angle $d\Omega$ localized in the vicinity of the direction given by unit vector $\underline{\ell}$ [3.51]. If $\Phi_b(\underline{\ell})$ is known, it is possible to calculate the flux of Burgers vectors of all dislocations cross- ing unit area of given orientations. This flux can be determined by the tensor density of dislocations (cf. [3.51])

$$\alpha_{ij} = \sum_{\underline{b}} \int b_j \Phi_{\underline{b}} (\underline{\ell}) \ell_i d\Omega \quad , \qquad (3.27)$$

where index i determines the orientation of the area, and index j that of the Burgers vector components.

In this case the scalar dislocation density $\bar{\rho}$ may be obtained by inte- grating and summing the orientation function $\Phi_b(\underline{\ell})$ over all directions $\underline{\ell}$ and over all possible Burgers vectors [3.51]

$$\bar{\rho} \sim \sum_{\underline{b}} \int_{\underline{\ell}} \Phi_b(\underline{\ell}) d\Omega \quad . \qquad (3.28)$$

Another important statistical characteristic of the dislocation ensemble is the distribution of internal stress fields (cf. Sect.5.7). There do not, unfortunately, exist any methods, at present, of direct experimental study of stress field distribution [3.51]. Only in some cases is it possible to estimate the dispersion of such distribution (cf. Sect.5.7).

The mean amplitude of internal stress $\bar{\sigma}_A$ for dislocations distributed at random with density $\bar{\rho}$, approximately corresponds to the stress caused by unit dislocation at distance $\bar{\lambda}$ [3.51,52]

$$\bar{\sigma}_A = \frac{\mu b}{2\pi} \cdot \frac{1}{\bar{\lambda}} = \frac{\mu b}{2\pi} \sqrt{\bar{\rho}} \quad . \tag{3.29}$$

The mean elastic energy per unit length of the dislocation line in the dislocational ensemble equals [3.51]

$$\bar{E} = \frac{\mu b^2}{4\pi} \ln \frac{\bar{\lambda}}{r_C} \quad , \tag{3.30}$$

where r_C is the radius of the dislocation core.

Various modifications of the above formulae for a dislocation ensemble have been suggested [3.51]. If, for instance, a dislocation net is formed in the crystal, $\bar{\lambda}$ in (3.30) should be replaced by the mean length of a link in the dislocation net. If dislocations aggregate into groups of n, such aggregates can be regarded as unit dislocations with summary Burgers vector $n\underline{b}$ located at a distance $\bar{\lambda}_{ef} = \sqrt{n/\bar{\rho}}$ (cf., e.g. [3.52]). It follows from (3.29) that for such a distribution of dislocations the amplitude of the mean internal stress field ought to be of the order [3.51]

$$\bar{\sigma}_A \approx \frac{\mu b}{2\pi} \sqrt{n\bar{\rho}} = \frac{\mu b}{2\pi} \cdot \frac{\sqrt{n}}{\bar{\lambda}} \quad . \tag{3.31}$$

As can be seen, in this case $\bar{\sigma}_A$ is $\sqrt{n}$ times larger than for a homogeneous distribution of dislocations. Hence, dislocation ensembles can form strong local fields of stress and, accordingly, give rise to relatively deep structural traps of polarization origin for charge carriers (cf. Sect.5.7).

It should be mentioned that at present our knowledge of the properties of dislocation ensembles and aggregations is still rather insufficient, both theoretically and experimentally (cf. e.g., review article [3.51]). However, the phenomenological correlations discussed above may be instrumental for qualitative and, in some cases, for semi-quantitative analysis of the statistical nature of distribution of structural local states in molecular crystals (cf. Sect.5.7).

3.8 Grain Boundaries, Their Energetic Characteristics

3.8.1 Energy of Grain Boundaries in Molecular Crystals

Vertical alignment of edge dislocations via polygonization leads to formation of boundaries separating crystal grains of different orientation

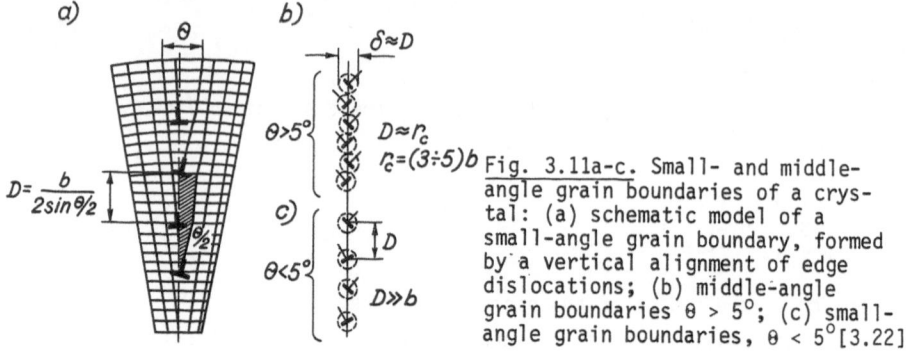

Fig. 3.11a-c. Small- and middle-angle grain boundaries of a crystal: (a) schematic model of a small-angle grain boundary, formed by a vertical alignment of edge dislocations; (b) middle-angle grain boundaries θ > 5°; (c) small-angle grain boundaries, θ < 5° [3.22]

(cf. Sect.3.7.2). The angle θ between disoriented crystal grains (see Fig.3.11a) is connected with distance D between dislocations and with the Burgers vector $\underline{b}$ by means of a well-known empirical formula [3.22]

$$D = \frac{b}{2 \sin\theta/2} \quad , \tag{3.32}$$

where θ is the angle.

At small values of θ an approximated expression may be used

$$D \approx \frac{b}{\theta} \quad , \tag{3.33}$$

where θ is the angle in radians.

This can be illustrated by Table 3.7 containing different values of interdislocation distances D, as dependent on boundary angle θ. The data refer to unit edge dislocation (001) [010] in naphthalene and anthracene crystals ($\underline{b}$ = 0.6 nm) which have been calculated according to (3.32). The Table

Table 3.7. Values of interdislocation distances D, as dependent on boundary angle θ for alignment of edge dislocations (001) [010] with unit Burgers vector ($\underline{b}$ = 0.6 nm) in naphthalene and anthracene crystals

θ	0.12°5'	0.5°30'	1°	2°	5°	10°	15°	20°	25°	30°
D[nm]	286.5	68.8	34.4	17.2	6.9	3.4	2.3	1.7	1.4	1.15

Linearized boundaries $D \gg \underline{b}$	Small-angle boundaries $D > \underline{b}$	Middle-angle boundaries $D \approx r_c = (3-5)\underline{b}$

presents also a conditional classification of subgrain and grain boundaries
of crystalline grains after MATARÉ [3.22] (cf. also Fig.3.11b,c).

Small-angle (subgrain) boundaries may be classified also according to
the manner in which various types of dislocations align [1.77]. Thus, a
symmetrical *tilt boundary* is composed entirely of edge dislocations of the
same type, having identical Burger vectors, whereas asymmetric tilt boundaries consist of alignments of two or more types of edge dislocations.

It follows from elasticity theory that the field of stress does not
spread to any extent beyond distance D from the angular boundary. It may
therefore be assumed that the width δ of an elastic distortion around the
boundary is of the order of D, i.e., $\delta \approx D$ [3.39].

In middle-angle boundaries edge dislocations are situated so closely
that their fields of stress completely overlap. D then becomes of the order
of the dislocation core radius: $D \approx r_c = (3-5)\underline{b}$ (cf. Table 3.7 and Fig.3.11).
The same applies to the width of the boundary, viz. $\delta \approx D$. In anthracene
single crystals linearized boundaries with values of θ lying between 0.6'
(D $\approx$ 2.8 μm) and 1°, as well as small-angle boundaries with values of θ
between 1° and 5° (D = 34-7 nm) have been observed [3.53] (cf. Table 3.7).
ROBINSON and SCOTT [3.48] considered a value of $\theta \approx 4^\circ$ (D $\approx$ 9 nm) as typical for small-angle grain boundaries in anthracene. The typical boundary
width $\delta \approx D$ must accordingly also be of the order of 9 nm, i.e., of sufficient dimensions for the formation of local states of polarization origin
in the respective regions of compression and dilation between dislocations.

Figure 3.12 presents a typical electron micrograph of a dislocation
alignment of a small-angle boundary in a p-terphenyl crystal belonging to
the same space group as anthracene (cf. also Fig.3.13,14).

The energy of grain boundaries $E_\perp$ (D) depends on dislocation density 1/D
at the boundary owing to superposition of stress fields (cf. Fig.3.6), and
may be evaluated by the expression [3.22]

$$E_\perp(D) = \frac{E_0}{D} (A - \ln\theta) = 2E_0 (A - \ln\theta) \sin \frac{\theta}{2} \quad , \qquad (3.34)$$

where

$$E_0 = \frac{\mu b}{4\pi(1-\nu)} \quad , \qquad (3.35)$$

A being a constant.

180

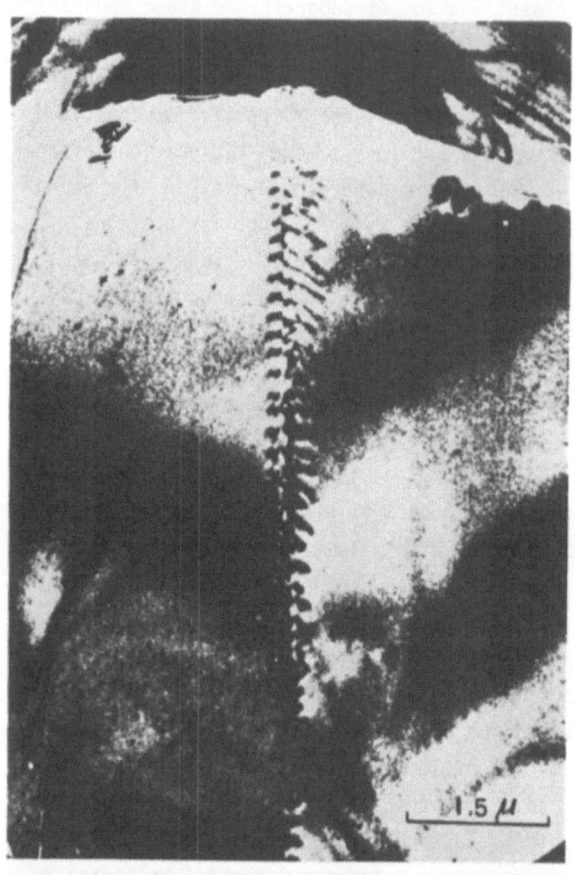

Fig.3.12.

Figure caption see opposite page

Fig.3.13.

Figure caption see opposite page

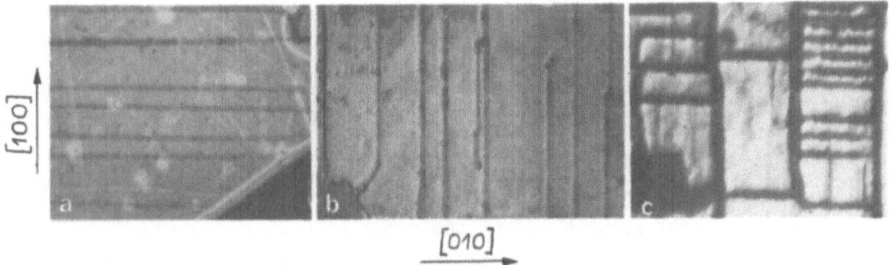

Fig.3.14. Optical micrographs of alignments of etch pits of thermally deformed anthracene crystals: (a) dislocation alignment of the [010] direction, producing the D_2 band of defect fluorescence (cf. Fig.4.11b); (b) dislocation alignment of the [100] direction, producing the D_1 band of defect fluorescence (cf. Fig.4.11a); (c) network of dislocation alignments of the [010] and [100] directions, producing both defect fluorescence bands D_1 and D_2 (cf. Fig.4.11c) [1.74]

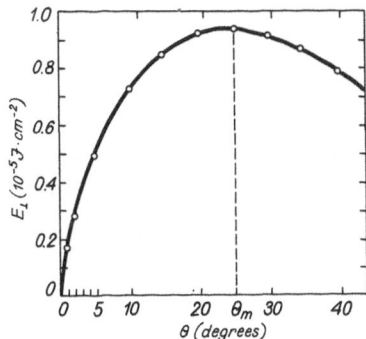

Fig. 3.15. Calculated dependence of grain boundary energy $E_\perp$ (D) on disorientation angle θ for an anthracene crystal

The $E_\perp(D) = f(\theta)$ curve, according to (3.34) shows a maximum at $\theta \approx \theta_m \approx 25°$ for the majority of materials [3.22]. Hence the following empirical expression may be recommended for practical calculations of $E_\perp(D)$ [3.22]

$$E_\perp(D) = 2E_0 \sin \frac{\theta}{2} (1 + \ln \theta_m - \ln \theta) \quad . \tag{3.36}$$

◄ Fig.3.12. Electron micrograph of an alignment of dislocations comprising a small-angle boundary in a p-terphenyl crystal [3.59]

◄ Fig.3.13. Optical micrograph of a (00ℓ) face of a cleaved anthracene crystal etched with oleum (300×). AB delineates a twist boundary, whereas CD is a row of etch pits of the emergence points of edge dislocations [1.77]

The calculated $E_\perp(D) = f(\theta)$ dependence for an anthracene crystal, according to (3.36), is presented in Fig.3.15. The calculation refers to a boundary formed by alignment of (001) [010]-type dislocations with $\underline{b}$ = 0.6 nm and θ_m = 25°; the other parameters of anthracene as in Table 3.5.

Figure 3.15 shows that at small θ values the grain boundary energy $E_\perp(\theta)$ is directly proportional to θ, i.e., to dislocation density 1/D [cf. (3.33)]. This yields an energy value $E_\perp(D)$ of the order of 0.5×10^{-5} J $\cdot$ cm^{-2} = 50 erg $\cdot$ cm^{-2} at $\theta \approx 5°$.

3.8.2 Relative Lattice Compression on Grain Boundaries of an Anthracene Crystal

According to MATARÉ [3.22] the resulting relative compression or pressure $\delta p/p$ on middle-angle tilt boundary ($\theta \gtrsim 5°$) may be calculated from the expression

$$\frac{\delta p}{p} = \frac{\nu}{b} \cdot \frac{E_\perp'}{\mu} \quad , \tag{3.37}$$

where $E_\perp'$ is the surface energy of the grain boundary in J $\cdot$ cm^{-2}. $E_\perp'$ can be estimated, assuming that during formation of the boundary the full dislocation energy $E_\perp$ is concentrated on the tilt boundary having width $\delta \approx D$. Hence,

$$E_\perp' = \frac{E_\perp}{\delta} \approx \frac{E_\perp}{D} \quad . \tag{3.38}$$

The results of calculated estimates of relative compression $\delta p/p$ on a middle-angle tilt boundary of an anthracene crystal, formed by alignment of dislocations of type (001) [010] and (001) [100], are given in Table 3.8 for different values of angle θ.

The total energy $E_\perp$ of the above-mentioned edge dislocations in the anthracene crystal, as used for calculation, was obtained by means of (3.19) at $r_1 \approx 10^{-3}$ cm, r_c = 3$\underline{b}$ and $\underline{b}$ = 0.6 nm and 0.855 nm, respectively. The values of anisotropic shear moduli μ and Poisson's ratio's ν were taken according to Table 3.4.

Table 3.8 shows that the value of $\delta p/p$ at middle-angle tilt boundaries of an anthracene crystal can be as high as 8-13%.

Table 3.8. Calculated estimates of relative compression $\delta p/p$ on middle-angle boundaries ($\theta > 5^\circ$) of an anthracene crystal for different values of disorientation angle θ

Type of dislocation forming the boundary	θ	5°	10°	15°	20°	25°	30°
(001) [010]	$\delta = D$ [nm]	6.9	3.4	2.3	1.7	1.4	1.15
	$\delta p/p$ [%]	2.1	4.3	6.3	8.5	10.3	12.6
(001) [100]	$\delta = D$ [nm]	9.8	4.9	3.3	2.5	2.0	1.65
	$\delta p/p$ [%]	1.3	2.6	3.9	5.2	6.5	7.8

It must, however, be borne in mind that $\delta p/p$ values, as presented in Table 3.8, are only average estimates obtained from linear elastic theory of dislocations. Inside the dislocation core where there is considerable divergence from linearity one may, naturally, expect local regions of stronger compression—as high as 20-30%.

Calculated values of increase in electronic polarization energy ΔP in regions of compression at middle-angle tilt boundaries of an anthracene crystal are given in Sect.5.2.

3.9 Stacking Faults in Molecular Crystals

3.9.1 General on Stacking Faults

A stacking fault is a planar defect bounded by a pair of partial dislocations of opposite sign (cf. Sect.3.5.1). This follows from the general rule that the Burgers contour of a partial dislocation must start and end in the plane of the stacking fault [3.34]. Stacking faults usually arise as a result of dissociation of a unit dislocation into two partial ones, if the following condition is fulfilled [3.34]

$$b_1^2 > b_2^2 + b_3^2 \quad , \tag{3.39}$$

where $\underline{b}_1$ is the Burgers vector of the full dislocation, $\underline{b}_2$ and $\underline{b}_3$ of the partial ones.

Stacking faults dominate in lamellar anisotropic structures, such as graphite and anthracene-type molecular crystals (cf. Sect.3.9.2). In these cases the plane of the defect is, as a rule, parallel to the basic slip plane of the crystal, e.g., the ab plane in anthracene.

The formation of a defect between partial dislocations leads to an increase in energy by a unit of length $\gamma_{sf}r$, where γ_{sf} is the stacking fault energy in erg/cm^2. This is accompanied by the appearance of an interaction force between the partial dislocations (per unit length), equal to

$$\delta(\gamma_{sf}r)/\delta r = \gamma_{sf} \quad .$$

The equilibrium distance between partial dislocations, i.e., the width of the stacking fault ribbon r_{sf} corresponds to the energy minimum at which the force γ_{sf} is equal and opposite to the elastic force. The value of r_{sf} for an isolated planar stacking fault ribbon can be found from the expression [3.34,54]

$$r_{sf} = \frac{\mu b^2 \ (2-\nu)}{8\pi\gamma_{sf} \ (1-\nu)} \ \left(1- \frac{2\nu \ \cos \ 2\varphi}{2-\nu}\right) \ , \tag{3.40}$$

where φ is the angle between the Burgers vector of the full edge dislocation and the direction of the ribbon.

As shown in [3.45], the energy minimum corresponding to the equilibrium width r_{sf} may be rather flat and subject to modulation by the periodic Peierls potential (cf. Fig.3.5a), causing the formation of a number of metastable local minima. The value of r_{sf} must, accordingly, have statistical distribution with a certain dispersion. One can therefore speak of the mean value of a typical width of a stacking fault ribbon in a given crystal.

Since r_{sf} is inversely proportional to stacking fault energy ($\sim 1/\gamma_{sf}$), the width of the ribbon may become extremely narrow at high values of γ_{sf}, as in the case of face-centred cubic crystals of some metals with energy γ_{sf} of more than 20 erg/cm^2. Such narrow ribbons are difficult to detect experimentally [3.34]. On the other hand, in the case of lamellar structures, e.g., in graphite, the value of γ_{sf} is small—of the order of 10^{-2}-10^{-1} erg/cm^2, and the corresponding value of r_{sf} can get as large as 10^{-5}-10^{-4} cm [3.34,54]. From this point of view anthracene-type crystals may probably be placed somewhere inbetween these extreme cases (see Sect.3.9.2).

3.9.2 Stacking Faults in Anthracene-Type Crystals, Their Energetic Characteristics

In their studies of mechanically deformed naphthalene and anthracene ROBINSON and SCOTT [3.48,55,56] showed that dislocational slip of [110] direction

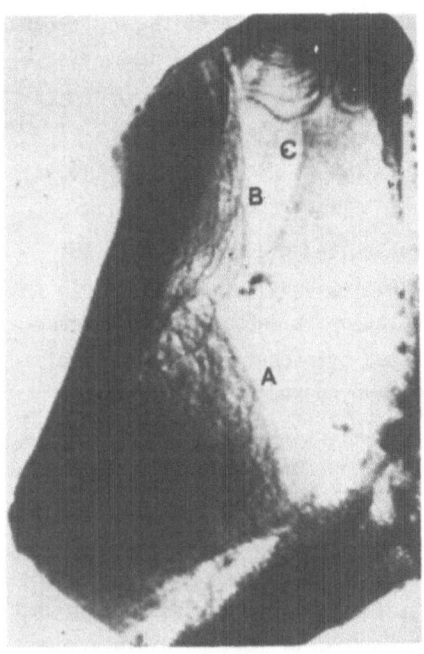

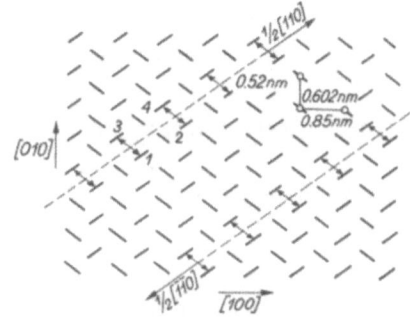

Fig. 3.16. X-ray topogram of an anthracene single crystal (× 15) reflection from ($\bar{1}$11) plane obtained
using monochromatic Cu K$_{\alpha1}$ radiation.
Regions A and B show dislocations
parallel to the [010] direction;
region C dislocations parallel to
the [110] direction [3.57]

Fig. 3.17. Schematic scalar diagram
of a stacking fault ribbon in the ab
plane of an anthracene crystal [1.78]

occurs in the ab plane of the crystals. They suggested these dislocations to
be, most likely, partial ones with Burgers vectors 1/2 [110] and 1/2 [$\bar{1}\bar{1}$0]
setting bounds to a stacking fault ribbon in the ab plane of the crystal.
The presence of a [110]-slip system in anthracene crystals was later directly confirmed by X-ray topography [3.57] (cf. Fig.3.16), as well as by optical [1.37,3.58] and electron micrographs [1.79,80].

A typical stacking fault ribbon in the ab plane of anthracene is shown
schematically, according to THOMAS et al. [1.78] in Fig.3.17. It can be seen
that the given stacking fault is bounded from two sides by partial dislocations of types (001)1/2 [110] and (001)1/2 [$\bar{1}\bar{1}$0]. The stacking fault thus
constitutes a ribbon of molecules displaced by a vector 1/2 [110], and positioned between layers of perfectly packed crystal lattice. In this configuration the specific position of molecules along the very line of partial dislocation is rather noteworthy. For instance, molecules 1 and 3 (cf. Fig.3.17)
are situated in parallel positions at a distance of 0.52 nm, forming a so-
called predimer state.

Partial dislocations of type (001)1/2 [110] in crystals of the anthracene space group are most likely formed through dissociation of the basal dislocation (001) [100] according to equation

$$(001) \ [100] \rightarrow (001)1/2 \ [110] + (001)1/2 \ [1\bar{1}0] \quad . \tag{3.41}$$

Such a dissociation is energetically favorable, in agreement with (3.39) ($b_1^2 = 73.3 \ \text{Å}^2$; $b_2^2 = b_3^2 = 27.6 \ \text{Å}^2$). For dominating basal dislocations of type (001) [010], on the other hand, such dissociation is energetically unfavorable. These considerations have been confirmed by recent full and partial dislocation self-energy estimates in anthracene by KOJIMA [3.58], which show that the self-energy of a pair of 1/2 [110] partial dislocations is intermediate between the self-energies of [100] and [010] full dislocations (cf. Table 3.6).

A typical fault width r_{sf} in anthracene space group crystals may be of the order of a hundred Å (see [1.25]).

An estimate of stacking fault energy γ_{sf} according to (3.40) for a planar ribbon in an anthracene crystal, bounded by a pair of partial dislocations 1/2 [110] and 1/2 [$\bar{1}\bar{1}0$] in the ab plane (cf. Fig.3.17), yields the following values: at $r_{sf} = 52$ Å (number of rows of molecules in ribbon n = 10) $\gamma_{sf} \approx 13$, at $r_{sf} = 104$ Å (n = 20) $\gamma_{sf} = 6.5$, and at $r_{sf} = 208$ Å (n = 40) $\gamma_{sf} \approx 3$ erg/cm^2. (Averaged μ_v value according to Table 3.4 and $\nu = 0.2$ was used in the calculations).

As may be seen, the specific energy of the stacking fault ribbon in anthracene is of the same order as the specific energy concentrated on the small-angle boundary (cf. Sect.3.8.1).

Some years ago THOMAS [1.80], and JONES et al. [3.59] reported the discovery of a new type of basal dislocation in p-terphenyl crystals which belong to the anthracene space group (cf. Table 3.11). The electron micrographs obtained by the method of diffraction contrast [3.59] showed that in the ab slip plane of p-terphenyl dislocations with Burgers vector [12ω] are formed, with ω, most likely, equal to zero. The authors of [1.80,3.59] suggested that these dislocations appear to be partial dislocations of type (001)1/2 [120], which may occur also in anthracene crystals. The existence of this type of partial dislocation in anthracene was later confirmed by SLOAN et al. [1.37] (see Sect.3.12.1).

A schematic drawing of a partial dislocation of (001)1/2 [120] type in an anthracene crystal is presented, according to THOMAS [1.80], in Fig.3.18.

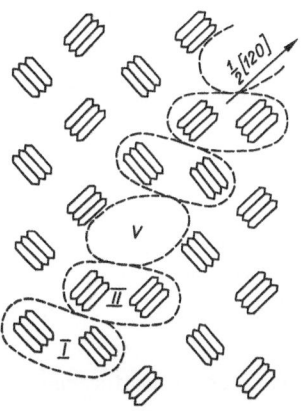

Fig. 3.18. Schematic picture of a partical dislocation of 1/2 [120] type in the ab plane of an anthracene crystal (by courtesy of J.M. THOMAS)

Fig.3.19. Electron micrograph of doubly-imaged dislocation segments [(a), (b) and (c)] in a p-terphenyl crystal, which could signify presence of partial dislocations separated by stacking fault ribbons [3.59]

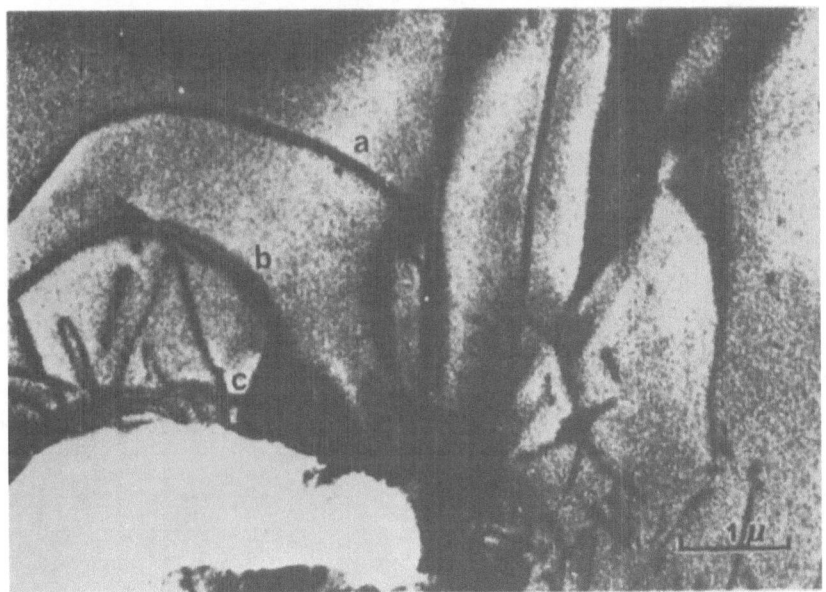

As may be seen from Fig.3.18, in this case two pairs of predimer states are formed along the line of partial dislocation separated by a vacancy between them.

Figure 3.19 presents an electron micrograph of segments of doubled dislocations in a p-terphenyl crystal which may be interpreted as partial dislocations enclosing a stacking fault ribbon [3.59].

3.9.3 Calculations of Equilibrium Configuration of Molecules in Stacking Faults of an Anthracene Crystal

Figure 3.17 presents an "idealized" schematic configuration of molecules in a stacking fault ribbon, a so-called zero-approximation configuration which does not take into account relaxation of molecules into a new equilibrium state. Considering the possible consequences of the specific configuration of molecules along the dislocation line 1/2 [110], according to the diagram Fig.3.17, the authors of [1.78] anticipated that there might be a further mutual approach between the parallel molecules of 1 and 3, 2 and 4 of the presumed "incipient" dimer states.

Such predimer states might serve as centers for excimer or photodimer formation (cf. Sect. 3.10).

A more realistic schematic diagram of molecule configuration along the dislocation line 1/2 [110] (see Fig.3.20) shows, however, that parallel molecules 1 and 3 of the presumed predimer pair cannot actually approach each other owing to strong overlap of the Van der Waals radii between molecules 1 and 7, 3 and 4. The opposite phenomenon ought therefore to be expected - not mutual approach, but repulsion of molecules 1 and 3 in opposite directions from the dislocation line.

In order to find the real molecular configuration along the dislocation line (cf. Fig.3.20), as well as inside the stacking fault ribbon, it is necessary to consider the process of relaxation of molecules into a new equilibrium state in the defect region. In this case it is advisable to use the atom-atom potential method (see Sect.1.2). Owing to a smaller amount of

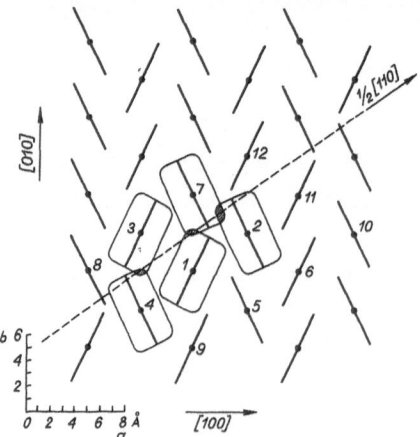

Fig. 3.20. Schematic diagram of molecular configuration along a partial dislocation line of (001)1/2 [110] type. The diagram shows an "idealized" picture of the given dislocation, disregarding relaxation of molecules to a new equilibrium state

variable parameters the problem is simpler as compared to that of edge dislocation (cf. Sect.3.6.2).

The first attempt to obtain a rough estimate of the configuration of the
equilibrium state of molecules along a partial dislocation (001)1/2 [110]
in an anthracene crystal was made by SWORAKOWSKI [1.21]. In this work the
method of atom-atom potentials (cf. Sect.1.2) was applied using the Buckingham function (1.4) with parameter A,B and C values after [1.17] (cf.
Table 1.1).

In [1.21] relaxation of 45 molecules was considered, positioned in several rows on both sides of the dislocation line 1/2 [110]. The molecules
concerned were allowed to relax freely away from the dislocation line. No
boundary conditions or quenching of the displacement of molecules were imposed. Only linear shifts of molecules were considered, assuming that the
molecules retain the angular parameters of perfect lattice.

The minimization procedure was carried out, starting from the ith molecule, all the other molecules remaining in fixed positions. When the minimum
for the ith molecule has been found, all other equivalent molecules are also
shifted to the corresponding minimum position. A similar procedure is then
performed with the (i + 1)-st molecule, and so on till complete self-consistency.

The first calculations performed by the author of [1.21] demonstrated
that the molecules of the presumed predimer pairs are actually displaced
away from the dislocation line in the relaxation process, whereas molecules
of the closest approach are to be found in the second and third rows from
the dislocation line.

The initial assumptions for calculations in [1.21], especially neglect
of possible changes in angular parameters of molecules, as well as the absence of boundary conditions yield a very rough description of the actual
relaxation process. However, the calculated configuration of relaxed molecules in [1.21] gives nevertheless a qualitatively adequate approach to
the picture obtained later from more accurate calculations.

SWORAKOWSKI's approach [1.21] was extended in the work of SILINSH and
JURGIS [1.22] in which the equilibrium postrelaxation configuration of molecules was calculated for a stacking fault ribbon bounded by partial dislocations 1/2 [110] and 1/2 [$\bar{1}\bar{1}0$] in the ab plane of an anthracene crystal
(cf. Fig.3.17). The variational procedure of energy minimization was carried
out applying the atom-atom potential method (cf. Sect.1.2). The Buckingham
function was used with parameters A, B and C after KITAIGORODSKY [1.2]
(cf. Table 1.1).

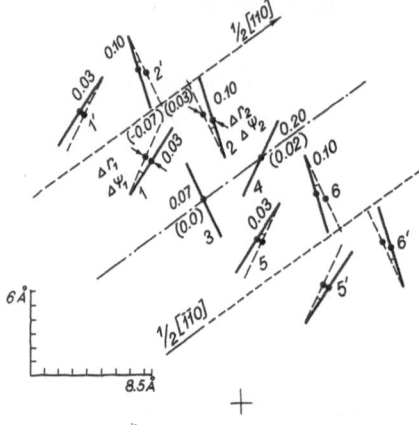

Fig. 3.21. Schematic diagram of calculated equilibrium configuration of molecules in a stacking fault ribbon
bounded by partial dislocations (001)
1/2 [110] and (001)1/2 [$\bar{1}\bar{1}$0] in the
ab plane of an anthracene crystal;
broken line - initial position of molecules; solid line - equilibrium position of molecules. Figures at molecule sites denote calculated ΔP values
for equilibrium configuration of molecules, figures in brackets - for initial configuration of molecues [1.22]

As initial zero approximation for minimization an "idealized" structure
of a stacking fault ribbon was taken, shifted by vector [1/2, 1/2, 0] with
respect to the remaining part of the ab plane of the crystal, and assuming
that all molecules of the ribbon retain their initial linear and angular
coordinates in the unit cell (cf. Fig.3.20).

Atom-atom interaction between pairs of adjacent molecules was considered
inside the stacking fault ribbon, as well as beyond the line of partial dislocation in the ab plane. The influence of molecules in the plane above the
stacking fault ribbon was also taken into account.

The following boundary conditions were applied. Firstly, it was assumed
that the partial dislocation lines do not shift in the relaxation process,
i.e., the width of the faulty stacking ribbon r_{sf} remains constant. This
means that molecular displacements are considered as symmetrical with respect to and in opposite directions from both dislocation lines 1/2 [110]
and 1/2 [$\bar{1}\bar{1}$0] (cf. Fig.3.21). Secondly, it was assumed that the molecules
situated in the middle row of the ribbon (e.g., molecules 3 and 4 in
Fig.3.21) retain their coordinates.

Such an approach obviously sets an upper limit to molecular displacements
for a separate isolated stacking fault ribbon. It follows that an upper
limit is thus set for the increase in energy of electronic polarization
ΔP in real stacking fault ribbons.

The assumption that repulsion forces act most strongly on molecules positioned directly along both sides of the dislocation line, has been confirmed by calculation. At "zero" configuration (cf. Fig.3.20) we get rather
strong repulsion between molecules 3 and 4, 2 and 7 and, especially, between

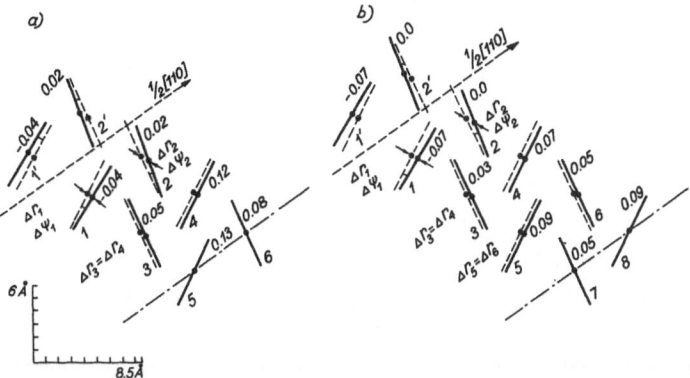

Fig. 3.22a,b. Schematic picture of calculated equilibrium configuration of molecules in stacking fault ribbon in the ab plane of an anthracene crystal: (a) stacking fault ribbon containing five rows of molecules (figure shows half of a ribbon, up to the middle row); (b) stacking fault ribbon containing seven rows of molecules (notation as in Fig.3.21) [1.22]

Table 3.9 Linear (Δr) and angular ($\Delta \psi$) displacements of molecules and intermolecular distance (d) after relaxation to equilibrium configuration in stacking fault ribbons of an anthracene crystal [1.22]

Figure showing configu-ration of stacking fault	Number of rows of mole-cules in ribbon	Width of ribbon, r_{sf} [Å]	Fixed molecules of middle row of ribbon	Displaced molecules	Δr, [Å]	$\Delta \psi$ (degress)	Mean intermolecular in-teraction energy in rib-bon (rel. units)	Molecules	d[Å]
Fig.3.21	3	15	3; 4	1	0.36	-5.5	-1.01	1-3	4.88
				2	0.62	8.4		2-4	4.62
Fig.3.22a	5	24	5; 6	1	0.50	-4.6		1-3	4.98
				2	0.66	5.3	-1.58	2-4	4.82
				3	0.24	-			
				4	0.24	-			
Fig.3.22b	7	34	7; 8	1	0.56	-4.2		1-3	5.04
				2	0.68	3.6		2-4	4.92
				3	0.36	-	-1.90		
				4	0.36	-			
				5	0.18	-			
				6	0.18	-			

1 and 7. Accordingly, these molecules get displaced into opposite directions form the dislocation line in the relaxation process (cf. Fig.3.21).

In the case of nonequivalent molecules of the first row — molecules 1 and 2, and 1' and 2' (cf. Figs.3.21,22) — two parameters were subjected to variation: linear displacements Δr_1 and Δr_2 in directions $[1\bar{1}0]$ and $[\bar{1}10]$, respectively, and rotation angles $\Delta\varphi_1$ and $\Delta\varphi_2$ around the long axis of the molecule. (It was found that rotation around other axes of the molecule is negligible compared to that around the long axis). For molecules of the internal rows of the ribbon only linear displacement Δr_i was considered (cf. Fig.3.22).

The calculation of the equilibrium configuration of molecules was carried out for stacking fault ribbons of width r_{sf} = 15, 24 and 34 Å and containing, respectively, 3, 5 and 7 rows of molecules each (cf. Fig.3.21,22).

As can be seen from Figs.3.21,22, and from Table 3.9, the molecules positioned in the first row along the dislocation line (molecules 1 and 2, 1' and 2') are most displaced. The angular displacement $\Delta\varphi$ of these molecules provides the most "convenient" configuration, with minimum mutual repulsion.

The absolute value of displacement Δr increases with increase in width r_{sf} of the ribbon. At the same time there is also an increase in distance d between the molecules of closest packing. In other words, compression of molecules diminishes with increase in width of the stacking fault ribbon. As expected, according to (3.40), there is also a corresponding diminution in mean interaction energy between neighboring molecules, i.e., in the stacking fault energy.

One should be rather cautious applying the atom-atom potential method for calculation of relaxation processes in structural defects in molecular crystals.

It should be recalled that the empirical Buckingham formula gives correct approximation only in the vicinity of the equilibrium state between interacting pairs. However, in the initial unrelaxed configuration separate atom pairs of neighboring molecules in the defect region may be at a distance considerably smaller than the equilibrium one, i.e., in the region of strong repulsion. The Buckingham formula (1.4) has a false peak in this region after which further approach seemingly leads to attraction, i.e., to negative values of φ_{ij}. The critical distance at which this takes place is ca. 0.6—0.7 Å for H...H and ca. 0.8—0.9 Å for C...C interaction. Our computing program especially provided limiting conditions taking into account the "dip" in the Buckingham function, if in some cases the interacting atoms approach each other to critical distance.

Calculations carried out in [1.21,22] may be regarded as the first attempt to build up a microscopic model for finding direct correlations be-

tween coordinates of molecular displacement in stacking faults, and the corresponding energy spectrum of local states in molecuuar crystals (cf. Sect.
5.3).

A different computational approach to the studies of stacking fault defects in molecular crystals, which may be characterized as a topological
one, was proposed by MIRSKY and COHEN [1.23,24], and RAMDAS et al. [1.25].
In the work of the former the atom-atom potential method was used to calculate the relative facility of slip in various directions of the known slip
planes of anthracene, including the slip along the unit cell diagonal in the
ab plane, viz. (001) [110] slip. In this case a uniform slip of one semi-
infinite half of the crystal against the second such half was considered.
The authors focused their attention on the calculation of the energy barriers that have to be surmounted to permit the given type of slip to occur,
i.e., they computed the interaction energy between the two half crystals as
a function of the slip vector along the [110] direction. The authors showed
that at slip vector values 1/4 and 3/4 of the unit cell diagonal Peierls'
maximums are formed while at slip vector value 1/2 [110] there exists a definite energy minimum. This means that stacking faults enclosed between 1/2
[110] type of partial dislocations actually form metastable defect structures in the anthracene lattice [1.24].

Since in [1.23,24] no allowance was made for relaxation the quantitative
data obtained may be regarded only as a rough approximation. In the work of
RAMDAS et al. [1.25] stacking fault calculations by atom-atom potential
method are illustrated with reference to a substituted anthracene, viz.
1,8-dichloro-9-methyl-anthracene. In this case the half crystal that had already experienced the slip of (100)1/2 [010] was allowed to "relax" along
the [$\bar{1}$00] direction by an amount x$\underline{a}$, where x is a small fraction of the unit
cell vector. It was found that the fault energy falls sharply and continu-
osly with increase in x, reaching a broad minimum at x $\approx$ 0.3. It was also
found that specific intramolecular distortion (rotation about a definite
C-C bond) may still further diminish the fault energy [1.25].

The results obtained so far by computational approach to the study of
molecular displacements in specific structural defects are rather encouraging and lend support for wider future applications of the atom-atom potential method in this field, so important for energy spectra investigations of
local states in molecular crystals (cf. Sect.5.3).

3.10 Formation of Predimer States in the Regions of Extended Structural Defects of Anthracene-Type Crystals

Anthracene crystals belong to the lattice type A. Accordingly, in a perfect anthracene crystal (unlike pyrene or perylene belonging to lattice type B) there are no parallel predimer configurations of nearest neighbor molecules in a unit cell (cf. Sect.1.3.4).

However, as shown by THOMAS et al. [1.78,3.10], such predimer states of molecules are formed in regions of dislocation defects. Experiments of an-thracene crystal photodimerization show that photodimerization proceeds se-lectively, preferably in the core of dislocation defects, demonstrating thus the existence of predimer configurations of molecules in these struc-tural defects of the crystal. Regions of strong local compression in the dislocation core seem to be the sites of pred.mer state formation (cf. Sect.3.11). Excimer luminescence observed by TANAKA et al. [3.60] on strong hydrostatic compression of an anthracene crystal, which disappears after subsequent thermal treatment, supports this view. It is not unlikely that inclusions of domains of other metastable modifications of anthracene pos-sessing, according to CRAIG et al. [1.87], predimer structure of B-lattice type (cf. Sect.1.3.1) may be formed in compressed regions of dislocation defects. It was suggested by SCHIPPER and WALMSLEY [3.27] that excition trapping at the compressed regions of dislocation defects in anthracene (cf. Sect.4.3) may initiate local dimerization. At the same time, compres-sive strains set up in the dimerization region may be responsible for fur-ther exciton trapping, leading to the formation of new dimers in the region. An effect of this type was recently experimentally observed by WILLIAMS et al. [3.28]. The authors reported that in ultra-pure anthracene crystals subjected to photodimerization a wide excimer luminescence band is observed at ca. 475 nm indicating formation of "incipient" predimer states in close proximity to the stable photodimer, which act as excition traps of ca. 4000 cm^{-1} depth (cf. Sect.4.3).

Excimer luminescence was also applied by MÜLLER and BAESSLER [3.61], and MÜLLER et al. [3.62] for detecting predimer states of molecules in quasi-amorphous and amorphous layers of tetracene (cf. Sect.5.12). These dimer states are possibly responsible for the appearance of trapping centers for charge carriers in these layers, possessing a depth of ca. 0.7 eV [3.63].

HOFFMANN et al. [3.64] have recently shown that predimer states are also formed in amorphous layers of anthracene prepared by vacuum sublimation onto a substrate cooled to 80 K. Such layers exibit three distinct excimer bands at 21900, 21500 and 19600 cm^{-1}.

3.11 Some More Complex Two- and Three Dimensional Lattice Defects in Molecular Crystals

In our discussion of topography and energetics of dislocations in previous sections we were using an idealized model based on the asumption of straight-line dislocations.

A strict, noncontroversial dislocation theory has so far been involved only for such a simplified dislocation model (cf., e.g. [3.34]). For real crystals, however, particularly for those possessing high dislocation densities, the application of such a model presents considerable difficulties.

When dislocations interact, they may form kinks, jogs, or vacancies according to their specific properties (Burgers vectors and slip planes) [1.77]. The manifold dislocations appear in the form of curved and tangled closed loops, building an extensive dislocation network, or bend-contour poles and other rather complicated defect configurations [1.79-81]. As a result complicated dislocation ensembles are created, as can well be seen from optical and electron micrographs and X-ray topograms (cf. Figs.3.23, 24,16) of aromatic hydrocarbon crystals.

At present there is a rather considerable gap between theory describing interaction between dislocations of simple geometrical configuration, and the possibility of applying this theory for the treatment of complicated dislocational formations in regions of strong deformations in real crystals [3.34]. Such a situation is rather typical for a theory using simplified models for describing various complex systems in solids. Two alternative approaches are, in principle, possible in these cases. The first uses simplified "idealized" models for which a mathematically strict and physically self-consistent theory can be evolved. Such models, however, provide only a very rough approximation in the description of complex real systems. An extrapolation of such a simplified model for the interpretation of results obtained beyond its possible validity limits involves considerable risks. Another alternative consists in building up phenomenological models, very often based on physical intuition, which provide a visual physical picture of the complex phenomenon or yield a statistically averaged description of the system.

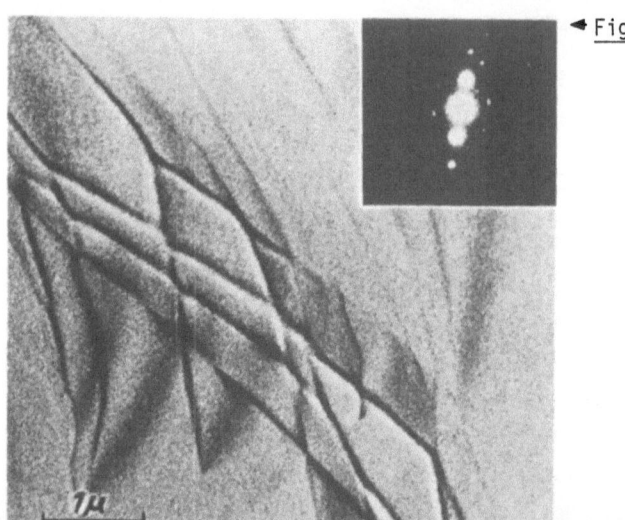

◄ Fig.3.23

▼Fig.3.24

Fig.3.23 Electron micrograph showing a network of extended dislocations in the ab plane of a p-terphenyl crystal [1.79]

Fig.3.24 Electron micrograph showing a bend-contour pole and extensive dislocation network in a p-terphenyl crystal [1.79]

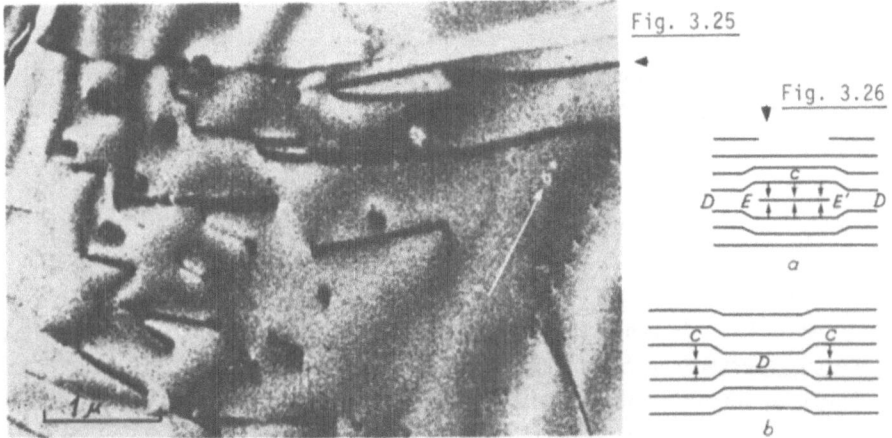

Fig. 3.25

Fig. 3.26

Fig. 3.25. Electron micrograph showing strong evidence for nonbasal dislocations in a p-terphenyl crystal [3.59]

Fig. 3.26a,b. Schematic picture of dislocation loops: (a) closed lopp of edge dislocation EE' forms a disk of excess molecules in the plane of the loop; (b) cross section of disc of vacancy loop formed as a result of vacancy condensation (C — region of compression; D — region of dilation)

This latter approach is usually employed in the analysis of the properties of dislocation ensembles and of defects with complicated structure (cf., e.g. [3.34]). some such models are discussed below and will be used later in considering randomizing factors of local state formation (cf. Sect.5.7).

Only part of the dislocations, mainly straight-line ones, emerge to the free surface of the crystal (cf. Figs.3.13,14,25). The majority of curved-line dislocations form closed loops or a dislocation network (cf. Figs.3.16, 23,24).

A closed loop of an edge dislocation forms a disk of excess molecules in the plane of the loop (Fig.3.26a). Such a dislocation disk is subjected to compression on the part of other crystal planes along its whole surface. The increase in electronic polarization energy ΔP thus depends on the diameter (or, generally, on the shape and size) of the disk. Such a disk forms an "island" for charge carrier trapping, with increased value of P.

If the Burgers vector is of opposite sign, then the loop presents a vacancy disk created by missing molecules (cf. Fig.3.26b). A vacancy disk may be formed as a result of vacancy condensation. Such a disk acts as an anti-trapping "island" for charge carriers having a reduced P value. In its turn, such an "island" is surrounded by a ring of compression with $\Delta P > 0$ (cf. Fig.3.26b).

A random distribution of diameter and shape of dislocation loops consti- tutes one of the randomizing factors in the energy distribution of local states in a molecular crystal (cf. Sect.5.7).

In Sect.3.9 we considered a model of a separate isolated stacking fault ribbon. Such simple planar defects can, however, aggregate in various col- lective configurations, such as a sequence [3.54] or a pile of ribbons, and form a three-dimensional inclusion of faulty stacking [3.34]. In crystals of aromatic hydrocarbons many cases have been observed when planes of faulty stacking intersect or form loops of faulty structure (cf. [1.79,80]).

Specific configurations of faulty stacking planes can be formed on inter- section of these planes, when so-called partial "pin" dislocations are cre- ated [3.20,34].

One of such configurations is called a Lomer-Cottrell barrier. It is formed by an immobile "pin" dislocation at a faulty stacking plane interac- tion line which, being unable to slip, presents an obstacle to other dis- location slips. Apart from the Lomer-Cottrell one, a number of other bar- riers have been described which are also formed by fixed "pin" dislocations from another slip plane (see for details [3.34,65]).

In a crystal of the anthracene space group we observe mainly slip in the basal plane ab in which, as a rule, basal unit dislocations are dominant (cf. Table 3.11).

ROBINSON and SCOTT [3.55], using methods of mechanical shear and compres- sion of the crystal, showed that starting stresses τ_R for nonbasal slip are approximately by an order of magnitude higher than for basal ones in an an- thracene crystal. Therefore typical nonbasal dislocations in anthracene-type crystals (cf. Table 3.11) arise mainly as a result of mechanical deforma- tion during growth or handling of the crystal.

It has been shown by ROBINSON and SCOTT [3.48] that nonbasal dislocations in molecular crystals may actually act as a "pin" for fixation of, say, small-angle boundaries and, quite likely, also of other dislocations or stacking faults. Dislocation aggregations shown in Fig.3.10b are likely to emerge near such barriers which create local regions of strong fluctuations of stress fields and, correspondingly, deep trapping centers of structural origin (cf. Sect.5.2). Basal and nonbasal dislocations in crystals of the anthracene space group are shown on the electron micrographs in Figs.3.23 and 25, respectively.

As we see, the scope of various structural defects and their interaction configurations is wide and diversified. This apparently is the main factor which determines the statistical nature of the energy spectrum of local trapping states of structural origin (cf. Sect.5.7).

An example of a typical three-dimensional defect in molecular crystals is an inclusion of another phase in the matrix of the parent crystal. This is due to pronounced polymorphism among a number of molecular crystals, particularly of heterocyclic and substituted aromatic compounds. The energy of different polymorphous phases can be so close that they may coexist at room temperature without losing stability.

In studies of polymorphic transformations in molecular solids MNYUKH et al. [3.66,67] have shown that, in general, no definite orientational relationship exists between the crystal lattices of two polymorphs.

An interesting divergence from this general rule was, however, recently observed by WILLIAMS et al. [1.88,89]. Electron and optical microscopic studies of 1,8-dichloro-10-methylanthracene, carried out by the authors, have revealed that the parent orthorhombic structure may, under certain circumstances undergo a stress-induced phase transformation into a daughter phase which is formed as thin laminar inclusions in the parent lattice.

It is rather noteworthy that these lamellae form a crystallographically coherent, definitely oriented phase boundary in the matrix of the parent crystal (see Fig.3.27). In other words, the phase boundaries do not destroy the monocrystalline structure of the parent crystal. Electron diffraction studies of the parent crystal and of the lamellae of the daughter phase show that the lamellae possess, just as the parent phase, orthorhombic structure oriented in a definite way in the matrix of the parent crystal [1.89].

Since the mean density of the daughter phase differs from that of the parent, one expects a corresponding change in the electronic polarization energy for charge carriers. Thus, such lamellae of second-phase inclusions, coherently linked with the matrix of the parent crystal, may form "tubes" of local states possessing definite crystallographic direction and causing considerable anisotropy of electrophysical parameters of the crystal.

Recently RAMDAS et al. [1.85] and PARKINSON et al. [1.86] have shown that stress- induced metastable second-phase inclusions may be produced also in anthracene crystals (cf. Sect.1.3.1). It was suggested earlier by CRAIG et al. [1.87] that in the defect regions of local compression inclusions of metastable structures of $P2_{1/c}$, $P2_{1/n}$ and $P\bar{1}$ symmetry may be formed in the parent anthracene crystal (see Sect.1.3.1). According to the authors these

Fig. 3.27. Electron micrograph of a crystallite of 1,8 dichloro-10-methyl anthracene showing plates of new phase within the parent matrix [1.88]

second-phase inclusions may form strings of 10 to 100 units, preferably along the [010] direction of an anthracene crystal.

Local photo-induced phase transitions may also occur in molecular crystals, e.g., such as photodimerization of predimer states found in the regions of extended structural defects of anthracene-type crystals (see Sect. 3.10).

The above-mentioned second-phase inclusions may create relatively deep trapping centers for charge carriers (cf. Sect.5.2).

3.12 Observation of Structural Defects in Molecular Crystals

At present there exist a variety of different experimental methods for direct observation and characterization of extended structural microdefects,
such as dislocations, tilt boundaries, planar faults, etc., in crystals.
Apart from conventional methods of optical microscopy, such up-to-date
techniques as X-ray topography, transmission electron microscopy and electron
diffraction studies ought to be mentioned. (This does not apply, however,
to point defects in solids which can be investigated only by indirect methods) (cf. Sect.3.3).

It may safely be said that modern experimental methods of investigating
extended microdefects in solids have reached a high stage of perfection and
have provided ample proof of the validity of the basic concepts of dislocation theory. These methods have, however, been developed from the extensive
experimental experience gained in studies of defects in anorganic solids-
metals, covalent and ionic crystals. As regards organic solids, and, in
particualr, molecular crystals, advances in the studies of microdefect topology are much less impressive, in comparison with anorganic crystals. This
is obviously due to various methodological difficulties, such as susceptibility of anthracene-type molecular crystals to electron-beam damage and lack
of thermal or chemical stability, quite apart from mundane difficulties in
preparing samples of acceptable dimensions [1.80,81,3.8]. Peculiarities of
molecular crystal structure have also impeded wide application of various
traditional optical and X-ray methods of microdefect research. It has been
very often necessary to evolve specific, modified, and occasionally rather
sophisticated methods for studying microdefects in organic crystals.

Pioneer work in investigating dislocation defects in organic molecular
crystals was done by ROBINSION and SCOTT [3.48,55,56,68] and THOMAS et al.
[1.77,78,3.7].

Applying methods of mechanical deformation and X-ray topography, ROBINSON
et al. were first to identify the dominant basal and nonbasal dislocations
in naphthalene and anthracene crystals, and to determine their critical
starting stresses of slip [3.48,55-57] (cf. Table 3.11). Significant progress was achieved at a later stage in the work of THOMAS and his group who
successfully applied a novel technique, particularly transmission electron
microscopy and selected area electron diffraction for studying dislocation
defects in a number of organic crystals (see review articles by THOMAS and
WILLIAMS [1.77,78,80,81,3.8]. It is rather evident that these extended

studies have laid the basic foundations of modern microdefect topography in organic solids demonstrating the vast impact of topological concepts upon novel approaches in organic solid-state physics and chemistry. As will be shown later, just microtopographic studies of structural defects form the real basis for the development of local state electronics in molecular crystals (see Chaps.4,5).

3.12.1 Optical Low Resolution Technique

Optical microscopic examination of the surface topography of cleaved, deformed or etched surfaces of organic crystals has become a long established method in microdefectoscopy of organic molecular crystals [1.77,80]. Optical microscopy does not require complicated apparatus and sophisticated experimental techniques. Information obtained by this method is, however, rather limited. It permits only indirect determination of "points" where dislocation lines emerge to the surface of the crystal. It does not, however, provide a possibility of studying the fine structure of dislocation defects and of determining their quantitative parameters. In general, only the direction (and not the magnitude) of the Burgers vector may be found. More correct identification of a specific dislocation and determination of its Burgers vector usually relies on more sophisticated methods such as electron or X-ray diffraction contrast techniques [1.81] (see Sects.3.12.2,3).

Optical microscopy is usually employed in combination with selective chemical etching of the surface of the crystal, or its mechanical deformation. Such studies in crystals grown by various techniques (see Sect.3.13) lead to the conclusion that high temperatures and strains during growth are responsible for the presence of most of the dislocations [1.81]. Similarly, deformation of the samples subesequent to growth, either by compression or tension at various temperatures, may produce additional specific types of dislocations in the crystal. This offers a unique opportunity of studying the role of specific types of dislocations in solid state processes [1.81].

Mechanical deformation of a crystal leads to a considerable increase in dislocation density (cf. Table 3.12). Dominant types of dislocations become clearly visible, as well as the prevalent directions of angular boundaries. Deformation with subsequent thermal treatment of the crystal leads also to the formation of polygonized vertical dislocation alignments. This extends the possibilities of optical observation of dislocation defects.

Deformation of the crystal can be effected in the process of cleavage, as well as at the stage of subsequent bending of the cleaved lamella [3.48,

55]. There are various methods of bending, the most widespread being that of three-point fixing [3.48].

Earlier studies by ROBINSON and SCOTT [3.48,55,56,68] as well as a recent paper by KOJIMA [3.58] convincingly demonstrate the excellent possibilities of optical microscopy of dislocation defects in deformed naphthalene and anthracene single crystals.

Dissolution and vaporization of molecules may be regarded as the reverse of crystal growth, and these two processes serve as a basis of the etch-pit technique for the revelation of the sites of emergence of dislocations at free crystal surfaces [1.81]. Both processes rely on the fact that extra free energy is associated with the dislocation cores which act as centers of increased reactivity [1.77,81]. Experimentally the "etch-pit" technique is connected with the exposure of cleavage or habit faces of a crystal to a particular reagent (etchant) or to evaporation and subsequent observation of the surfaces by methods of optical microscopy [1.81].

Dislocational etch pits, unlike those nucleated at point defects, point defect clusters or impurity centers, are almost invariably of pyramidal form with the apex of the pyramid lying on the dislocation line and the center of the base located at the original intersection of the dislocation with the surface [1.77].

In addition to the regular geometrical form there also exists another criterion for identifying dislocational etch pits, viz. mirror symmetry of pit images for corresponding cleavage faces of the crystal [1.77].

Shape and internal structure of dislocation etch pits depend mainly on the nature of etchant, temperature, crystal symmetry, crystal-etchant inter-facial energy, direction of the dislocation line, Burgers vector value, presence of impurity molecules in the dislocation core, and on other factors [1.77].

If the etched crystal face is isotropic, a conic etch pit is obtained, whilst a pyramidal one is formed in the case of an anisotropic crystal face. The base of the pyramid reflects to a certain extent the symmetry of the crystal. If an imagined straight line is drawn from the center of the base of the pyramid through its summit, it is possible to determine almost pre-cisely the direction of the dislocation line from the shape of the pyr-amid. It is not so simple, however, to link this direction with the crys-tallographic axes of the crystal, and there are rather controversial data on this problem in the literature, depending, most likely, on etching methods used (see, e.g., [1.73,77, 3.53,69,70]).

A typical optical micrograph of etch pits on a cleavage face (00ℓ) of an anthracene crystal is presented in Fig.3.13. The micrograph clearly shows isolated dislocation etch pits, as well as the row CD of etch pits of edge dislocation alignment. AB represents the twist boundary of the crystal. The micrograph of etch pits conveys not only a qualitative picture of geometric distribution of dislocations, but permits also a quantitative estimate of mean dislocation density per unit area of the given crystal surface (cf. Table 3.12).

In the case of a vertical dislocation alignment forming small- or middle-angle boundaries it is possible to estimate the boundary angle θ by measuring the mean distance D between dislocation pits and using the empirical formula (3.32).

Unlike anorganic crystals, for which there exists a wide range of etchants, as well as different etching techniques (see, e.g. [3.21]), the choice of suitable etchants for organic crystals is rather limited. Strong acids or their mixtures are usually employed, preferably a mixture of concentrated sulphuric acid with oleum [1.77]. A more detailed description of the etching technique for anthracene-type crystals may be found in [1.37,73,3.68-70].
For anthracene an effective etching method has been evolved only for the ab cleavage face of the crystal [3.70]. This sets serious limitations to possibilities of using etching methods, since only nonbasal dislocations emerge on the ab face of the anthracene crystal (cf. Table 3.12).
Optical methods of studying the degree of perfection of a crystal by etch-pit technique usually yield a two-dimensional picture only. For obtaining direct information on dislocation defects in the bulk of the crystal the so-called decoration technique may be employed [3.21,71]. Its essence consists in chemical treatment of a crystal, with the aim of depositing "decorating impurity" along the dislocation line which may be observed directly or indirectly. This technique can, however, be applied to molecular crystals only in some special cases. Thus, photochemical methods of "decorating" a molecular crystal have been described, using selective photodimerization of "predimer" molécules along dislocation lines [1.77,78,3.72] (see also Sect.3.10).

An original method of selective thermal deformation of thin anthracene crystals grown from gaseous phase was recently applied by LISOVENKO et al. [1.73,74]. Anthracene crystal layers were put on optical contact with a quartz or plexiglass substrate and cooled to 4 K. Owing to difference in linear expansion coefficients, cooling produced considerable plastic deformation causing a large number of excess dislocations. The mean density of nonbasal dislocations, which was about $10-10^2$ cm^{-2} in a nondeformed crystal, reached a value of up to 10^6-10^7 cm^{-2} after such thermal deformation [1.73]. In the experiment the conditions were chosen selectively, the extent and nature of thermal deformation being dependent on the nature of the substrate, as well as on the final temperature of cooling. In this way the authors of [1.74] succeeded in obtaining specific nonbasal edge dislocation alignments in the [100] or the [010] direction of the crystal, or a network of alignments of both directions. Optical micrographs of etch-pit rows of such selectively deformed anthracene crystals are shown in Fig.3.14. Using defect fluorescence technique LISOVENKO et al. [1.74] demonstrated that for both types of dislocation alignments shown in Fig.3.14, there exist corresponding excition trapping states of definite depth (see Sect.4.3).

In recent years there has been considerable progress in the evolution of novel optical methods for studying microdefects in molecular crystals. Various special techniques are used for raising resolving power and image contrast [1.37,3.9]. For illustration a NOMARSKI [3.73] interference contrast optical micrograph is presented in Fig.3.28 obtained by SLOAN et al. [1.37] for (20$\bar{1}$) as-grown face of a vapor-grown anthracene crystal (cf. Fig.3.31). Examination of the micrograph reveals prominent slip traces running along the [010] direction. The fact that the slip traces are observed on the (20$\bar{1}$) face indicates that there is a component of the Burgers vector of the responsible dislocations in the [100] direction. Near the edge of the crystal marked curvature of the slip traces can be observed (see Fig.3.28). These data support the suggestion that slip preferably occurs in (001) planes by means of dislocations with Burgers vector in [010] or 1/2 [110] and/or 1/2 [120] directions which are the dominant slip vectors for basal dislocations in anthracene-type crystals (cf. Table 3.11). These con-

Fig. 3.28. Nomarski interference contrast optical micrograph of (20$\bar{1}$) as-grown face of vapor-grown anthracene crystal showing prominent slip traces in [010] direction ($\times$ 320) [1.37]

clusions were later confirmed by independent evidence obtained by diffraction contrast transmission electron microscopy for p-terphenyl crystals (see Sect.3.12.2).

3.12.2 Electron Microscopy and Diffraction Techniques

Transmission electron microscopy (TEM) and especially its more sophisticated modifications such as diffraction contrast techniques and selected area electron diffraction has recently become the most powerful tool for the studies of extended microdefects in organic molecular crystals [1.80,81,3.8,59]. In principle, structural imperfections of any material may be investigated by means of TEM technique as long as one can prepare samples of 100-500 nm thickness, transparent for an electron beam of ca. 100 keV energy.

Application of electron microscopes with high resolving power makes direct lattice resolution possible and thus, in principle, yields direct information about closely spaced shear planes in small areas of ca. 10^{-6}-10^{-4} cm in diameter [1.81]. By high resolution electron microscopy the periodicity of crystals may be directly imaged, and so aperiodicities in the stacking sequences of planes may be readily detected [1.81,3.8]. As early as at the end of the fifties, when electron microscope techniques came into general use (see [3.74]), MENTER [3.75] obtained direct electronic micrographs of molecular planes of platinum and copper phthalocyanine crystals with a distinctly discernible "excess" molecular plane of edge dislocation. However, attempts to widen the scope of objects of investigation applying electron microscopy in organic crystal studies proved to be rather disappointing. It was found, for instance, that aromatic hydrocarbon crystals, unlike phthalocyanine crystals, are rather susceptible to electron beam damage and undergo irreversible destruction in the course of taking the electron micrograph, i.e., the degree of perfection of the crystal changes beyond control in the very process of investigation.

In order to surmount these difficulties JONES et al. [3.59,76] conducted special experiments and determined approximate lifetimes of various aromatic crystals at room temperature in an electron microscope operated at 100 keV. These experiments showed that each kind of aromatic hydrocarbon crystal possesses its own specific critical doses of irradiation, i.e., has its own specific mean "lifetime" in an electron microscope beam. For instance, the mean lifetime of a p-terphenyl crystal in an electron microscope at 100 keV and 3.8×10^{-4} A cm^2 is 22 ± 3 min, that of anthracene only 5 ± 1 min, i.e., the critical dosis of irreversible destruction is 5 times lower in the case of anthracene, as compared to p-terphenyl. Hence, the time of exposure must be under 10 min for p-terphenyl and under 2 min for anthracene crystals under the given experimental conditions [3.59,76]. This sets rather strict conditions for obtaining an electron micrograph of anthracene. Special techniques should therefore be used for obtaining electron micrographs of anthracene-type crystals, such as low temperatures of the order of 100 K, high-sensitivity photographic material, electron-optical amplification of the image, special methods of raising the contrast of the image (diffraction contrast method), etc.[1.80,3.10,76].

These methodological difficulties may be to some extent surmounted using a carefully selected model system. Thus, much can be learned about the behavior of anthracene-type crystals using p-terphenyl instead of anthracene, the former being more electron-beam stable and belonging to the same space group as anthracene, viz. $P2_{1/a}$, [1.80].

A detailed electron microscopic study of thin p-terphenyl crystals, using diffraction contrast technique, was performed by THOMAS [1.80], and JONES et al. [3.59].

The diffraction contrast technique is based on the fact that the Bragg condition may be satisfied at the distorted regions of the lattice when no diffraction occurs elsewhere at perfect crystalline regions. The situation for an edge dislocation is schematized, according to [1.81], in Fig.3.29.

The theory governing the imaging of dislocations and other extended defects has been adequately discussed in a monograph by HIRSCH et al. [3.77]. In good approximation the conditions for the disappearance of dislocation contrast in an isotropic material may be given by

$$g \cdot b = 0 \qquad\qquad (3.42)$$

$$g \cdot b \wedge u = 0 \quad , \qquad\qquad (3.43)$$

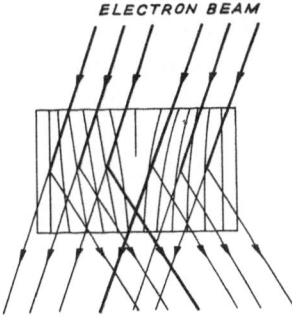

ELECTRON BEAM

Fig. 3.29. Schematic picture of diffraction of electron beam at the region of edge dislocation core, caused by distortion of crystal planes [1.81]

where $\underline{g}$ is the operating diffraction vector, $\underline{b}$ is the Burgers vector and $\underline{u}$ is a unit vector along the dislocation line. For $\underline{g} \cdot \underline{b} \wedge \underline{u} \neq 0$ and for strongly anisotropic material the $\underline{b} \cdot \underline{g} = 0$ disappearance condition is not absolute but only distinguishable [3.59].

Burgers vector determination involving such images relies on the fact that diffraction contrast from a dislocation line of Burgers vector $\underline{b}$ disappears when conditions (3.42,43) are satisfied [1.81]. In essence, therefore, the method employed here is to take a series of bright-field micrographs using various degrees of tilt of the sample, chosen so as to reveal dislocations (under one set of diffracting conditions) and then to render them invisible (under one or more sets of conditions) [3.59]. The characterization of stacking faults utilizes similar procedures [1.81].

For preparing thin samples of anthracene and p-terphenyl crystals of 100-500 nm thickness for electron microscope studies special methods are employed. One of them consists in preparing a dilute solution of anthracene [3.10] or p-terphenyl [3.59,76] in xylol and allowing it to evaporate slowly on the water surface. In this procedure small crystallites of ca. 4 mm^2 area are produced [3.59], with the (001) face oriented parallel to the water surface [3.76]. These thin layers are then transferred on a metal grid and placed into the electron microscope. Thin layers of p-terphenyl crystals can also be obtained by sublimation on a cooled substrate [3.59]. Both solution- and vapor-grown crystallites display a considerable density of grown-in dislocations and faults of others kinds [3.59].

Figures 3.12,19,23-25 present some typical electronic micrographs of p-terphenyl crystals obtained by THOMAS et al. [1.79,3.59] using diffraction contrast technique.

The micrograph in Fig.3.23 shows a fairly extensive network of basal dislocations in the ab plane of a p-terphenyl crystal part of which is brought into contrast by means of [110] bend contour. If this particular bend contour is swept, by tilting the sample across the network, the other dislocations may be imaged, provided conditions (3.42,43) are not statisfied [1.79, 3.59].

The micrograph in Fig.3.24 shows the bend contour network associated with the establishment of a (001) pole in a p-terphenyl crystal [1.79,3.59].

The micrograph in Fig.3.25 presents relatively short straight dislocations which probably run from the top to the bottom surface of the crystal and apparently are nonbasal [3.59].

The micrograph in Fig.3.19 demonstrates doubly imaged dislocation segments (a), (b) and (c) which could be interpreted as partial dislocations separated by stacking faults [3.59].

The micrograph in Fig.3.12 visualizes an array of dislocation alignments comprising a small-angle boundary in a p-terphenyl crystal.

An analysis of electron micrographs obtained by the diffraction contrast method made it possible for THOMAS et al. [1.79,80,3.59] to present direct evidence of the existence of dominant basal dislocations of (001) [010] and (001) [110] types, of nonbasal dislocations of (100) [010] and (100) [001] types (cf. Table 3.11), as well as of previously unknown basal dislocations of type (001) [120] (cf. Fig.3.18) in thin p-terphenyl crystals.

Electron microscopy data on thin crystals are usually complemented by electron diffraction patterns [1.79,80,3.59]. Figures 3.23,24, for instance, present insets of electronographs of the crystals under study, confirming the corresponding (001) orientation of the crystals.

Selected area electron diffraction was also successfully applied by THOMAS et al. [1.78] for microtopographic identification of photodimerization products along the dislocation lines in an anthracene crystal.

The detailed studies carried out by THOMAS and his group on the microtopography of dislocation defects in molecular crystals [1.79,80,81,3.59,76] have formed a basis for the development of real physical models of formation of polarization-caused local trapping states. For example, this work has led to quantitative calculations of the local state energy spectrum for stacking fault ribbons in anthracene crystals (cf. Sect.3.9.3).

In recent years electron diffraction techniques have been widely used for studying the crystalline state of thin evaporated layers of organic crystals, e.g., for investigating the structure of oriented, quasi-amorphous and amorphous layers of tetracene and pentacene [1.104,2.153,3.63] (cf. Sect.5.14). The applicability of the replica technique of electron microscopy was also demonstrated for studying surface structure of oriented crystalline layers of tetracene and pentacene [1.104,2.153] (cf. Sect.5.14).

In the course of the last ten years the scanning electron microscope has become a particularly effective instrument for studying the topography of the crystal surface (cf. e.g., [1.79,3.78,79]). This method may also be successfully applied for surface studies of oriented crystalline layers (see, e.g. [2.107,3.80]). Its wider application for organic crystal studies is at present also impeded by effects of probe damage by the electron beam. The method may, however, become applicable for organic molecular crystals in the near future by using a more sophisticated technique, such as, e.g., preliminary coating of the sample with a layer of metal, as practised in case of biological objects [3.79].

Promising results in the surface study of organic molecular crystals can be expected also from such modern methods as Auger spectroscopy and the method of low-energy electron diffraction (LEED) (cf. [3.81]).

Thus FIRMENT and SOMORJAI [3.82] have recently surmounted a number of methodological obstacles and first obtained surface-structural information on vapor-grown naphthalene crystal surface using the LEED technique (cf. Sect.5.14). Later BUCHHOLZ and SOMORJAI [3.83] used LEED measurements for studying monolayer surface structures of crystalline phthalocyanine films vapor-deposited on single-crystal copper substrates. The LEED technique according to [3.82] is applicable to studies of a wide variety of molecular crystal surfaces and opens a novel field of structural surface chemistry of organic solids.

3.12.3 X-ray Methods

The feasibility of X-ray technique for studying dislocation defects in anthracene crystals was clearly demonstrated at the end of the sixties by MICHELL et al. [3.57]. X-ray methods did not, however, become generally accepted for studying structural defects in other molecular crystals. According to THOMAS and SHERWOOD [1.79,80,3.23], this is mainly due to insufficient resolving power for identification of individual dislocations in small samples. X-ray topography has considerable advantages in studying dislocations in the bulk of comparatively large-size crystals and at relatively low dislocation density ($n_{dis} \leq 10^5$ cm^{-2}) [3.23]. WEBB [3.84], however, is of the opinion that resolution of separate dislocations by means of X-ray topography is possible at dislocation density $\lesssim 10^6$ cm^{-2} in the case of thin crystals, and $\lesssim 10^4$ cm^{-2} in the case of surface layers and of thick crystals. Resolution in these cases reaches 5 μm, approaching the value obtained by the etch-pit method, but making possible the detection of dislocations in the bulk of the crystal without its destruction. The limits imposed by crystal size are also not so rigid. In the above-mentioned work by MICHELL et al. [3.57] an anthracene crystal grown form the vapor phase and of size ca. $5 \times 3.5 \times 0.5$ mm^3 was used. It is therefore to be expected that new advances in the technology of obtaining perfect organic crystals will, most likely, enhance further development in the techniques of X-ray microtopography.

Theoretically the range of reflection angles ought to be of the order of a few angular seconds in the case of large perfect crystals, and reflection intensity must be proportional to the structure factor. For most crystals the reflection angle range is, however, of the order of angular minutes, reflection intensity being proportional to the square of the structure factor [3.21]. These discrepancies are mainly due to the structural imperfection of most crystals which consist of mosaic domains, disoriented with respect to each other owing to dislocation alignments and small- and medium-angle boundaries.

Thus X-ray diffraction provides a general method of assessing the degree of perfection of a crystal [3.21,84]. Among such methods of rough estimation of perfection of a crystal the popular method of GAY et al. [3.85] might be mentioned, based on observing distortions in the diffraction pattern. With increase in dislocation density diffraction spots widen and form diffuse arcs along the Debye-Sherrer rings.

X-ray topography of almost perfect crystals (at dislocation densities $\lesssim 10^6$ cm^{-2}) is based on the theory of X-ray diffraction on perfect crystals (see [3.86]). Dislocations and other defects are considered as local faults in a perfect crystal, causing the appearance of intensity contrast [3.84].

Modern X-ray methods of studying defects in crystals, as well as their main features and fields of application are summarized in Table 3.10. These methods may be conditionally divided into three main groups:

a) methods of roughly estimating the degree of perfection of a crystal [Schulz method (I)];
b) topographic methods of studying nearly perfect crystals [Berg-Barret (II, III), Borman (IV), Lang (V) methods];
c) methods of studying crystals of high degree of perfection [two-crystal spectrometry (VI) and X-ray interferometry (VII) techniques].

The essence of the methods of the second (b) group is schematically outlined in Fig.3.30. These methods are varieties of topographic mapping of dislocations in near-perfect crystals of different thickness. According to the value of limiting dislocation density ($n_{dis} < 10^6$ cm^{-2} or $n_{dis} \lesssim 10^4$ cm^{-2}, respectively) these methods are applicable for studying high-purity anthracene crystals grown from vapor or from solution, but in some cases even from melt (cf. Table 3.12). The second (b) group of topographical methods permits observation of separate dislocations, as well as drawing maps of dislocation networks and provides sufficient spatial resolution of the order ~ 5 μm [3.84].

The third (c) group comprises methods of studying single crystals of a very high degree of perfection [3.86]. These crystals are going to be an object of great interest in the near future owing to rapidly improving techniques.

Table 3.10 contains also references quoting review articles and original papers in which the reader will find particulars on experimental techniques.

Table 3.10. X-ray methods of studying structural imperfections in crystals

Method	Essence	Basic features	Fields of application and peculiarities
I Schulz [3.21,84, 87,88]	Registration of r flec-tion lauegrams frc the crystal, using conJinu-ous X-ray spectrum.	Laue pattern spots are observed, of size comparable to that of sample. Each spot is formed as result of reflection from weakly disoriented sites in the crystal.	Simplest X-ray method for studying degree of perfection of crystals. Can be used for studying small-angle boundaries and disorientation within range from several minutes to several degrees. Method provides low spatial and angular resolution.
II Berg-Barret back reflection [3.21,84]	Registration of Bragg reflection from the crystal surface using monochromatic X-rays (cf. Fig.3.30a).	Dislocations are observed as darker spots, owing to stronger reflection of X-rays from deformation regions.	Method particularly useful for studying dislocation pattern of surface regions of crystal at dislocation densities $\leq 10^4$ cm^{-2}; absorption criterion $\mu_0 t > 1$ *.
III Berg-Barret direct transmission (diffraction) [3.21,84, 88]	Registration of diffracted monochromatic X-rays after passing through thin crystal, so-called Laue position (Fig.3.30b).	Dislocations are observed as dark spots, since diffraction is more pronounced in these sites.	Method permits study of internal dislocation pattern of sufficiently thin crystals; resolution of separate dislocations possible at dislocation densities $< 10^6$ cm^{-2}; corresponding absorption criterion $\mu_0 t \ll 1$ *.

* μ_0 is the linear coefficient of effective absorption of X-rays; t is the thickness of the crystal in the direction of the incident beam. Condition $\mu_0 t \ll 1$ holds for a thin, $\mu_0 t \gg 1$ for thick and $\mu_0 t > 1$ for a crystal of medium thickness (cf.Fig.3.30a-c).

Table 3.10. (continued)

Method	Essence	Basic features	Fields of application and peculiarities
IV Borman anomalous transmission [3.21,84,88].	Registration of monochromatic X-rays passing through thick crystal, as well as of diffracting rays (Laue position)(cf. Fig.3.30c)	Regions of dislocations weaken both beams more strongly and are observed as lighter spots.	Method permits study dislocations in bulk of thick crystal, if dislocation density $< 10^4$ cm^{-2}; corresponding absorption criterion $\mu_0 t \gg 1$*.
V Lang technique [3.21,84,89,90]	Registration of Bragg reflection of monochromatic X-rays according to Berg-Barret direct transmission method (cf. Fig.3.30b).	Simultaneously moving film and sample, a full "map" of dislocation pattern can be obtained; changing angular orientation of crystal we get a three-dimensional topogram of dislocation network (cf.Fig.3.16).	Method permits mapping internal dislocation pattern of thin crystals; absorption criterion $\mu_0 t \ll 1$*.
VI Two-crystal spectrometry [3.21,86,91]	Characteristic monochromatic X-rays are reflected at Bragg angle from reference crystal to crystal under study.	Registration of intensity of X-rays reflected from crystal under study at various angles. Integral intensity and width of peak indicate degree of perfection of crystal.	Method permits detection of disorientation of order of 0.01 angular second.
VII X-ray interferometry [3.21,86,92]	Moire fringe interference from perfect and faulty structures.	Used in combination with Borman or Lang method.	The most sensitive method of detecting small deformations in crystals. May be used for studying crystals of high perfection: detects displacements of the order of 0.1 Å and disorientation of order of 0.001 angular second.

* μ_0 is the linear coefficient of effective absorption of X-rays; t is the thickness of the crystal in the direction of the incident beam. Condition $\mu_0 t \ll 1$ holds for a thin, $\mu_0 t \gg 1$ for thick and $\mu_0 t > 1$ for a crystal of medium thickness (cf.Fig.3.30a-c).

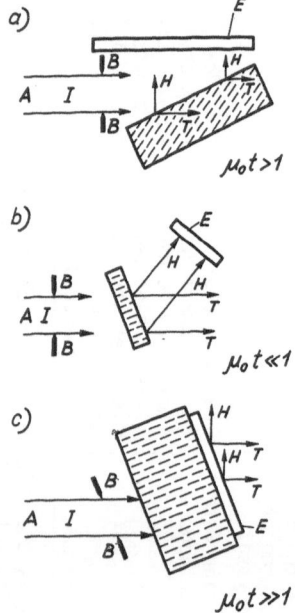

a)

b)

c)

Fig. 3.30a-c. Schematic diagrams of basic methods of X-ray diffraction topography, according to [3.84]: (a) Berg-Barret method of back-reflection; (b) Berg-Barret method of direct transmission (diffraction); (c) Borman method of anomalous transmission. A — source of monochromatic X-rays; B — limiting slits; I — incident beam; H — diffracted beam; T — transmitted beam; E — photographic film

Fig.3.16 shows a typical X-ray topogram of an anthracene crystal, obtained by MICHELL et al. [3.57] using the Lang technique. The topogram presents reflection from the $(\bar{1}10)$crystal face. It displays groups of dislocations in regions A and B, with Burgers vector components parallel to the [010] direction, as well as dislocations in region C with Burgers vector component in direction [1$\bar{1}$0]. Reflection from other faces revealed also dislocations in the [110] direction. The topographical studies of dislocations in the bulk of the crystal show, in its turn, that the given dislocations lie in the basal plane (001) of the crystal. In this way, studying the reflection topograms form various faces of the crystal, different types of dislocations can be visualized and identified using the exclusion method [3.57].

Figure 3.16 clearly illustrates the large potential possibilities of X-ray topography for direct observation and characterization of dislocations in organic molecular crystals.

Many types of imperfections in crystals change such a parameter as the lattice period in the local regions of their appearance. These changes in lattice period can be determined by the BOND method [3.93] which permits measurements of interplane distances with an accuracy of ca. 10^{-4} %. Since the bulk in which the X-rays diffract is very small—of an order of

$1 \times 0.025 \times 0.005$ mm^3 — it becomes possible to study local lattice period changes from point to point in the defect region [3.94]. One can say with certainty that this method of studying structural imperfections in molecular crystals is extremely promising from the point of view of local state formation.

The present rapid development of automatic diffractometry in X-ray structure analysis opens many new and promising possibilities for studying the degree of perfection of molecular crystals.

In earlier days a determination of so-called temperature factors B from an X-ray diffraction pattern was connected with a large amount of work and was carried out rather as an exception (cf. e.g. [1.30]). At present, however, measurements of these parameters on an automatic diffractometer have become a routine procedure. It must be borne in mind that the temperature factor B is determined not only by termal oscillations of molecules in the lattice, as might be expected from the direct meaning of the term. As was shown in [Ref.1.2, p.139], the total mean square deviation of molecules from equilibrium position in a crystal, $\overline{u_{tot}^2}$, which determines the value of the temperature factor B/8π^2 in X-ray diffraction patterns, can be a sum of two components, viz. the mean square value of the dynamic deviations $\overline{u_d^2}$ and the mean square value of the static deviations $\overline{u_s^2}$

$$\overline{u_{tot}^2} = \frac{B'}{8\pi^2} = \overline{u_d^2} + \overline{u_s^2} \quad . \tag{3.44}$$

The dynamic component $\overline{u_d^2}$ of the temperature factor is actually determined by oscillations of molecules in the lattice, depending on temperature. The static component $\overline{u_s^2}$, on the other hand, is determined by static distortion of the lattice due to the presence of structural or impurity defects in the crystal.

An estimate of the contributions of the dynamic component $(\overline{u_d^2})$ and the static one $(\overline{u_s^2})$ was made in [1.66]. Date on anisotropy of the total temperature factor $\overline{u_{tot}^2}$ for anthrahence were taken from [1.30]. The estimate showed that in the anthracene crystal studies in [1.30] the static component of the temperature factor is actually approximately twice as large as the dynamic one.

It should be noted that modern techniques of X-ray methods for determining the anisotropy of B factors make it possible to quench the dynamic component $\overline{u_d^2}$ in low-temperature measurements. A comparison between the $\overline{u_{tot}^2}$ values at room and at low temperature provides an experimental estimate of the contribution of the static lattice distortions. It may be expected that such studies on molecular crystals will be carried out already in the nearest future.

3.13 Main Characteristics of Dislocation Defects
in Some Model Molecular Crystals

3.13.1 Dominant Types of Dislocations in Anthracene Space Group Crystals

The most elaborate studies on dislocation defects have been performed on anthracene-type crystals. In this field a large number of experimental data have been accumulated from optical [1.37,77,3.58], X-ray [3.57] and electron microscopy [1.77,79-81,3.8,59] work. Studies have been performed on crystals obtained by various methods of growth from gaseous phase, melt and solution. Special investigations were carried out on crystals that were subjected to mechanical [1.72,77,3.55,58,95] or thermal deformation [1.73,74], to uniform hydrostatic compression [3.60,96], etc. Anthracene and its analogues are therefore rightly considered as the main model compounds in the study of dislocation defects in organic molecular crystals.

Table 3.11 presents literature data on the dominant dislocation slip systems of anthracene, p-terphenyl, naphthalene and anthraquinone crystals, all belonging to space group $P2_{1/a}$.

Table 3.11 shows that of all so-called basal dislocations gliding in the (00ℓ) planes of the crystal the (001) [010] dislocation is dominant. This dislocation has the largest probability of appearing under normal conditions of crystal growth, due to small starting slip stress value τ_R. (For instance, in a naphthalene crystal we have for the given dislocation $\tau_R \sim (1-3) \times 10^{-5} \mu$ [3.56], where μ is the shear modulus).

As proved by calculations, the energy $E_\perp$ of a (001) [010]-type basal dislocation in an anthracene crystal is almost half that of a (001) [100]-type dislocation (cf. Table 3.6). Therefore its formation is energetically most favorable. Among dominating dislocations in anthracene, p-terphenyl and naphthalene crystals basal partial dislocations of (001)1/2 [110] type are of particular importance. They have been observed independently by optical [1.37,3.48,55,58] (cf. Fig.3.28), electron microscopy [1.79,80,3.59] and X-ray techniques [3.57] (cf. Fig.3.16).

Table 3.11. Dominant systems of dislocation slip in anthracene space group crystals

Crystals	Lattice parameters	Dominant types of dislocation slip	References	Notes
Anthracene $P2_{1/a}$, Z = 2	a = 8.56 b = 6.04 c = 11.16 β = 124.7°	Basal dislocations (001) [010] (001) [100]	[1.37,77,3.53, 55,57,58]	Dislocation of type (001) [010] is dominant in the ab plane of the crystal. The calculated energy values of the (001) [010] - and (001) [100]- -type dislocations see Tables 3.5 and 3.6.
		1/2 [110] 1/2 [120]	[1.37,77,79,80, 3.23,48,55,57]	Partial dislocations 1/2 [110] and 1/2 [120] cause formation of stacking fault ribbons in the ab plane of the crystal (cf. Sect.3.9).
		Nonbasal dislocations (100) [010] (010) [100] (100) [010] (100) [001] (010) [001]	[1.77,79,3.55] [1.73,74,3.55] [1.73,74] [1.79,3.55] [1.77,3.55]	Nonbasal dislocations are mainly formed as a result of crystal deformation.
		(20Ī) [010] (12Ī) [2Ī0] (110) [1Ī0]	[1.37,77,3.55] [1.37] [1.37]	Observed in crystals grown from gaseous phase [1.37]
p-Terphenyl $P2_{1/a}$, Z = 2	a = 8.08 b = 5.60 c = 13.59 β = 91°55'	Basal dislocations (001) [010] (001) [11ω] or 1/2[110] (001) [12ω] or 1/2[120]	[1.79,3.59] [1.79,3.59] [1.79,80,3.59]	Dominating basal dislocation. Dislocations of types [11ω] and [12ω], most likely, have ω = 0. The corresponding partial dislocations 1/2 [110] and 1/2 [120] cause formation of stacking fault ribbons in the ab plane of the crystal (cf. Sect.3.9).
		Nonbasal dislocations (100) [010] (100) [001] (201) [010] (1Ī0) [110]	[1.77,79,3.59] [1.79,3.59] [1.77] [1.77]	

Table 3.11. (continued)

Crystals	Lattice parameters	Dominant types of dislocation slip	References	Notes
Naphthalene $P2_{1/a}$, $Z = 2$	$a = 8.24$ $b = 6.00$ $c = 8.66$ $\beta = 122.9°$	Basal dislocations (001) [010]	[1.77]	Dominating basal dislocation
		(001) 1/2 [110]	[1.77,3.56]	Partial dislocations of type 1/2 [110] causes formation of stacking fault ribbons in the ab plane of the crystal.
		Nonbasal dislocations (010) [001]	[1.77]	
Anthraquinone $P2_{1/a}$, $Z = 2$	$a = 15.85$ $b = 3.98$ $c = 7.92$ $\beta = 102°43'$	Basal dislocations (001) [010]	[1.77]	Dominating basal dislocation
		Nonbasal dislocations (100) [010] (20$\bar{1}$) [102]	[1.77]	Causes formation of stacking faults in the (20$\bar{1}$) plane of the crystal.

Since these partial dislocations cause the formation of stacking fault ribbons in the ab plane of the crystal (cf. Sect.3.9), their study is of great importance for elucidating the conditions under which trapping centers for charge carriers are formed in an anthracene crystal (cf. Sect.3.2 and 5.2).

An interesting empirical correlation was discovered by ROBINSON and SCOTT [3.56], concerning the formation of (001) [110]-type dislocations in the process of mechanical deformation of a naphthalene crystal. They found that the starting stress τ_R of slip of the basal dislocation of type (001) [110] is lower in highly purified naphthalene crystals than in crystals containing impurities. This means that the conditions for the formation of stacking fault ribbons in high-purity naphthalene and anthracene crystals are more favorable than in impurity-containing crystals.

Similar correlations have been reported in [3.56], also concerning (001) [010]-type basal dislocations. In this case the value of τ_R in naphthalene crystals containing 0.4-1.5% impurities is 1.5-2.5 times higher than in high-purity single crystals. This means that ultra-high purification of the crystal removes impurity-created local centres, but, at the same time, produces favorable conditions for the formation of dislocation-caused local centers. This peculiarity of molecular crystals is not always sufficiently taken into account. It appears that it is just the nature of weak intermolecular interaction which makes it almost impossible to obtain dislocation-free organic molecular crystals.

The value of τ_R for creation of nonbasal dislocation slips in anthracene-type crystals is by an order or more higher than in the case of basal dislocation slips. Thus, for a nonbasal dislocation of the (100) [010] type the starting stress of slip must be of the order of $\tau_R = 10^{-3} \mu$ [3.55]. Hence, nonbasal dislocations can, as a rule, be formed only under nonequilibrium conditions, e.g., under mechanical or thermal deformation of the crystal (cf. e.g. [1.73,74,3.55]). As it was demonstrated by LISOVENKO et al. [1.74], under special conditions of thermal deformation of thin anthracene crystals specific types of nonbasal dislocations may be made dominant (cf. Sect.3.12.1 and Fig.3.14).

On the other hand, THOMAS [1.79] has recently reported that after photooxidation of transstilbene crystals a number of new, unusual dislocation slip systems have been observed, not present prior to photooxidation. The author is of the opinion that, as a result of photooxidation, and formation of oxidation products in the bulk of the stilbene crystal, local fields of stress are created, producing new dislocations.

As regards dislocational defects in other organic crystals, only rather fragmentary data are available, insufficient for obtaining any general conclusions.

3.13.2 Density of Dislocations in Anthracene Crystals, Its Dependence on Crystal Growth and Treatment

Density of dislocations largely depends on the method of growing the crystal, on its thermal treatment, and mechanical deformations.

Anthracene single crystals can be obtained from melt, solution, or gaseous phase.

Typical density values of nonbasal dislocations in an anthracene crystal, as dependent on growing technique and treatment, are presented in Table 3.12.

The mean density of nonbasal dislocations, emerging in the (001) face of an anthracene crystal, is usually determined by counting the number of etch pits per unit area of the surface.

As may be seen from Table 3.12, the most perfect anthracene crystals can be obtained from the vapor phase. However, this method involves other difficulties. One obtains, as a rule, crystals in the form of facet-like thin

Table 3.12. Values of mean nonbasal dislocation densities in anthracene for different methods of crystal growing and treatment

Method of crystal growing and treatment	Mean dislocation density on (00ℓ) face of crystal $[\text{cm}^{-2}]$	Ref.
From gaseous phase:	10^2-10^3	[1.37,77,3.53]
	5×10^3-5×10^4	[3.69,97]
undeformed	10 -10^2	[1.73]
after thermal deformation, on cooling to 4.2 K	10^6-10^7	[1.73]
From solution	10^3-10^4	[1.77,3.53]
	5×10^3-5×10^4	[3.69,97]
From melt:	10^5-10^6	[1.77,3.53,97]
	5×10^5	[3.69]
deformed	10^7-10^8	[1.77]
without thermal treatment	10^6-10^7	[3.70]
after thermal annealing during 200 hrs at 200°C	10^5-10^6	[3.70]
anthraquinone-doped	10^8	[3.70]

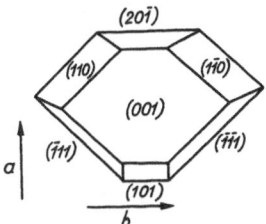

Fig. 3.31. Morphology of a vapor-grown anthracene single crystal [1.37]

lamellae which are difficult to remove from the substrate without mechanical damage and deformation [3.23].

Figure 3.31 illustrates the morphology of crystals of ultra-pure anthracene grown by SLOAN et al. [1.37] from the vapor phase in the presence of an inert gas (such as N_2, He, Ar or Xe). Figure 3.31 shows prominent as-grown faces of the obtained crystals having typical dimensions ca. $0.5 \times 0.5 \times 0.2$ cm^3. A practically identical morphology was obtained by ROBINSON and SCOTT [3.68] for anthracene crystal grown from the gaseous phase in vacuo.

Anthracene crystals are colorless, and their fluorescence lies in the violet region, being easily quenched even by minute quantities of tetracene. A triplet exciton in high-purity anthracene single crystals has a mean lifetime of 10-20 ms at 300 K [3.95].

Attempts to cleave, polish and subject to chemical etching faces of an anthracene crystal other than the basal (001) face have not been successful up to now with large single crystals obtained from melt [1.37]. Therefore there is practically no information available on dislocation density on nonbasal faces. The authors of [1.37] were the first to succeed in obtaining etch pits on the as-grown ($20\bar{1}$) and (110) faces of an anthracene crystal, apart from the (001) face (see Fig.3.31). The basal cleavage face (001) of these vapor-grown crystals showed a dislocation density of ca. 1×10^3 cm^{-2}. The corresponding value for faces ($20\bar{1}$) and (110) was 1×10^5 cm^{-2}, i.e., the same as in crystals obtained from melt (cf. Table 3.12). This demonstrates the considerable anisotropy of dislocation distribution and of the corresponding anisotropy of structure-caused charge carrier traps in the crystal.

The best specimens of anthracene single crystals of average perfection can be grown from solution. This method, however, always involves the risk of impurity and solvent inclusions [3.98].

High-purity anthracene single crystals may be obtained from melt. In this case, however, one always gets a high density of nonbasal dislocations (cf. Table 3.12) although a lot depends on the crystal-growing technique employed.

According to SHERWOOD [3.23], the most perfect crystals are obtained by the Czochralski method, whilst those grown by the traditional Bridgeman method usually contain considerable dislocation densities.

Despite the high density of structural defects, single crystals obtained from melt are most widely employed in physical experiments. This is due to the fact that the method provides a possibility of obtaining large high-purity anthracene crystals which are convenient for processing and experiment. For this and other, perhaps, purely psychological reasons, the investigator frequently prefers an ultra-pure, even if imperfect, crystal.

Mechanical deformation of the crystal leads to an increase in dislocation density by an order of magnitude and more (cf. Table 3.12). This value can rise to 5-6 orders of magnitude in the case of thermal deformation [1.73]. It is highly probable that there should also be a corresponding rise in average density of structural charge carrier trapping centers.

It must be borne in mind that Table 3.12 contains data on the average density of only nonbasal dislocations which emerge at the ab plane of the anthracene crystal. The density of basal dislocations, in accordance with Sect. 3.12.1 is, most probably, considerably higher. Unfortunately there is little experimental evidence at present on basal dislocation density in anthracene-type crystals.

For more detailed information on methods of crystal growing the reader is referred to special papers on anthracene [1.35-37,77,3.68,69], as well as to general review articles and monographs [3.21,99].

4. Local Trapping Centers for Excitons in Molecular Crystals

*They are ill discoverers that think
there is no land, when they can see
nothing but sea.*

Francis Bacon

*Research is to see what others have
seen and to think what no else has
thought.*

Albert Szent-Györgyi

The neutral excited state energy $E_\mu(\underline{k})$ [cf. Sect.2.1] is determined by the topology of molecules in the crystal, since both dispersional (D) and resonance $[\mathscr{E}_\mu(\underline{k})]$ interaction depend on the distance R between the excited molecule and the surrounding molecules of the lattice. Hence, the terms D and $\mathscr{E}_\mu(\underline{k})$ will be unequivocally set only for a perfect crystal, and only in this case will the exciton bands be strictly fixed and equal for the whole crystal.

If, however, there are local deviations from ideal geometry in the crystal, for instance, in the region of deformation around structural defects, then the energy E_μ of the excited state also assumes corresponding local values. In regions of compression the terms D and $\mathscr{E}_\mu(\underline{k})$ are increased, thus lowering the self-energy value of the excited state and forming a local exciton trapping center. On the other hand, in regions of dilation the interaction terms D and $\mathscr{E}_\mu(\underline{k})$ would be diminished and would form local centers of increased excited state energy which act as "antitrapping" centers for excitons.

One and the same structural defect of a crystal may form local centers both for excitons and charge carriers, i.e., they can create both neutral and ionized local states. However, the forces which give rise to these local states, their physical nature, R dependence, and energy values differ in both cases.

Local trapping centers for charge carriers are formed (cf. Sect.3.1) as a result of local change in electronic polarization energy ΔP by an excess charge carrier located in a structural defect of the lattice. Local trapping centers for excitons, on the other hand, are created, as can be seen from an analysis (cf. Sect.2.1), through local change of dispersion and resonance interaction between the excited molecule and surrounding molecules in the region of the structural defect.

In other words, the forces which create local ionized states (trapping centers for charge carriers) are of classical electrostatic nature, representing charge-induced dipole interaction in the crystal (cf. Sect.2.4). On the other hand, the forces giving rise to neutral local states (trapping centers for excitons) are of intrinsically quantum mechanical nature, viz. dispersion (London) and resonance interaction forces between the excited molecule and the molecular environment (cf. Sect.2.1). Common in both cases is the dependence of the respective forces on lattice geometry and their high sensitivity to changes in this geometry.

Only a comparatively small chapter in the present work deals with neutral local centers of structural origin. This is due to the fact that our main interest, as repeatedly stressed, is focused on ionized states. Neutral states are discussed only for the sake of maintaining a certain completeness in our discussion on the energetics of electronic states in molecular crystals. Another circumstance is the comparatively small amount of available theoretical and experimental data concerning neutral local states of structural origin in molecular crystals. However, existing data are instrumental in better understanding the problem of the formation of ionized local states. This is all the more so owing to the general crystallographic approach applied in the interpretation of the formation of both kinds of local states in one and the same defect region of the crystal.

4.1 Theory of Exciton States in a Deformed Molecular Crystal

Molecular exciton theory usually considers excited states in an ideal crystal (cf., e.g. [1.53,2.2,4,5,8-10,4.1-3] and Sect.2.1). In addition, local exciton states in molecular crystals, primarily connected with the presence of guest impurity molecules, are also extensively studied, both theoretically and experimentally (cf., e.g. [2.4,4.1,4-10]).

The first detailed theoretical study of the formation of local exciton states in a structurally deformed molecular crystal was carried out by SCHIPPER [4.11]. He showed that exciton theory parameters are comparatively sensitive to changes in crystal lattice parameters and that lattice deformations produce changes in the spectrum of excited states in the crystal. The expression on p.48 may be presented, according ot SCHIPPER [4.11] in the

following form for an excited state in a crystal of the anthracene space
group, with two molecules in a unit cell:

$$E_{\pm}^{S}(\underline{k}) = \left[(E^S - E^0) + D^{SS} + I_{11}^{SS}(\underline{k})\right] \pm I_{12}^{SS}(\underline{k}) \quad , \tag{4.1}$$

where E^0, E^S are, respectively, the ground and the excited states of the
molecule; D^{SS} is the term of dispersion (Van der Waals) energy shift due
to interaction between the excited molecule (S) with the surrounding mole-
cules; $I_{11}^{SS}(\underline{k})$, $I_{12}^{SS}(\underline{k})$ are the terms of resonance interaction (exciton transfer) $[I_{11}^{SS}(\underline{k}) = I_{22}^{SS}(\underline{k})$ for $\underline{k}$ in the region of $\underline{k} = 0]$.
The term D^{SS} may be represented in the following simplified form:

$$D_{mn}^{SS} = \frac{A^S}{R_{mn}^6} \quad , \tag{4.2}$$

where A^S is an empirical constant for the given excited S state which may
be determined from crystal spectra.

The term I_{mn} may be described, in dipole-dipole approximation, as follows:

$$I_{mm}^{st} = \frac{d^S d^t}{R_{mn}^3} \ F_{mn}^{[kk']} \quad , \tag{4.3}$$

where d^S and d^t are the dipole lengths of the corresponding s and t transitions; F_{mn} characterizes the relative orientation of the molecules m and n.

As may be seen from (4.2,3), the terms of dispersion- and resonance-
caused energy shifts exhibit different dependence on intermolecular distance
R_{mn}, proportional to $1/R_{mn}^6$, and $1/R_{mn}^3$ respectively. These dependences con-
stitute the main factors determining the effect of lattice deformation on
the energy spectrum of neutral states of the crystal.

The shift in energy levels of the excited states of a deformed anthracene-type crystal may be described, according to [4.11] by the equation:

$$\Delta E_{\pm}^{S}(\underline{k}) = \left[\Delta D^{SS} + \Delta I_{11}^{SS}(\underline{k})\right] \pm \Delta I_{12}^{SS}(\underline{k}) \quad . \tag{4.4}$$

In order to simplify the problem certain approximations are used in
[4.11] concerning the nature of deformation of the crystal.
First, uniform deformation of the whole crystal is assumed. This means
that translational symmetry of an ideal crystal remains unchanged in the

deformed one, and the wave vector $\underline{k}$ and vector $\underline{T}_{mn}$ connecting molecules m and n in the undeformed crystal change into the corresponding vectors $\underline{k}'$ and $\underline{T}'_{mn}$:

$$\underline{k}\underline{T}_{mn} = \underline{k}'\underline{T}'_{mn} \quad . \tag{4.5}$$

This makes the fundamental concepts of exciton theory of an ideal crystal applicable to a deformed crystal.

Secondly, only so-called external deformation is considered, i.e., changes in size and shape of the unit cell. Possible effects of internal de-formation, i.e., changes in position and orientation of molecules inside the unit cell, are disregarded. Under these conditions any point inside the crystal with initial position $\underline{r}$ is shifted into a new position $\underline{r}'$ where

$$\underline{r}' = \underline{r} + \underline{d} \tag{4.6}$$

with coordinates $\underline{r}(x,y,z)$ and $\underline{d}(u,v,w)$ in (x,y,z) system of coordinates.

After corresponding transformations of the initial expressions for ener-gy and accounting for deformation of the crystal SCHIPPER [4.11] obtained the following general formulae for calculating D_{mn}^{SS} and I_{st}:

a) the deformation-caused energy shift of dispersion term is

$$\Delta D^{SS} = \sum_{n(\neq m)} D_{mn} = \sum_{n(\neq m)} \left[\sum_i D'_{mn}(U_i)\delta U_i \right. $$
$$\left. + \sum_j D'_{mn}(x_j^{mn})\delta x_j^{mn} \right] \quad , \tag{4.7}$$

where δx_j^{mn} is the matrix element of deformation displacement.

The derivatives D'_{mn} assume the following form [cf. (4.2)]:

$$D'_{mn}(U_i) = \frac{6A^S}{R_{mn}^7}\left(\frac{\partial R_{mn}}{\partial U_i}\right)_0 \quad ; \tag{4.8}$$

b) the deformation-caused energy shift of the resonance term is

$$\Delta I_{mn}^{st} = d^s d^t \Delta G_{mn}^{[kk']} \quad , \tag{4.9}$$

where

$$G_{mn}^{[kk']} = \frac{F_{mn}^{[kk']}}{R_{mn}^3} \quad ; \tag{4.10}$$

$$\Delta G_{mn}^{[kk']} = \sum_i G_{mn}^{'[kk']}(U_i)\delta U_i + \sum_k G_{mn}^{'[kk']}(W_k)\delta W_k \quad . \tag{4.11}$$

The derivatives $G_{mn}^{'}$ assume the following form:

$$G_{mn}^{'[kk']}(U_i) = R_{mn}^{-3}\left[\left(\frac{\partial F_{mn}^{[kk']}}{\partial U_i}\right)_0 - \frac{3F_{mn}^{[kk']}}{R_{mn}}\left(\frac{\partial R_{mn}}{\partial U_i}\right)_0\right] \quad . \tag{4.12}$$

It might be of interest to mention that the problem of pressure effect on bathochromic shift of molecular crystal absorption spectrum, disregarding specific effects of deformation of the crystal, was approached in an earlier paper by RICE and JORTNER [4.12]. The authors suggested the following approximated formula for estimating the dispersion term D^{SS}:

$$D^{SS} \approx \sum R_i^{-6}\alpha\left(\mu^2 + \frac{1}{9}\alpha\Delta E\right) \quad , \tag{4.13}$$

where α is the isotropic molecular polarizability; μ is the transition dipole moment; ΔE is the energy of the excited state.

The term D^{SS}, as can be seen, is proportional to the polarizability α of the molecule which regularly increases in the sequence anthracene-tetracene-pentacene, with corresponding increase in the value of dispersion term D^{SS}.

The author of [4.11] calculated the energy level shift for an excited single state of a homogeneously deformed anthracene crystal. The following approximations were employed.

Absorption of light was considered for a beam falling normally upon the ab plane of the crystal, the corresponding wave vector of the exciton being $\underline{k} = 0$. Accordingly, the expression for $E_{\pm}^{S}(\underline{k})$ in (4.1) assumes the following form:

$$E_{\pm}^{S}(0) = \left[\left(E^S - E^0\right) + D^{SS} + I_{11}^{SS}(0)\right] \pm I_{12}^{SS}(0) \quad . \tag{4.14}$$

Davydov splitting in this case is

$$E_D = 2I_{12}^{SS}(0) \quad , \tag{4.15}$$

and the mean absorption band shift of the crystal with respect to the isolated molecule equals

$$E_S = \frac{1}{2} \left[D^{SS} + I_{11}^{SS}(0) \right] \quad . \tag{4.16}$$

Davydov splitting, according to (4.15) is determined by the resonance term only, whilst the mean absorption band shift of the crystal — both by the dispersion and the resonance terms [cf. (4.16)].

For calculating the vibronic components of the S_1 absorption band of anthracene, (4.14) has to be transformed in the following way:

$$E_{\pm}^{Sv} = \left(E^S + vh\nu - E^0 \right) + D^{SS} + \xi^2(v) \left[I_{11}^{SS} \pm I_{12}^{SS} \right] \quad , \tag{4.17}$$

where v is the vibrational quantum number (v = 0,1,2,3,4); ξ (v) is the Frank-Condon factor.

It is further assumed in the calculation that by means of hydrostatic pressure p homogeneous deformation of the anthracene crystal is effected. The pressure p is divided into corresponding stress components σ_1, σ_2, σ_3.

Using given σ_i values and experimentally determined elastic constants c_{ij} for anthracene (cf. [3.43] and Table 3.3), the corresponding deformation displacements may be calculated [cf. (4.7,9)], using the well-known matrix formula connecting deformation displacement ε_i with corresponding stress σ_i (cf. e.g. [3.34]

$$\sigma = c\varepsilon \quad , \tag{4.18}$$

where σ and ε are column matrices with elements σ_i and ε_i, respectively; c is the elastic constant matrix c_{ij}.

After evaluation of the respective derivatives of the dispersion and the resonance terms, according to (4.8,12), it is possible to obtain simple expressions for estimating the shift E_S^{Sv} of the vibronic levels of the first single band of anthracene, as well as the increase in Davydov splitting E_D^{Sv} as functions of the hydrostatic pressure p [4.11]

$$\Delta E_S^{Sv} = -k_S^v p; \tag{4.19}$$

$$\Delta E_D^{Sv} = k_D^v p \quad . \tag{4.20}$$

Calculated values of coefficients k_S^v and k_D^v are given in Table 4.1. These data show that pressure p produces observable bathochromic shift of excited S_1^v levels of anthracene. This shift is of the order of 80 cm^{-1}/kbar, i.e., about 0.01 eV/kbar and is due to diminution in intermolecular distances and corresponding increase in dispersion and resonance interaction terms between the excited molecule and its environment.

Table 4.1. Calculated values of coefficients k_S^v—mean shifts of vibronic levels of the first singlet band, and k_D^v—increase in Davydov splitting due to hydrostatic compression of an anthracene crystal, according to [4.11]

Vibrational quantum number v	$k_S^v = \Delta E_S^{Sv}/p$ $[cm^{-1}/kbar]$	$k_D^v = \Delta E_D^{Sv}/p$ $[cm^{-1}/kbar]$	$E_D, [cm^{-1}]$
0	79	4.0	132
1	79	4.0	123
2	77	2.8	82
3	75	1.2	32
4	-	0.7	15

Note: E_D is the calculated value of Davydov splitting at normal pressure [4.11].

An increase in the resonance term raises also the value of Davydov splitting (cf. Table 4.1).

Results of calculation [4.11] show that the contribution of the dispersion term to the energy level shift considerably exceeds that of the resonance term. Thus, the contribution of the resonance term constitutes ca. 10-20 % of the total for S_1 vibronic bands of anthracene with v = 0 and v = 1. It becomes still lower for higher vibronic levels (v = 2, 3 and 4).

According to (4.19,20) ΔE_S^{Sv} and ΔE_D^{Sv} depend linearly on hydrostatic pressure p. This linearity holds, according to SCHIPPER [4.11], up to pressures of 40 kbar.

The effect of hydrostatic pressure upon the crystal actually manifests itself in small microregions around the excited molecules. The derivatives of the dispersion term, as obviously follows from (4.8), converge rather quickly within the sphere of the nearest neighbors. The resonance term derivatives have a longer range of interaction: convergence to 2 % is attained within a 300 Å radius in the ab plane of anthracene, whilst in the c' direction only the adjacent planes contribute to any extent [4.11].

It follows that all the basic conclusions obtained in [4.11] may be used in qualitative analysis of shifts of excited state energy levels in regions of local compression of the crystal. It must, however, be borne in mind, in considering local regions of lattice compression or dilation around dislocational defects of the crystal, that these pressures are by no means homogeneous (cf., e.g. Fig.3.4). The pressures have a considerable gradient in these regions, and one can safely speak only of some average effective pressure in the given local region. It is therefore hardly likely that translational symmetry of the lattice [cf. (4.5)] and the wave $\underline{k}$ of the exciton are conserved in such regions of local compression of the crystal. It appears that trapping of an exciton takes place in regions of local compression of the crystal, with subsequent localization of the excited state on a new, energetically more favorable local level.

All basic dependences which are characteristic of the effect of dispersion and resonance interaction between an excited molecule and the surrounding ones apparently hold for such a localized excited state. This is borne out by a comparison between excited state level shifts in a crystal uniformly compressed by hydrostatic pressure and conforming with the idealized Schipper model, and corresponding level shift effects in real crystals in which dislocation defects of local compression regions are produced by thermal or mechanical deformation. These experimental effects will be discussed in the two following sections.

4.2 Electron Level Shifts in Hydrostatically Compressed Molecular Crystals

The theoretical results of SCHIPPER [4.11] concerning excited state energy level shifts in a uniformly deformed anthracene crystal (cf. Sect.4.1) have been confirmed experimentally by observations of bathochromic shifts of absorption and fluorescence bands at hydrostatic compression of anthracene and other aromatic hydrocarbon crystals.

Results obtained by means of high-pressure techniques for studying optical and electronic properties of a number of molecular crystals are discussed in some detail in the review article by OFFEN [4.13]. We should like to note especially the pioneer work in this field, carried out by DRICKAMER et al. [3.96,4.14-16], as well as interesting studies performed by TANAKA et al. [3.60], SHIROTANI et al. [1.103,4.17,18], and ARNOLD et al. [4.19, 20] which we are going to discuss in some detail.

RICE and JORTNER [4.12] already pointed out that high-pressure technique might provide a very promising approach for the studies of electronic states in molecular crystals. Linear and bulk compressibility of anthracene-type crystals is rather high, as compared to covalent and ionic crystals (cf. Sect.5.6). This leads to considerable changes in optical and electronic properties already within the range of comparatively moderate pressures. Thus, for instance, a red shift of ca. $50-100$ cm^{-1}/kbar can be observed in the singlet absorption band $S_0 \rightarrow S_1$ in the case of linear polyacenes [4.13]. In other words, pressures of a few kilobar are sufficient for easy experimental observation of these phenomena.

The spectral shift in a molecular crystal depends on dispersion and resonance interaction between the excited molecule and the surrounding ones (cf. preceding Section). These interactions differ in magnitude for different excited states. Hence, a different kind of bathochromic shift dependence on pressure ought to be expected for singlet ($S_0 \rightarrow S_1$) and triplet ($S_0 \rightarrow T_1$) transitions. Triplet level shifts must be smaller at equal pressure than the singlet ones, owing to the low value of oscillator strength in $S_0 \rightarrow T$ transitions [4.20].

These considerations are confirmed by experimental data, as can be seen from Table 4.2. The specific long-wave shift $\Delta E^T/p$ of the triplet level T_1 at hydrostatic compression of anthracene and tetracene crystals is 4 to 6 times smaller than that of the first singlet absorption band $\Delta E_a^S/p$.

As a result, hydrostatic compression of a crystal changes the mutual position of the T_1 and S_1 levels. This effect was successfully used by WHITTEN and ARNOLD [4.20] who applied pressure for modulating the decay constant of a single exciton into two triplet ones ($S_1 \rightarrow 2T_1$) in tetracene crystals.

Table 4.2. Reported experimental values of specific bathochromic shift of absorption and fluorescene bands at hydrostatic compression of anthracene (Ac) and tetracene (Tc) crystals

Crystal	$\Delta E_a^S/p$ [cm^{-1}/kbar]	$\Delta E_f^S/p$ [cm^{-1}/kbar]	$\Delta E^T/p$ [cm^{-1}/kbar]
Ac	45 [4.14] 70 [4.15]	80 [3.60]	$\sim$ 11 [4.19]
Tc	110 [1.103,4.17] 92 [4.30]	165 [4.20]	15 ± 7 [4.20]

Note: $\Delta E_a^S/p$ is the specific shift of $S_0 \rightarrow S_1$ absorption;

$\Delta E_f^S/p$ is the specific shift of the $S_1 \rightarrow S_0$ fluorescence;

$\Delta E^T/p$ is the specific shift of the triplet level.

Table 4.2 clearly demonstrates another interesting empirical dependence of optical properties of molecular crystals upon hydrostatic compression. The specific shift of the fluorescence band $\Delta E_f^S/p$ for anthracene, as well as for tetracene, is about twice as large as that of the singlet absorption band. This is most likely due to a difference in the pressure dependence of mutual positions of potential curves of the ground and the excited states of the molecules.

The theoretical estimate of the specific shift of the singlet 0-0 level of an anthracene crystal, as found by SCHIPPER [4.11] ($K_S^{V=0} = 79$ cm^{-1}/kbar) (cf. Table 4.1), is in close agreement with the experimental value of $\Delta E_f^S = 80$ cm^{-1}/kbar (cf. Table 4.2) for fluorescence according to TANAKA et al. [3.60], but is somewhat larger if compared to the corresponding values for absorption (cf. Table 4.2). The values for K_D^V and E_D, on the other hand, as given in [4.11], appear to be somewhat underestimated (cf. Tables 4.1 and 3) apparently owing to the limitations of dipole-dipole approximation in resonance interaction [cf. (4.3)].

At high hydrostatic pressures [3.96] the bathochromic shift of absorption bands E_a^S of aromatic hydrocarbon crystals can reach extremely high values. Thus, a relative decrease in volume $\Delta V/V$ of a crystal to ca. 0.30 (corresponding to ca. 17 kbar, cf. Fig.5.3) causes a shift of ΔE_a^S of the order of 5000 cm^{-1}. At pressure of about 100 kbar ΔE_a^S is as high as 9000 cm^{-1}, i.e., 1.1 eV [3.96]. This illustrates the high compressibility of molecular crystals of the anthracene space group.

Another rather curious effect was observed by TANAKA et al. [3.60] in the study of pressure effects on shift ΔE_f^S of the fluorescence band ($\lambda = 450$ nm) of anthracene. At high pressures of the order of 50 kbar residual fluorescence was observed in form of a wide band with a peak at 475 nm. The authors interpreted this phenomenon as excimer luminescence. This means that at high pressures predimer configurations can be formed in defect regions (cf. Sect. 3.10). These give rise to comparatively deep trapping centers of excimer nature for excitions. It is noteworthy that these centers continue to exist in the crystal after cessation of pressure, but can be removed by thermal treatment of the crystal [3.60].

SHIROTANI et al. [1.103] studied also the effect of pressure on absorption spectra of oriented and quasi-amorphous tetracene (Tc) and pentacene (Pc) layers: This work is of particular interest with respect to Davydov splitting which is rather pronounced in oriented Tc and Pc polycrystalline layers and practically unnoticeable in quasi-amorphous layers [1.102,2.30]. Such layers therefore make it possible to study separately the pressure effect on Davydov splitting and on the general spectrum shift. (For details of oriented and quasi-amorphous layers of tetracene and pentacene see Sect.5.14).

SHIROTANI et al. [4.18] investigated also the effect of pressure on optical properties of layers of tetrathiotetracene (TTT), a rather exotic derivative of tetracene.

Some reported experimental data on the shift of excited state energy levels of Tc, Pc and TTT layers under hydrostatic pressure can be found in Table 4.3. In polycrystalline Tc and Pc films obtained by sublimation in vacuo at room temperature, the ab plane of crystallites being preferably oriented parallel to the surface of the substrate, pronounced Davydov splitting was observed in unpolarized light [1.102,103,2.130]. In such oriented pentacene films (Fig.4.1) one can clearly discern the a component at 630 nm (E_-^S), corresponding to absorption of light in a monocrystal parallel to the a axis, as well as the b component at 670 nm (E_+^S), corresponding to polarization parallel to the b axis.

The Davydov a and b components show considerable long-wave shift in oriented Tc and Pc films at hydrostatic compression: the $\Delta E_+^S/p$ value reaches values of -110 and -130 cm^{-1}/kbar, and that of $\Delta E_-^S/p$ -75 and -50 cm^{-1}/kbar for Tc and Pc, respectively (cf. Table 4.3). The Davydov splitting itself increases with pressure: $\Delta E_D/p$ becomes as high as 37 and 70 cm^{-1}/kbar for Tc and Pc, respectively (cf. Table 4.3 and Fig.4.1). These effects are in full agreement with the model proposed by SCHIPPER [4.11]: pressure-caused mutual approach of molecules raises both the dispersion D_{mn}^{SS} and the resonance I_{mn}^{st} terms of interaction between the excited molecule and the surrounding ones [cf. (4.2,3)] thus lowering its energy levels.

Table 4.3. Reported experimental values of the specific shift of the singlet 0-0 absorption band at hydrostatic compression of oriented ($\Delta E_+^S/p$) and quasi-amorphous ($\Delta E^S/p$) layers of tetracene (Tc) and pentacene (Pc), as well as of polycrystalline tetrathiotetracene layers

Compound	$\Delta E_+^S/p$ [cm^{-1}/kbar]	$\Delta E_-^S/p$ [cm^{-1}/kbar]	E_D, [cm^{-1}]	$\Delta E_D/p$ [cm^{-1}/kbar]	$\Delta E^S/p$ [cm^{-1}/kbar]
Tc	-110 [1.103,4.17]	-75 [1.103,4.17]	575 [1.103]	37 [1.103]	-130 [1.103]
Pc	-130 [1.103]	-50 [1.103]	950 [1.103]	70 [1.103]	-140 [1.103]
TTT					-32 [4.18]

Note: $\Delta E_+^S/p$ and $\Delta E_-^S/p$ are the specific shifts of the b and a components, respectively; E_D is the value of Davydov splitting at normal pressure (E_D = 230 cm-1 for anthracene [1.43]); $\Delta E_D/p$ is the specific increase in Davydov splitting at hydrostatic compression.

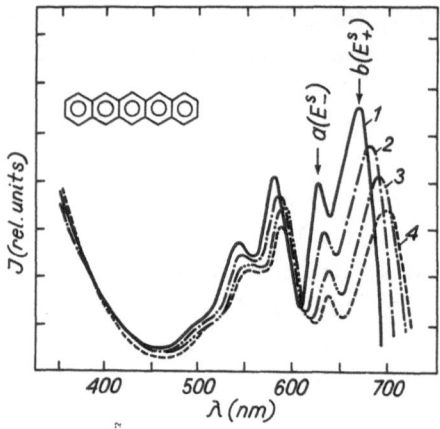

Fig.4.1. Electronic absorption spectra of pentacene layers with oriented crystallites at various hydrostatic pressures p. Curves (1) p = 0; (2) p = 1.8; (3) p = 3.6; (4) p = 5.4 kbar. a(E^S_-) and b(E^S_+) are, respectively, the Davydov a and b components of the 0-0 transition in the S_1 absorption band [1.103]

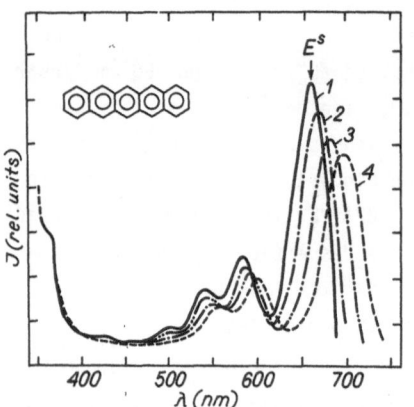

Fig.4.2. Electronic absorption spectra of quasi-amorphous pentacene layers at various hydrostatic pressures p. Curves (1) p = 0; (2) p = 1.8; (3) p = 3.6; (4) p = 5.4 kbar. E^S is the 0-0 transition in the S_1 absorption band [1.103]

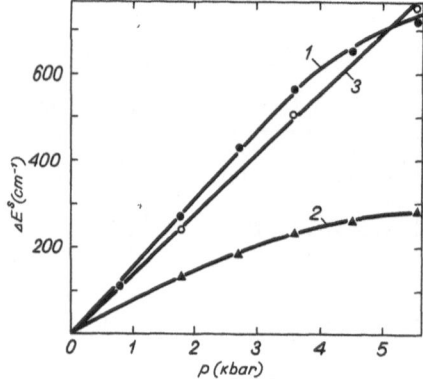

Fig.4.3. Shift ΔE^S of the 0-0 transition in the S_1 absorption band for thin layers of oriented (1,2) and quasi-amorphous (3) pentacene, as dependent on hydrostatic pressure p. (Curve 1) Davydov b component, (Curve 2) Davydov a component for oriented pentacene layers, (Curve 3) 0-0 transition of a quasi-amorphous pentacene layer [1.103]

It ought to be noted that in the case of quasi-amorphous Tc and Pc films obtained by sublimation on a cooled substrate (cf. Sect.5.14), when Davydov splitting is hardly observable (Fig.4.2), the hydrostatic compression-caused bathochromic shift of the first singlet absorption band $\Delta E^S/p$ is just as marked and approximately of the same magnitude as in the case of oriented layers (cf. Table 4.3 and Figs.4.2,3). It can be seen from Table 4.3 that the $\Delta E^S/p$ value (-130 and -140 cm^{-1}/kbar) is even somewhat higher than that of $\Delta E^S_+/p$ for the b component of oriented films. This indicates that packing density and diminution of intermolecular distance with pressure in quasi-amorphous films is of the same order as in oriented films, and the corresponding energy level shifts are approximately equal in both structures.

Table 4.3 shows that for tetrathiotetracene, which possesses considerably lower compressibility, the $\Delta E^S/p$ value is also considerably lower than in the case of its analogues — tetracene (Tc) and pentacene (Pc).

Theoretical (cf. Sect.4.1) and experimental studies of the hydrostatic pressure effect on excited state energy levels clearly show that in local regions of compression of real molecular crystals local states can be formed which may act as exciton trapping centers. Since the specific depth of these states, depending on pressure, equals $80 - 140$ cm^{-1}/kbar (cf. Tables 4.2 and 3), i.e., ca. $0.01 - 0.02$ eV/kbar, we may expect formation of exciton trapping centers of $0.03 - 0.1$ eV depth in regions of local compression at pressures of 3 to 5 kbar.

It will be shown in the next section that such centers of structural origin have actually been observed, and their depth can serve for an approximate estimate of local pressures in dislocation defects.

In addition, predimer configurations of molecules (see Sect.3.10) can be formed in regions of structural defects. These configurations may act as deeper exciton trapping centers and create excimer states.

4.3 Formation of Local Exciton Trapping Centers in Structural Defects of a Crystal

In the course of studying absorption or luminescence spectra of molecular crystals one frequently observes weak "satellite" bands beyond the long-wave edge of intrinsic absorption or luminescence of the crystal. In some cases these bands can be attributed to impurity molecules or to local states created by perturbation of the surrounding molecules of the host crystal by impurity molecules [1.71,4.10].

However, even after most thorough purification of a crystal, such as anthracene, one still observes in its luminescence spectrum residual long-wave bands which are attributed to structural defects of dislocation type.

The structural origin of these local bands is borne out by the fact that their intensity and spectral distribution may be altered by mechanical deformation or thermal treatment of the crystal. In addition, they are strongly dependent on the crystal growing techniques.

The role of the first factor was already clearly demonstrated by SCHNAITHMANN and WOLF [4.21]. The authors found that plastic deformation of a naphthalene crystal leads to the appearance of quasi-discrete traps in its fluorescence spectrum, which are situated ca. 165 cm^{-1} below the S_1 level of the crystal. In addition, traps with practically continuous spectral distribution appear in the longer-wave region. Subsequent thermal annealing considerably lowers the intensity of fluorescence from these defect levels. Similar results were also reported by LIPSETT and MacPHERSON [4.22] after various kinds of external treatment causing structural changes in a naphtalene crystal.

The influence of crystal growing techniques was studied in detail by HELFRICH and LIPSETT [1.70]. They investigated structural traps for singlet excitons in high-purity anthracene single crystals obtained from melt and from gaseous phase after complete removal of traces of impurities.

The local excited state levels were studied from fluorescence spectra under excitation in the singlet absorption band $\nu_{ex} > \nu_{0-0} = 25078$ cm^{-1} and under direct excitation in absorption bands of defect structures at $\nu_{ex} < \nu_{0-0}$.

The fluorescence spectra of an anthracene crystal obtained from the vapor phase [Ac(v)] and those obtained from melt [Ac(m)] considerably differ from each other (cf. Figs.4.4 and 4.5a).

Ac(v) crystals possess a discrete fluorescence spectrum (cf. Fig.4.4) in which, apart from the high intensity peak of the 0-0 transition, typical vibronic bands can be observed, corresponding to transition to higher (v > 0) vibrational levels of the electronic ground state. These vibronic bands reveal characteristic double structure and may be attributed to corresponding intramolecular vibrations at 394, 1013, 1168, 1253, 1407, 1555, and 1803 cm^{-1}, respectively [1.70]. The only exception is the band 260 cm^{-1} below the 0-0 transition. This band is more diffuse than other bands, does not reveal any doublet structure and thus cannot be attributed to any Raman-active

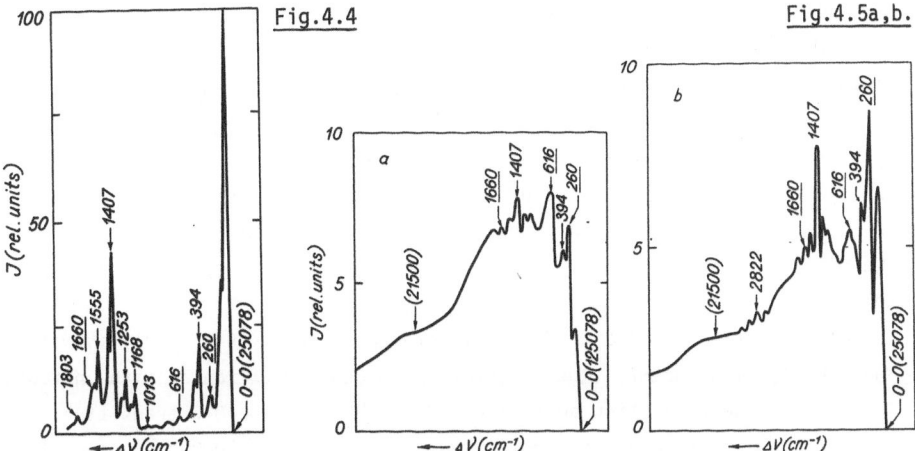

Fig.4.4. Fluorescence spectrum of a vapor-grown anthracene crystal [Ac(v)]. Δν shows the bathochromic shift of the bands with respect to the 0-0 transition. The underlined bands at 260, 616, 1660 cm⁻¹ are attributed to fluorescence of excitions trapped in structural defects of the crystal [1.70]

Fig.4.5a,b. Fluorescence spectra of a melt-grown anthracene crystal [Ac(m)]: (a) before thermal treatment; (b) after thermal treatment. The underlined bands at 260, 616 and 1660 cm⁻¹ are attributed to fluorescence of excitons trapped in structural defects of the crystal [1.70]

vibration of the anthracene molecule. The authors of [1.70] suggested that this band is due to fluorescence from structural defects of the crystal. Similar views have also been expressed in later work by BRIDGE and VINCENT [4.23].

The structural nature of the band 260 cm⁻¹ below the 0-0 transition was finally confirmed by Helfrich and Lipsett in their studies of thermal treatment of the crystal. After thermal annealing of an Ac(v) crystal the intensity of this band increased about tenfold, exceeding in intensity the vibrational band at 394 cm⁻¹. [Thermal annealing of Ac(v) and Ac(m) crystals proceeded at 130°C in argon atmosphere during 3 days]. In addition, two new bands of structural origin became distinctly visible in the fluorescence spectrum of the crystal after thermal treatment, viz. the 616 and 1660 cm⁻¹ bands which may be noticed in the spectrum already before thermal treatment of the Ac(v) crystals (cf. Fig.4.4).

Ac(m) crystals obtained from melt reveal a wide, almost continuous fluorescence band below the 0-0 transition with hardly discernible most intensive vibration peaks at 394 and 1407 cm⁻¹. This continuous fluorescence spectrum "background", uniformly decreasing towards longer wave-lengths, constitutes ca. 70 % of the total intensity. At the same time there are pronounced bands of structural defect fluorescence at 260, 616 and 1660 cm^{-1} below the 0-0 transition band in the spectrum of Ac(m) crystal (cf. Fig. 4.5a).

Thermal treatment of the Ac(m) crystal leads to a considerable increase in intensity of the 260 cm^{-1} band and to decrease in intensity of the 616 cm^{-1} band (cf. Fig.4.5b). In general, there is also a drop in intensity of the continuous background fluorescence, and the discrete spectral bands become more clearly discernible.

The excitation spectra of defect fluorescence (Fig.4.6) clearly show that the density of structural local centers monotonously decreases with increase in depth of their position below the exciton band of the 0-0 transition. The deepest centers of defect fluorescence are situated ca. 2000 cm^{-1} below the 0-0 transition. A continuous excitation spectrum of defect centers has been observed both for Ac(m) crystals (cf. Fig.4.6, Curve 3), as well as for Ac(v) crystals (cf. Fig.4.6, Curve 1). And this despite the fact that in the fluorescence spectra of Ac(v) crystals (cf. Fig.4.4) the continuous "background" is considerably less pronounced than in Ac(m) crystals (cf. Fig.4.5a). It is rather interesting that, upon thermal treatment of Ac(v) crystals, simultaneously with intensity increase of discrete bands of structural origin at 260, 616, and 1660 cm^{-1} there is a concurrent decrease in the density of centers possessing continuous spectral distribution (cf. Fig.4.6, Curves 1 and 2).

Local centers of exciton trapping with a continuous energy spectrum and maximum depth of the order of 2000 cm^{-1} are, most likely, formed as a result of statistical distribution of local fields of lattice compression in regions of dislocation ensembles (cf. Sect.3.7.3). The spectra in Figs.4.4,5 convincingly confirm the accepted view that structural defect densities in melt-grown crystals, as a rule, considerably exceed defect densities in vapor-grown crystals (cf. Sect.3.13).

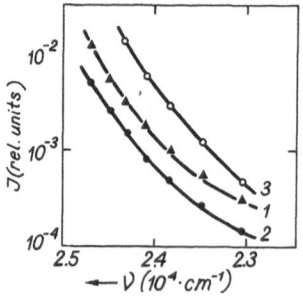

Fig.4.6. Intensity J of continuous defect fluorescence as dependent of excitation frequency (excitation spectra of defect fluorescence). (Curve 1): for vapor-grown anthracene crystal Ac(v) before thermal treatment; (Curve 2): the same after thermal treatment; (Curve 3): for melt-grown anthracene crystal Ac(m) [1.70]

In the process of thermal treatment of the crystal the random distribution of dislocation ensembles undergoes rearrangement (e.g., polygonization), and as a result various specific dislocation alignments and aggregations of definite structure are formed (cf. Sect.3.7). Such processes of thermal rearrangement of dislocation defects are, naturally, accompanied by a decrease in the density of local states with continuous distribution of the energy spectrum (cf. Fig.4.6). Simultaneously, there is a concurrent increase in intensity of characteristic bands in the defect fluorescence of "quasi-discrete" states. Sometimes thermal treatment is accompanied by a redistribution of the density of various quasi-discrete centers (cf. e.g., Fig.4.5a,b).

It was suggested by the authors of [1.70] that deep structural exciton trapping centers at 1660 cm^{-1} may be connected with predimer configurations of anthracene, which enhance formation of excimer states. Such predimer states cannot, however, be regarded as point defects of a crystal, as assumed by HELFRICH and LIPSETT [1.70]. As was demonstrated in Sect.3.10, predimer states of molecules in anthracene-type crystals are, most likely, formed only in regions of strong lattice compression associated with dislocations, stacking faults, and other extended lattice defects.

After the detailed research by HELFRICH and LIPSETT [1.70] a number of other papers appeared confirming the basic conclusions of the aforementioned authors. Thus, LYONS and WARREN [4.24] observed a rich spectrum of local states of exciton trapping in anthracene crystals at 194, 534, 700, 933, and 1221 cm^{-1} below the singlet level. These local states were also interpreted by the authors as being of structural origin.

These and other earlier reported data on structural defect luminescence in organic crystals, up to 1974, are discussed in the review articles by WILLIAMS and THOMAS [1.71] and by BIRKS [1.41,46].

We shall additionally discuss here a number of papers published in the 1974-1978 period, which have made a significant contribution to the studies of exciton trapping phenomena in structural defects of an anthracene crystal [1.72-74,3.28,95,4.25,26].

GOODE et al. [3.95] applied triplet excitons as probes in their studies of structural defects in purposefully deformed and subsequently thermally treated anthracene crystals. Ultra-pure anthracene single crystals were used, grown from melt and from vapor (the mean lifetime of triplet excitions in purified crystals was about 8 to 24 ms at 300 K). Structural traps of triplet excitons were investigated by delayed fluorescence technique within the 4 to 400 K temperature range, as well as using the method of long-wave phosphorescence band shift at low temperature.

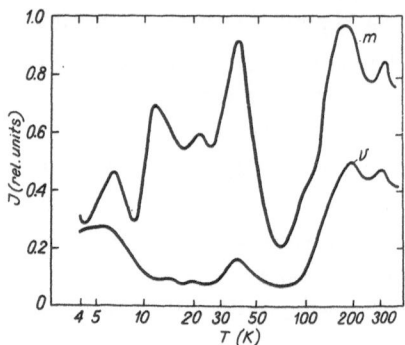

Fig.4.7. Delayed fluorescence spectra J = f(T) of anthracene crystals. (m) for a melt-grown crystal; (v) for a vapor-grown crystal [3.95]

The delayed fluorescence temperature spectra of anthracene present a rich variety of local trapping states for triplet excitons (Fig.4.7). As may be seen from Fig.4.7, the spectra of melt-grown anthracene [Ac(m)] differ considerably from those of vapor-grown crystal [Ac(v)] (cf. Fig.4.7, spectra m and v). The spectra of Ac(m) consist of wide overlapping bands with characteristic peaks at 6, 11, 20, 30, 38, 85, 170, and 320 K.

The spectra of Ac(v) contain a flat band in the 4-10 K range, small peaks at 38 and 85 K, and stronger bands at 170 and 320 K. In other words, they possess a poorer spectrum of traps than specimens of Ac(m).

The picture of delayed fluorescence spectra undergoes radical change upon deformation with subsequent thermal treatment of the crystal. Thus, for instance, bending of an Ac(m) crystal on a knife edge along the [100] and [110] directions in the (001) plane in helium atmosphere produces a strong intensification of bands at 320 and 6 K, whereas the intensity in the 10-50 and 100-200 K ranges decreases. This indicates a redistribution of structural defects at the given mechanical deformation. Subsequent thermal annealing of the crystal partly restores the initial structure of the spectrum, the decrease in intensity being particularly strong at 320 and 6 K, with narrowing of bands and intensity redistribution in other spectral ranges.

Such behavior of the spectra speaks in favor of structural origin of the given local centers and demonstrates the effect of external influence (deformation, temperature) on the dynamics of distribution of their energy spectra. These effects are, naturally, due to the dynamics of interaction between dislocation defects in the crystal (cf. Sect.3.7).

Studies of phosphorescence spectra of Ac crystals, as reported in [3.95] convincingly complement the general picture of structural defect formation. In this work the trap depth was determined from phosphorescence band shift

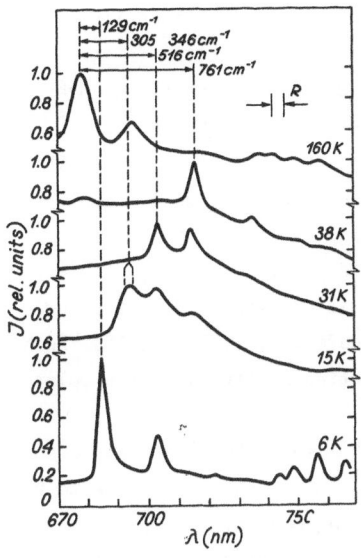

Fig.4.8. Normalized phosphorescence spectra of a melt-grown anthracene Ac(m) crystal at various temperatures. The top part shows corresponding trap depths for triplet excitions. (R) spectral slit width of instrument [3.95]

at low temperatures. The phosphorescence spectra of an Ac(m) sample are shown in Fig.4.8. These data are in good agreement with corresponding peaks in the temperature spectrum of delayed fluorescence (cf. Table 4.4).

A long-wave shift of Ac phosphorescence caused by triplet exciton trapping takes place only at temperatures below 100 K. Above 150 K the phosphorescence spectrum of free triplet excitons is observed (cf. Fig.4.8).

Table 4.4. Mean depth of triplet exciton traps E_T in an anthracene single crystal obtained from melt [Ac(m)], as determined from long-wave shifts in phosphorescence spectra [3.95]

E_T		$T_{DF}max$	Characteristic of traps
[cm^{-1}]	[meV]	[K]	
129	15	6	Traps of structural origin, due to local deformation of crystal
305	38	11	
346	43	20	
516	64	30	
761	94	38	Most likely, X center in the environment of an oxidized Ac molecule
2500	301	302	Most likely, a center of excimer origin

Note: $T_{DF}max$ are the corresponding peaks of delayed fluorescence temperature spectrum bands (see Fig.4.7).

An analysis of the effect of deformations and of thermal annealing leads the authors of [3.95] to the conclusion that the trap spectrum in the $130-500$ cm^{-1} range, created by local regions of deformation of an anthracene crystal, is most likely of dislocation origin (cf. Table 4.4).

The behavior of traps of depth 760 cm^{-1} provides the authors of [3.95] with reasons to suggest that they are connected with structural X centers in the environment of impurity molecules of oxidized anthracene.

As regards traps for triplets of 2500 cm^{-1} depth (ca. 0.3 eV), the authors of [3.95] try to link them with the formation of predimer structures in the compressed region of extended lattice defects (cf. Sect.3.10). Such "incipient dimers" may form excimer states, acting as "traps" for triplet excitons.

Recently WILLIAMS and ZBOINSKI [4.26] have measured the temperature spectra of delayed fluorescence of melt-grown high-purity Ac crystals, with the aim of investigating the triplet exciton trapping characteristics. The data obtained are interpreted by the authors in terms of structural triplet exciton traps associated with polymorphic inclusions (cf. Sect.1.3) and "incipient dimers" created in the compressed regions of extended lattice defects (cf. Sect.3.10). These structural defects yield a quasi-continuous energy spectrum of triplet exciton trapping levels extending downwards for ca. 4700 cm^{-1}. The deepest traps are removed by prolonged ultra-violet irradiation of the crystal under conditions that lead to formation of photodimers, whereupon other types of shallower traps appear [4.26].

WILLIAMS et al. [1.72] employed the method of prompt fluorescence for studying the kinetics of structural trap formation for singlet excitons upon mechanical deformation of an anthracene single crystal. In this work ultrapure anthracene single crystals were used, grown from melt and subjected to thermal annealing in vacuo during a week at a temperature of a few degrees below melting point. Figure 4.9 (Curve 1) presents the spectrum of prompt fluorescence of a well-annealed undeformed anthracene crystal excited in the 27400 cm^{-1} region at 4 K.

The fluorescence spectrum in Fig.4.9, Curve 1, shows the characteristic vibronic bands, as in [1.70] (cf. Figs.4.4,5), as well as the characteristic band of structural origin at ca. 1660 cm^{-1} below the 0-0 transition.

Deformation of the anthracene crystal was effected by means of the three-point bending method in the (00ℓ) planes at 373 K in nitrogen atmosphere. The axis of bend was parallel to the [100] direction. Under such deformation conditions basal (00ℓ) dislocation slips are dominant, as well as, possibly,

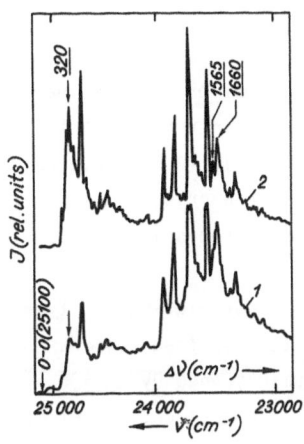

Fig.4.9. Prompt fluorescence spectrum at 4·K of a melt-grown anthracene single crystal after careful annealing. (Curve 1) before mechanical deformation; (Curve 2) after mechanical deformation (ν_{ex} = 27400 cm^{-1}). The underlined bands at 320, 1565 and 1660 cm^{-1} are attributed to fluorescence of excitions trapped in structural defects [1.72]

partial dislocation of the 1/2 [110] and 1/2 [120] type in the ab plane of the crystal (cf. Sect.3.13 and Table 3.11).

After such deformation a conspicuous wide band appears in the fluorescence spectrum at ca. 320 cm^{-1}, as well as a band ca. 1565 cm^{-1} below the 0-0 transition (cf. Fig.4.9, Curve 2). Traces of these bands of structural origin can be noticed already in the spectrum of an undeformed crystal (cf. Fig.4.9, Curve 1).

As regards the structure-caused fluorescence band at 1660 cm^{-1} depth, its intensity remains practically unchanged at the given deformation of the crystal. This band belongs, most likely, to a comparatively stable configuration of structural defects. It is particularly pronounced in the fluorescence spectra of anthracene single crystals obtained from melt (cf. Fig. 4.5a,b).

The assumption that the 320 and 1565 cm^{-1} bands actually act as single exciton traps is supported by the fact of their disappearance with temperature rising to 40 K [1.72]. Simultaneously, there is also a sharp drop in the intensity of the 1660 cm^{-1} band.

WILLIAMS et al. [3.28] have studied fluorescence spectra of ultra-pure anthracene subjected to photodimerization and photo-oxidation. Photodimerization of samples produces two kinds of singlet exciton traps having depth of ∼4000 cm^{-1} and ∼300 cm^{-1}, respectively. The deep trapping centers are interpreted as "incipient" dimer states formed in close proximity to the stable photodimer molecules (cf. Sect.3.10). The traps of ∼300 cm^{-1} depth with distribution extending to around ±50 cm^{-1}, are attributed to displaced and compressed anthracene molecules in the proximity of photodimerization nuclei.

It is interesting to notice that photo-oxidation of anthracene, in contrast to the process of photodimerization, does not introduce any singlet exciton traps in the anthracene lattice [3.28]. The oxidation products have S_1 states well above that of anthracene and, remarkably, introduce little distortion into the host structure [3.28]. GAIEVSKII et al. [4.27] have recently observed defect luminescence from anthracene single crystals with photoinduced shallow local centers of structural origin, red-shifted 32 cm^{-1} below the S_1 excition band.

LISOVENKO et al. [1.73,74] were first to succeed in correlating two typical defect luminescence bands of an anthracene crystal, shifted towards the long-wave region from the 0-0 band of exciton luminescence by $\Delta E = 220-240$ (D_1) and $\Delta E = 260-280$ (D_2) cm^{-1}, with specific dislocation defects of the lattice. The D_1 and D_2 bands have been frequently observed by other authors in imperfect or deformed anthracene crystals (cf., e.g., Figs.4.4,5). Both bands were usually attributed to structural lattice defects. Thus the D_2 band practically always appears in the luminescence spectra of anthracene single crystals grown from melt.

LISOVENKO et al. [1.73,74] have made a successful attempt to identify the nature of the structural defects leading to the appearance of these bands. The authors of [1.73,74] studied the effect of plastic deformation on prompt fluorescence of anthracene crystals. The crystals were grown by means of sublimation from the gaseous phase, on glass, quartz or plexiglas substrates, and their plastic deformation was effected thermally by cooling the substrates to 4 K (cf. Sect.3.12.1).

Figure 4.10 presents fluorescence spectra of an undeformed (a) and of a thermally deformed (b) anthracene crystal at 4.2 K.

The fluorescence spectra of an undeformed crystal (cf. Fig.4.10a) are rather similar to those obtained from an anthracene crystal grown from the gaseous phase as reported in [1.70] (cf. Fig.4.4).

In a thermally deformed anthracene crystal (cf. Fig.4.10b) a new series of bands appears, the strongest one belonging to the 0-0 band at 24840 cm^{-1}. This new series of defect fluorescence is shifted towards longer waves by $\Delta E = 240$ cm^{-1} from the 0-0 band of the exciton fluorescence 0-0 band at $\nu = 25080$ cm^{-1}. It can therefore be classified as a D_1 band.

Figure 4.11 demonstrates fluorescence spectra (T = 4.2 K) obtained under various conditions of thermal deformation of anthracene crystals. These spectra contain separately either a D_1 band of defect fluorescence (a), or only a D_2 band (b), or both bands at the same time (c).

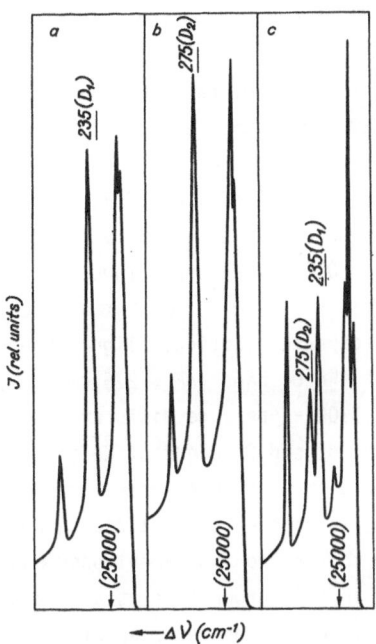

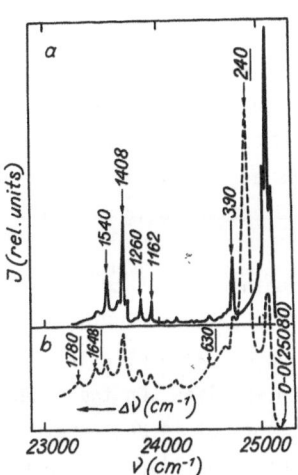

Fig.4.10. Fluorescence spectra of anthracene crystals (T = 4.2 K, thickness of crystals ~ 1 μm) (a) thermally undeformed crystal; (b) after thermal deformation of the crystal. Bands at 390, 1162, 1260, 1408 and 1540 cm^{-1} correspond to intramolecular vibrations of anthracene. Defect fluorescence band is underlined [1.73]

Fig.4.11. Initial parts of fluorescence spectra of anthracence crystals (T = 4.2 K) (a) after thermal deformation of specimen deposited on a quartz substrate, with only the D_1 band of defect fluorescence appearing; (b) the same on polymethylmetacrilate substrate, with only the D_2 band of defect fluorescence appearing; (c) after thermal deformation, at conditions when both D_1 and D_2 bands of defect fluorescence appear [1.74]

After registration of the low-temperature fluorescence the crystals were etched in an oleum-sulphuric acid mixture and studied under an optical microscope. It was found that the crystals showing a D_1 band of defect fluorescence (cf. Fig.4.11a) exhibit etch-pit rows mainly in the [100] direction (cf. Fig.3.14b) but crystals having a D_2 band (cf. Fig.4.11b) - in the [010] direction (cf. Fig.3.14a). In specimens possessing both bands, D_1 and D_2 (cf. Fig.4.11c), etching produces a picture of etch-pit network having etch-pit rows in the [100] and [010] directions (cf. Fig.3.14c).

Initially, in [1.74], the directions of the crystallographic axes were determined from polarization of the luminescence spectrum. X-ray analysis showed, however, that the initial determination of direction of the axes was incorrect. The corrected data of [1.74] are in agreement with [3.53,70]

An analysis of data obtained in [1.74] showed that the etch-pit rows of the [100] direction are, most likely, vertical alignments of nonbasal edge dislocations of the (100) [010] type which may be formed in the process of polygonization (cf. Sect.3.7). Accordingly, the etch-pit rows of the [010] direction are likely to be vertical alignments of nonbasal edge dislocations of the (010) [100] type. Thus the authors of [1.74] concluded that the D_1 band ($\Delta E = 235$ cm^{-1}) in thermally deformed anthracene crystals is caused by [100]-directed vertical alignment of (100) [010]-type edge dislocations, whilst the D_2 band ($\Delta E = 275$ cm^{-1}) by [010]-directed vertical alignment of (010) [100]-type edge dislocations.

Both D_1 and D_2 bands of defect fluorescence, of depth $\Delta E = 235$ cm^{-1} and $\Delta E = 275$ cm^{-1}, respectively, were later observed by WILLIAMS and ZBOINSKI [4.26] in vapor-grown anthracene crystals. These bands may also be attributed to the two above-mentioned nonbasal dislocation families. On the other hand, in melt-grown Ac crystals, according to the authors of [4.26], there is only one pronounced peak at 275 cm^{-1} which is rather sensitive to mechanical deformation of the crystal. Apart from this D_2-type of band in melt-grown Ac crystals shallow singlet exciton traps in the region 20 to 120 cm^{-1} have been detected [4.26], which may be attributed to so-called X traps of displaced Ac molecules (see, e.g. [1.71]), as well as a broad structureless background emission with maximum at ~23750 cm^{-1} which may be attributed to "excimeric" fluorescence from "incipient dimer" pairs formed in compressed regions of the lattice (cf. Sect.3.10), or to defect fluorescence from polymorphic inclusions (cf. Sect.1.3). It is interesting to mention that WILLIAMS and CLARKE [4.28] have recently observed a broad set of singlet exciton structural traps with maximum at ~275 cm^{-1} in carbazole- and acridine-doped Ac single crystals. These traps are considered to be caused by displaced Ac molecules at internal surfaces where segregation of impurity molecules takes place.

LISOVENKO et al. [4.25] have recently observed a set of singlet exciton traps with maximum at 130 cm^{-1} in solution-grown Ac single crystals. This set of structural traps is attributed to basal dislocations of (001) [010] type. After thermal treatment of the crystal at 398 K basal dislocations vanish and nonbasal dislocations of (100) [010] or (20$\bar{1}$) [010] type appear, giving rise to a D_2-type fluorescence band. This phenomenon is partially reversible: at room temperature nonbasal dislocations gradually relax to basal ones, which is confirmed by the fact that the D_2 band decreases in intensity and the D_{130} band slowly reappears.

OSTAPENKO et al. [4.29] have discovered, by studying the temperature dependence of luminescence intensity of naphthalene crystals, a family of singlet exciton traps with peaks at 160, 270, and 340 cm^{-1} depth. The density of these traps is negligible in naphthalene single crystals, but it increases considerably in polycrystalline samples and particularly in deformed single crystals.

LISOVENKO et al. [1.73] have proposed a rather ingenious method of estimating the mean pressure $\bar{p}$ in defect regions of the crystal, based on the long-wave shift of defect fluorescence bands. As already mentioned in Sect. 4.2, it was found by TANAKA et al. [3.60] that hydrostatic compression of an anthracene crystal causes linear long-wave shift of the fluorescence band, with specific shift value $\Delta E^S = -80\ cm^{-1}/kbar$ (cf. Table 4.2). This correlation of spectral shift under hydrostatic compression of the anthracene crystal can be used for estimating the mean pressure values $\bar{p}$ in the regions of local compression of extended structural lattice defects.

Thus for [100]- and [010]-directed dislocation alignments of a thermally deformed anthracene crystal (cf. Fig.3.14a,b), giving rise to D_1 and D_2 defect fluorescence bands (cf. Fig.4.11a,b) such an estimate yields the following mean values of local pressures: $\bar{p}_{D_1} = -235/-80 \approx 2.9$ and $\bar{p}_{D_2} = -275/-80 \approx 3.4$ kbar.

Values of the same order ($\bar{p} \approx 3$ kbar) were obtained also from calculated relative pressures $\delta p/p$ on middle-angle boundaries, formed by basal dislocations of (001) [010] and (001) [100] type (cf. Table 3.8), at relative pressures ca. 10 %.

We used the specific fluorescence band shift value $\Delta E^S = -80\ cm^{-1}/kbar$ [3.60] for estimating the mean pressures $\bar{p}$ in local compression regions of structural defects in anthracene crystals, based on reported data of characteristic defect fluorescence band values. The results of such an estimate are presented in Table 4.5.

Table 4.5 shows that the mean local pressures in structural defects of an anthracene crystal, obtained by the "fluorescence probe" method, are of values from several kilobar up to 20 kbar in case of defects with a quasi-discrete spectrum, and from zero to 25 kbar for defects with a continuous spectrum.

The data on mean $\bar{p}$ values from Table 4.5 will be used in Chap. 5 for evaluating electronic polarization energy changes for charge carriers localized in the given defect regions of an anthracene crystal (cf. Sect.5.2).

Table 4.5. Reported characteristic depths E_S of structural traps for singlet excitons and estimated mean pressures $\bar{p}$ in local compression regions of corresponding structural defects of an anthracene crystal

Structural defect fluorescence band [cm^{-1}]	Band shift with respect to 0-0 transition ΔE_S [a] [cm^{-1}]	E_S [eV]	$\bar{p}$ [kbar]	Characteristics of band	References
24950	130	~0.016	1.6	Observed in solution-grown crystals; attributed to basal dislocations of (001) [010] type.	[4.25]
24840 24845	240 (D$_1$) 235 (D$_1$)	~0.030	3.0 2.9	D$_1$ bands observed in vapor-grown crystals upon thermal deformation [1.73,74] or without it [4.26]; attributed to nonbasal (100) [010] type dislocations.	[1.73] [1.74,4.26] (cf. Figs. 4.10,11)
24805	275 (D$_2$)	~0.034	3.4	D$_2$ bands observed in vapor-grown crystals upon thermal deformation [1.74] or without it [4.26]; attributed to non-basal (010) [100] type dislocations	[1.74,4.26] (cf. Fig.4.11)
24818	260 263 275	~0.03	~3.3	Apparently D$_2$ type of bands. Observed in melt-grown [1.70,4.26] and vapor-grown crystals [1.70] without special treatment; also in doped crystals [4.28]	[1.70] [4.23] [4.26,28] (cf. Figs.4.4,5]
24780	320	~0.04	4.0	Appears under mechanical deformation of crystal, most likely, connected with stacking faults in (00ℓ) planes.	[1.72] (cf. Fig.4.9)
23535 24462	1565 616	0.19 0.08	19.6 7.7	Observed in melt-grown crystals (cf. Fig.4.5); appears in vapor-grown crystals after thermal treatment (cf. Fig.4.4).	[1.70,72] (cf. Figs. 4.4,5,9)
23420	~1660	0.20	20.8	Wide stable band of structural origin, particularly characteristic of melt-grown crystals.	[1.70,72] (cf. Figs.4.4,5,9)
Continuous fluorescence band in region 25000-23000 cm^{-1}	0-2000	0-0.25	0-25	Particularly pronounced in melt-grown crystals (cf. Fig.4.6, Curve 3); weaker in vapor-grown crystals (cf. Fig.4.6, Curve 1); diminishes under thermal treatment (cf. Fig.4.6, Curve 2).	[1.70,4.26] (cf. Fig.4.6)

[a] The position of the singlet S_1 D-D transition band of an anthracene crystal is, according to various reported data (cm^{-1}): 25078 [1.70]; 25093 ± 3 [4.23]; 25100 [1.72]; 25080 [1.73].

It should be noted that, according to [1.73], the bands of the exciton fluorescence series of a thermally deformed anthracene crystal are some three to four times wider than the corresponding bands of a non-deformed crystal (cf. Fig.4.10a,b). In addition, upon deformation of the crystal, the whole spectrum shifts by some 30 cm^{-1} towards shorter waves, and the short-wave edge of the 0-0 band becomes diffuse. Such widening is, apparently, caused by quasi-continuous shallow structural traps at the very bottom of the exciton band. These shallow traps are, most likely, formed by slightly displaced molecules with random distribution.

One might suggest that a statistical distribution of defect regions of compression and dilation is also the basic reason for the usually observed widening of electronic absorption bands of molecular crystals as compared to solution spectra, whilst the local regions of considerable compression are the cause of a long-wave "tail", characteristic of absorption bands of organic solids.

5. Local Trapping States for Charge Carriers in Molecular Crystals

Man hat nur Bausteine, kein Gebäude,
so lange man nicht die verwickelten
Erscheinungen einem Princip unter-
wurfig gemacht hat.

C.F.Gauss
Das Erfinden ist kein Werk des logi-
schen Denkens,
wenn auch das Endprodukt an die lo-
gische Gestalt gebunden ist.

A.Einstein

Our discussion on the role of structural defects in molecular crystals has come to a certain climax. After a long and arduous ascent along the slippery slopes of dislocational and other structural defects we have, at last, reached the summit, from the height of which we can now clearly see the main outlines of the whole problem of local states in molecular crystals.

In the present chapter we shall attempt to formulate this problem more precisely and, using the knowledge obtained from reading Chaps.3 and 4, to expand and substantiate our point of view on the physical nature and statistical origin of local states in molecular crystals. To this end three fundamental aspects of the problem have to be discussed.

First of all, it is necessary to show that in regions of compression or dilation of a molecular crystal local centers of ionized states, with changed energy of electronic polarization, are actually formed. It is, subsequently necessary to estimate the self-energy of charge carriers localized on these centers.

Secondly, it is necessary to demonstrate, applying our knowledge from Chap.3 on the wide variety of different structural defects in molecular crystals, the statistical nature of the energy spectrum of the local states of structural origin. We must, further, show the validity of Gaussian distribution for describing the energy spectrum of these local states.

In the third place, a phenomenological theory of space charge limited currents (SCLC) must be worked out for Gaussian distribution of local charge trapping states. And, finally, experimental evidence of the adequacy of this approach has to be provided.

5.1 Electronic Polarization Energy of a Compressed Anthracene Crystal

According to the phenomenological model proposed by SILINSH [1.66] and SWORAKOWSKI [3.5] (cf. Introduction to Chap.3), local charge carrier trapping states in molecular crystals are formed as a result of increased electronic polarization energy ΔP in local compressed regions of specific structural defects, such as dislocations, stacking faults, etc.

In order to make quantitative estimates of polarization energy changes ΔP and of respective trap depths E_t in structural defects of the crystal, it is essential to find, primarily, correlations between pressure p or relative volume change $\Delta V/V$ of a uniformly compressed crystal and the corresponding increase in polarization energy ΔP.

Such correlations were estimated by SILINSH et al. [1.62,5.1] for a uniformly (hydrostatically) compressed anthracene crystal. In [1.62] the relative volume change $(-\Delta V/V)$ up to 10.6 % compression was obtained from extrapolated values of temperature dependence of anthracene lattice parameters according to KITAIGORODSKI [1.2].

In [5.1] the pressure dependence of unit cell parameters of a uniformly compressed anthracene crystal was obtained by means of the following empirical expressions evolved by SCHIPPER [4.11] from experimental values of elastic constants for anthracene (cf. Sect.4.1)

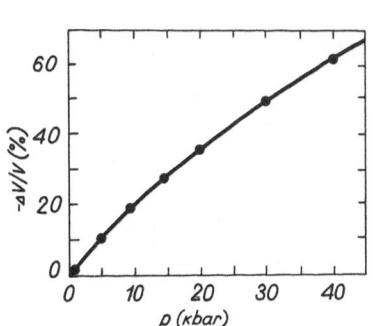

Fig. 5.1. Calculated relative lattice volume change $-\Delta V/V$ of an anthracene crystal as dependent on uniform hydrostatic pressure p

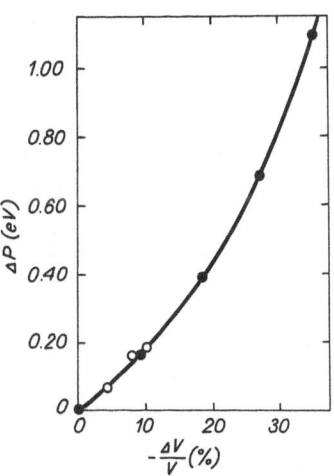

Fig. 5.2. Calculated increase of electronic polarization energy ΔP in an anthracene crystal, as dependent on relative lattice compression $-\Delta V/V$ under uniform hydrostatic pressure. (o) parameters of compressed lattice according to [1.62]; (•) according to [5.1]

$$\delta a = -0.67 \times 10^{-2} p, \quad \delta b = -0.31 \times 10^{-2} p,$$
$$\delta c = -0.10 \times 10^{-2} p, \quad \delta \beta = 0.8 \times 10^{-2} p, \tag{5.1}$$

where δa, δb, δc are the changes in parameters a,b,c of an anthracene unit cell in nm; $\delta \beta$ is the change in monoclinic angle β of an anthracene unit cell in rad; p is the uniform hydrostatic pressure, in kbar.

Pressure dependences of anthracene unit cell parameters, calculated according to (5.1), are presented in Table 5.1. Corresponding pressure dependences of relative lattice compression $-\Delta V/V = f(p)$ are shown in Fig.5.1. The relative lattice volume changes were evaluated up to a pressure of 40 kbar, i.e., up to the applicability limits of (5.1), according to [4.11].

Table 5.1 shows that under uniform hydrostatic compression the unit cell undergoes changes not only in size, but also in shape. The c parameter changes very little, whereas the monoclinic angle β changes to a much more considerable extent. This leads to a change in height of the cell, i.e., to a change in parameter c'.

The increments of polarization energy ΔP for a compressed anthracene crystal were calculated by the method of self-consistent polarization field (SCPF) (cf. Sect.2.4.2) within a region including 160 molecules.

The calculated dependences $\Delta P = f(\Delta V/V)$, as reported in [1.62,5.1], are presented in Fig.5.2. As may be seen from Fig.5.2, both methods of determination of lattice parameters for a compressed crystal provide concurrent values of ΔP.

Fig.5.2 shows that even a small (5 to 10 %) relative lattice compression of an anthracene crystal produces an increase of ΔP of the order of 0.1-0.2

Table 5.1. Calculated values of unit cell parameters of compressed anthracene crystal, as dependent on uniform hydrostatic pressure p [5.1]

Unit cell parameter (cf.Fig.1.5)	Normal pressure [1.2]	p [kbar]			
		5	10	20	30
a [nm]	0.855	0.822	0.788	0.721	0.654
δa [nm]	-	-0.0335	-0.067	-0.134	-0.201
$\delta a/a$ [%]	-	3.92	7.81	15.67	23.51
b [nm]	0.603	0.587	0.572	0.541	0.510
δb [nm]	-	-0.0155	-0.051	-0.062	-0.093
$\delta b/b$ [%]	-	2.57	5.14	10.28	15.43
c [nm]	1.117	1.112	1.107	1.097	1.087
δc [nm]	-	-0.005	-0.010	-0.20	-0.030
$\delta c/c$ [%]	-	0.45	0.90	1.79	2.69
β [°]	124.584	126.88	129.169	133.748	138.333
$\delta \beta$ [rad]	-	0.04	0.08	0.16	0.24

eV. In other words, such local compression of the anthracene lattice leads to the formation of shallow charge carrier traps of 0.1-0.2 eV depth.

We wish to draw attention to the superlinear nature of ΔP increase with growing relative lattice compression -ΔV/V. Thus, for -ΔV/V ca. 10 % we have an increase of ΔP = 0.20 eV, whereas at -ΔV/V = 30 % the increase of ΔP reaches the value of 0.9-1.0 eV. This means that in the regions of local lattice compression of such an order of magnitude, comparatively deep charge carrier traps of structural origin can be formed.

For comparison, the calculated pressure dependence of charge carrier trap depth $E_t \equiv \Delta P$ in a compressed anthracene crystal $E_t = f(p)$ according to reported data by JURGIS and SILINSH [5.1], and corresponding calculated pressure dependence of singlet exciton trap depth $E_s = f(p)$ according to SCHIPPER [4.11] are presented in Fig.5.3. These correlated dependences will be used later in Sect.5.2 for estimating the mean trap depth E_t for charge carriers from the data on singlet exciton trap depth E_s in defect regions of local compression in anthracene crystals.

It can be seen from Table 5.1 that anthracene-type crystals possess a pronounced anisotropy of lattice compressibility. It is maximal in the direction of the a axis of the crystal and minimal along the c axis.

The linear compressibility β of anthracene in the directions of the main axes [100], [010] and [001] of the crystal, as reported by DANNO and INOKUCHI [3.42], is presented in Table 5.2.

SWORAKOWSKI [3.12] used the data of linear compressibility (Table 5.2) for estimating ΔP at linear deformation of an anthracene crystal. The results of this estimate are shown in Fig.5.4. As one might expect, the largest contribution towards increase of ΔP is due to compression along the a axis, i.e., in the [100] direction of the maximum compressibility of the crystal.

The compressibility of anisotropic molecular crystals of the anthracene type can be illustrated by cross section of hydrostatic compressibility

Table 5.2. Linear compressibility β of an anthracene crystal [3.42]

Compressibility direction	$\beta \ [cm^2 dyn^{-1}]$
[100]	12.7×10^{-12}
[010]	4.44×10^{-12}
[001]	3.59×10^{-12}

254

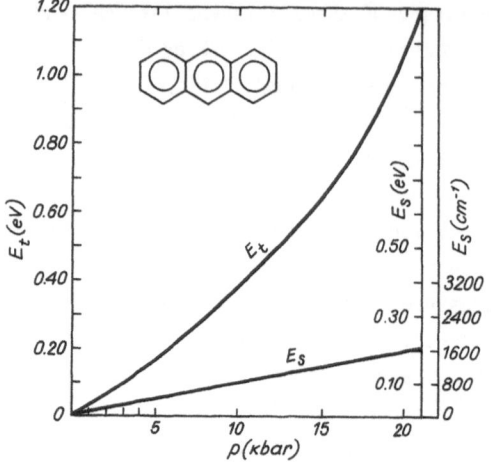

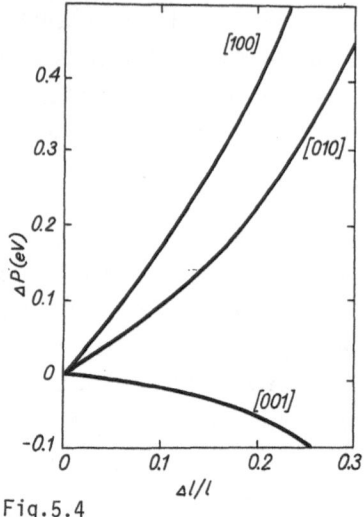

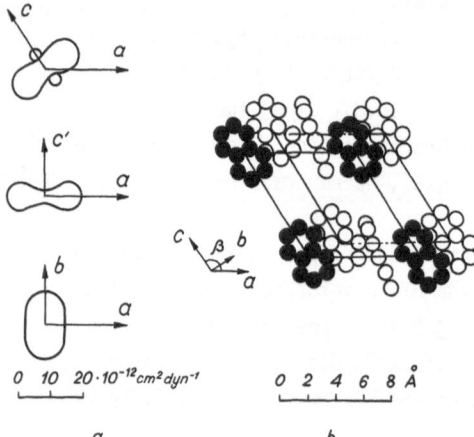

Fig.5.3

Fig.5.4

Fig. 5.3. Calculated pressure dependences of charge carrier trap depth $E_t = \Delta P$ in compressed anthracene crystal $E_t = f(p)$, according to [5.1], and corresponding singlet exciton trap depth $E_s = f(p)$, according to [4.11]

Fig. 5.4. Change of electronic polarization energy ΔP in anthracene as dependent on relative linear deformation $\Delta l/l$ along the main axes of the crystal [3.12]

Fig. 5.5a,b. Cross section of hydrostatic compressibility figures of a naphthalene crystal (a) and positions of naphthalene molecule in unit cell (b) ($\beta = 122°$) [1.2]

figures (cf. Fig.5.5). In naphthalene, for instance, as may be seen from Fig.5.5, the molecules are situated in the unit cell in such a fashion (cf. Fig.5.5b) that the directions of maximum compressibility (a) correspond to the mutual approach of the planar molecules by slipping along the ab plane of the crystal [1.2].

It must be borne in mind, at the same time, that, despite maximum compressibility of the anthracene crystal along the a axis, the direction along the b axis still remains the dominating one for dislocational slip, owing to lower formation energy of basal dislocation (001) [010], as compared to that of (001) [100] (cf. Sect.3.5.3 and Table 3.6).

5.2 Formation of Local Trapping Centers for Charge Carriers in Structural Defects of a Real Molecular Crystal

As discussed above (cf. Introduction to Chap.4), one and the same structural defect of a molecular crystal can form local trapping centers both for excitons, as well as for charge carriers. One should remember, however, that the physical nature of the forces creating these local states in the defect regions of lattice compression differs considerably. Trapping of an exciton is the result of an increase in dispersion (ΔD) and resonance (ΔI) terms of interaction between the excited molecule and its molecular environment (cf. Sect.4.1), whereas trapping of a charge carrier is the result of an increase in electronic polarization energy ΔP of the crystal in the defect region of local compression of the lattice (cf. Sect.3.1).

Common in both cases is only the pronounced dependence of these interactions on lattice geometry (cf. Introduction to Chap.4).

This allows one to use experimental data on singlet exciton trap depth E_s and mean pressure $\bar{p}$ estimates in defect regions of local compression of an anthracene crystal, obtained from defect fluorescence measurements (cf. Sect.4.3 and Table 4.5) for determining corresponding charge carrier trap depth E_t in the given structural defect. For this purpose the calculated pressure dependences $E_s = f(p)$ and $E_t = f(p)$ (cf. Fig.5.3), obtained under conditions of uniform hydrostatic compression of an anthracene crystal (cf. Sects. 4.1 and 5.1) can be used. [It should be recalled here that the $E_s = f(p)$ dependence, as calculated by SCHIPPER [4.11], is in good agreement with the experimental one, as determined by TANAKA et al. [3.60] (cf. Sects.4.1,2)].

Experimental trap depths of single excitons E_s and their corresponding counterparts — charge carrier traps E_t, estimated according to the correlated pressure dependence (cf. Fig.5.3), are presented for various typical structure defects in anthracene crystals in Table 5.3. (On the possible origin and nature of these structural defects see comments in Table 4.5).

As can be seen from Table 5.3, the same structural defects of an anthracene crystal form 3 to 5 times deeper traps for charge carriers than for singlet excitons, i.e., $E_t > E_s$.

Table 5.3. Singlet exciton trap depth E_s and mean pressure $\bar{p}$ in local compression regions, as determined by method of defect fluorescence (cf. Table 4.5) and the corresponding estimated trap depth of charge carriers E_t in various structural defect of an antrhacene crystal

E_s		$\bar{p}$	E_t
[cm^{-1}]	[eV]	[kbar]	[eV]
130	0.016	1.6	$\sim$0.05
240	0.03	$\sim$3.0	$\sim$0.10
275	0.034	3.4	0.13
320	0.04	4.0	0.15
616	0.08	7.7	0.30
1565	0.19	19.6	1.05
1660	0.20	20.8	1.1
1600-2000[a]	0.20-0.25	20.8-25.0	1.2-1.4

[a] the deepest structural traps of the "tail end" of the quasi-continuous sprectral distribution (cf. Fig.4.6).

As may be seen from Fig.5.3, the E_t = f(p) dependence, due to nonlinearity, increases more rapidly with pressure than the corresponding E_s = f(p) dependence. The nonlinearity region of the latter, lies, apparently, beyond the pressure values of p > 40 kbar.

The deepest structural singlet exciton trapping centers E_s, of 1500-2000 cm^{-1} (0.18-0.25 eV) depth (cf. Tables 4.5 and 5.3) are usually present, as already shown in Sect.4.3, in single crystals obtained from melt, or are formed after mechanical or thermal deformation of the crystal. Formation of such deep traps is, most likely, due to aggregation of dislocations in alignments and ensembles in the process of nonequilibrium growth or deformation of the crystal (cf. Sect.3.7 and 4.3). They may also be caused by local compressed inclusions of a different metastable phase in the matrix of the parent crystal and other more complex three-dimensional lattice defects, e.g., stacking fault alignments, etc. (cf. Sect.3.11). Corresponding charge carrier trapping states E_t, formed in the given structural defects, have a depth of ca. 1 eV (cf. Table 5.3).

The question about deep structural traps in molecular crystals is rather controversial. It is frequently discussed, whether it is at all possible, using general concepts of dislocation theory, to obtain structural defects with local stress fields of the order of 10 to 20 kbar, creating deep structural charge carrier traps E_t of 0.5-1.0 eV depth (cf., e.g. [3.12,13]).

It will be shown later that such deep local trapping centers actually cannot be created by stress fields of a single dislocation of a tilt boundary.

However, such strong stress field fluctuations can take place in cases of aggregations of dislocations (cf. Sects.3.7 and 5.7). For instance, if dislocations aggregate in a group consisting of n dislocations, then the amplitude $\bar{\sigma}_A$ of the mean internal stress field of such an aggregation is of the order of [cf. (3.31) in Sect.3.7.3]

$$\bar{\sigma}_A = \frac{\mu b}{2\pi} \frac{\sqrt{n}}{\bar{\lambda}} \quad . \tag{5.2}$$

If the aggregation of dislocations is of high density, with $\bar{\lambda} \approx r_c \approx 3b$, we get for anthracene ($\mu = 3.2 \times 10^{10}$ dyn $\cdot$ cm^{-2}, cf. Table 3.4) the following approximate values for $\bar{\sigma}_A$: for n = 10, $\bar{\sigma}_A$ = 5.4 kbar, and for n = 100, $\bar{\sigma}_A$ = 17 kbar.

Using the pressure dependence E_t = f (p) from Fig.5.3, it is possible to estimate the depth of the charge carrier traps E_t formed by the given aggregations.

Thus for a group of 10 dislocations (n = 10) $E_t \approx 0.2$ eV, and for n = 100, E_t = 0.85 eV. It follows that similar general considerations, based on the theory of dislocation ensembles (cf. Sect.3.7.3) eventually confirm the possibility of formation of structural charge carrier traps of depth E_t = 0.5-1.0 eV in real anthracene-type molecular crystals.

The existence of such deep charge carrier traps of depth E_t = 0.7-1.0 eV in anthracene and tetracene single crystals has been confirmed by many investigators (cf., e.g. [1.68,5.2-5]). These data are based on studies using different experimental techniques, such as methods of thermo- and photostimulated conductivity and of isothermic decay currents (cf. Sects.5.12,13), as well as studies of temperature dependence of space charge limited currents (cf. Sect.5.9.5).

Earlier such deep charge carrier trapping states were usually identified with impurity centers. At present, however, their structural origin seems proved beyond any doubt.

First, these trapping centers are observed in ultra-high-purity anthracene single crystals after repeated and meticulous purification, excluding traces of possible impurities [5.3,5].

Secondly, the existence of such deep trapping centers is characteristic of single crystals obtained from melt or mechanically deformed (cf. Sects. 4.3 and 5.12). Such crystals usually possess a high density of basal and nonbasal dislocations (cf. Sect.3.13 and Table 3.12). On the other hand, thin layers of molecular crystals, obtained by evaporation *in vacuo*, do not,

as a rule, display such deep trapping centers for charge carriers, but mainly have sets of shallow traps (cf. Sect.5.10 and Table 5.8).

Finally, the structural nature of these deep charge carrier traps is
borne out by the bipolarity of trapping. In other words, they act as trapping centers both for electrons and for holes. Such bipolarity is characteristic of stuctural local states of polarizational origin.

These deep trapping centers for charge carriers can be quasi-discrete,
of $G_g(E)$ type of distribution, with small dispersion parameter σ value (cf.
Sect.5.12), as well as quasi-continous, with a large value of σ, like corresponding centers for exciton trapping (cf. Sect.4.3).

Regular dislocation alignments, such as tilt boundaries (cf. Sect.3.8),
form shallower trapping centers for charge carriers. Thus, for example, tilt
boundaries of [100] and [010] direction, formed by vertical alignment of
nonbasal dislocations of the (100) [010] and (010) [100] type (cf. Fig.3.14
and Table 4.5) ought to form, according to correlation diagrams on Fig.5.3,
shallow charge carrier trapping states of depth $E_t \approx 0.10\text{-}0.13$ eV (cf. Table
5.3).

An independent approach of estimating local increase in polarization energy ΔP and of the corresponding depth of charge carrier traps E_t in compression regions of vertical dislocation alignments can be provided using
calculated data of relative pressure $\delta p/p$ on middle-angle tilt bondaries in
an anthracene crystal (cf. Sect.3.8.2 and Table 3.8). For instance, middle-
angle tilt boundaries of an anthracene crystal, created through polygonization of basal dislocations of (001) [010] type and having $\delta p/p$ values from
2 to 13 %, form charge carrier trapping states of mean depth E_t from 0.04 to
0.17 eV, depending on disorientation angle θ.

We see thus that both approaches of estimating E_t on tilt boundaries
yield values of the same order of magnitude. It stands to reason therefore
that middle-angle tilt boundaries in an anthracene crystal form local charge
carrier trapping states E_t of 0.05 to 0.20 eV in depth. However, whilst the
depth of singlet exciton trapping states E_s on tilt boundaries of nonbasal
dislocations of the (100) [010] and (010) [100] type in a thermally deformed
anthracene crystal has been determined experimentally with sufficient accuracy (cf. Table 4.5), the corresponding charge carrier trapping states E_t
have so far not been experimentally ascertained.

In compression regions of linearized and small-angle tilt boundaries, in
which dislocations are widely separated (D >> b) (cf. Table 3.7), as well as
in compression regions of single isolated edge dislocations a rough estimate

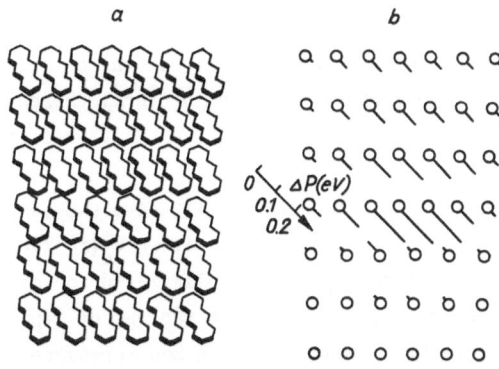

Fig. 5.6a,b. Schematic picture of edge dislocation of the (001) [010] type in an anthracene crystal (a) and corresponding changes of polarization energy ΔP at the dislocation core (b) [3.13]

of the mean depth of charge carrier trapping states based on elastic dislocation theory, yields E_t values of 0.03-0.05 eV.

On the other hand, SWORAKOWSKI [3.13] has made a rough estimate of local change of P in the core of the edge dislocation of (001) [010] type of an anthracene crystal which yields ΔP values of ca. 0.09-0.15 eV for molecules directly above the dislocation line, and ΔP ≈ -0.05 eV below the dislocation line (see Fig.5.6). This calculation is based on an "idealized" molecular dislocation core model, as shown on Fig.5.6a (cf. also Fig.3.3), without allowing for relaxation of displaced molecules. Therefore, these results, as pointed out by the author of [3.13] appear to be overestimated. Since the elasticity theory approach gives, as a rule, the lower limit of values under consideration, it stands to reason that the average value of ΔP = E_t actually is of the order of 0.05-1.0 eV in local compression regions of a single isolated dislocation.

It will be shown in Sect.5.3 that single stacking fault ribbons in anthracene crystals also form a complicated set of shallow traps for charge carriers of depth not exceeding 0.1 eV.

It was also found by defect fluorescence studies of molecular crystals (see, e.g. [4.10]) that shallow local states appear in the vicinity of impurity centers as well, owing to lattice deformations around the guest molecule. These shallow states do not, however, affect the charge carrier trapping directly since the final trapping is actually determined by a deeper level of the impurity molecule (cf. [1.45,135]). The existence of such local states can be ascertained indirectly, by their influence on the effective depth of the trapping level of the impurity molecule, as determined by measuring temperature dependence of charge carrier mobility in different crystallographic directions [1.45,135]. We see thus that structural defects in molecular crystals form a widely ranged energy spectrum of local charge carrier trapping states of polarization origin, spreading from E_t values of kT order at room temperature up to 1 eV.

Thus, in real molecular crystals the discrete conductivity levels E_e and E_h are enveloped by a distribution "bell" of local states with a $G_e(E)$ type of energy spectrum (cf. Fig.3.2). These local states of polarization origin are most likely to arise in local compression and dilation regions of separate edge dislocations and small-angle tilt boundaries. The mean pressure $\bar{p}$ in the compression regions of an edge dislocation is of the order of several kbar, and the corresponding charge carrier trapping states E_t have a depth, as stated above, of 0.05-0.1 eV. The existence of such trapping states with $G_e(E)$ type of distribution has been most clearly demonstrated for evaporated layers of some molecular crystals in which there are no deep trapping states with $G_g(E)$ type of distribution (cf. Sect.5.9.5 and Fig.5.29).

More deeply in the forbidden energy gap are positioned the charge carrier trapping states formed in the regions of various dislocation aggregations and ensembles, as well as created by more complex three-dimensional lattice defects, characteristic of melt-grown or deformed molecular single crystals.

Experimental evidence of the existence of different sets of charge trapping states with Gaussian distribution of the energy spectrum in real molecular crystals will be presented later in Sects.5.10,12.

The various randomizing factors determining the Gaussian character of the statistical distribution of these trapping states of structural origin will be discussed in detail in Sect.5.7.

It must be emphasized once more, in conclusion, that the local centers of structural origin under consideration are not point defects. They comprise a local microregion of deformed lattice in the crystal. A local charge carrier trapping center of polarization origin is actually formed by an assembly of displaced molecules containing several tens, even hundreds of molecules. The term "density N_t of local states" is, as a rule, used for denoting the number of such local assemblies of displaced molecules. The corresponding E_t value, in turn, denotes the maximum depth of the polarization well when the charge carrier is localized on a molecule with maximum increase in electronic polarization energy ΔP in the given defective microregion of the crystal.

A more rigourous approach demands accounting for the complete spectrum of local state sublevels upon localization of the charge carrier on any of the

molecules of the given assembly. In the case of such an approach each local
trapping center would contain n local sublevels equal to the number of molecules forming the center. Localization of a charge carrier on these sublevels, in turn, would be determined by the Boltzmann distribution.

Such an approach, however, is possible only if all the coordinates of
displaced molecules in the defect core are known. Unfortunately, such information is not possible at present. Traditional experimental methods (cf.
Sects.5.10,12) give only integral values for characterization of defect
centers, such as N_t, E_t, and the mean trapping cross section q_t.

The next section contains some calculated data on the energy spectrum of
sublevels for charge carrier trapping in stacking fault ribbons of an anthracene crystal. These results are based on calculated coordinates of displaced
molecules in the faulty lattice region, obtained by the method of atom-atom
potentials (cf. Sect.3.9.3).

Future progress in this field depends mainly on the possibility of obtaining reliable calculated data on the coordinates of displaced molecules
by the atom-atom potential method in various types of structural defects,
using molecular models of the defect core (cf. Sect.3.6.2).

5.3 Energy Spectrum of Local States of Polarization Origin in Stacking Faults of an Anthracene Crystal

The ΔP spectrum of local states for a charge carrier localized on displaced
molecules in a stacking fault ribbon of an anthracene crystal, was calculated by SILINSH and JURGIS [1.22].

The self-consistent polarization field (SCPF) method was used in the calculations (cf. Sect.2.4.2). The ΔP value was determined as the difference
between the electronic polarization energy of an ideal crystal and that of
a molecular configuration in the given defect structure. A lattice region
comprising 29 molecules was considered.

The calculated ΔP values are presented in Figs.3.21,22 both for relaxed
as well as unrelaxed configurations of molecules in the given stacking fault
ribbons.

Figure 3.21 shows that in a ribbon consisting of three rows of molecules
the increase in electronic polarization energy ΔP lies between 0.03 and 0.20
eV, the highest ΔP value falling into the middle row where mutual approach

of molecules is the greatest after relaxation (cf. Table 3.9). It is interesting to note that for the initial configuration of the unrelaxed molecules
in the ribbon the change in electronic polarization energy ΔP is inconsiderable (cf. Fig.3.21). Only molecules 1 and 1' of "predimer configuration"
show a decresase in P, apparently owing to the low value of the molecular
polarizability component α_N perpendicular to the plane of the molecule (cf.
Table 2.2).

With increase in width of the stacking fault ribbon r_{sf} the mutual approach of molecules in the middle rows becomes smaller after relaxation (cf.
Fig.3.22 and Table 3.9), and the maximum value of electronic polarization
energy increase ΔP becomes correspondingly smaller. It is rather characteristic that the increase ΔP in the middle of the ribbon is accompanied by a
decrease in the value of P ($\Delta P < 0$) along the dislocation line (cf. Fig.
3.22a,b).

We see thus that the stacking fault ribbon forms a potential "valley" for
charge carrier trapping of a rather complicated shape of the surface, exhibiting zig-zag-formed "channels" with equal or very close values of ΔP (cf.
Fig.3.22a,b). Therefore, it is possible, in principle, that a charge carrier,
trapped in such a stacking fault potential "valley", can travel by means of
hopping transfer along the middle rows of molecules of the ribbon in the
[110] direction of the lattice.

It is essential to note that the method of calculating equilibrium configuration of molecules, as performed in [1.22] (cf. Sect.3.9.3) (choice of
boundary conditions and limitation of number of variable parameters), yields
an upper limit for intermolecular approach in the ribbon after relaxation.
In this case we get, accordingly, the upper limiting ΔP values. One can
therefore conclude that in isolated stacking fault ribbons of anthracene-
type crystals a large number of shallow trapping states for charge carriers
are formed, possessing a complicated set of energy spectra of depth not exceeding 0.1 eV.

5.4 Local Surface States of Polarization Origin in Molecular Crystals

The free surface of an ideal crystal constitutes a two-dimensional structural defect (cf. Sect.3.2).

It may be shown that the near-surface molecules of a crystal form local
states of polarization origin for charge carriers. If a charge carrier ap-

proaches a free surface, its energy of electronic polarization P decreases, as can be seen from expression (2.12) for charge-induced dipole interaction W_{i-d} (cf. Sect.2.4.1). The terms of the sum in (2.12) disappear for molecules which are replaced by vacuum outside the boundaries of the surface, and the total polarization energy for a charge carrier localized on molecules in the near-surface region P_s is smaller than the total polarization energy P_v inside the bulk of the crystal, i.e., $P_s < P_v$.

As a result, a near-surface barrier of polarization origin is created for charge carriers. This barrier was first studied in microscopic approximation by SILINSH [1.66,5.6].

It was of considerable interest to get more accurate calculated data on the height φ_p and depth of penetration ℓ_p of such a "polarization barrier" in a model anthracene crystal. In the case of relatively high values of φ_p and, especially, of ℓ_p we can expect a strong influence of such a "polarization barrier" on photoemission, as well as on charge carrier transfer processes through the organic crystal-metal interface.

Such calculations of φ_p and ℓ_p values for an anthracene crystal were later performed by SILINSH and JURGIS [1.62,3.11]. In these studies the polarization energy P_s was calculated for a charge carrier localized in the surface layer near faces bounded off by ab-, ac- and bc planes of the crystal. The SCPF method was used (cf. Sect.2.4.2), according to (2.17), for a hemisphere containing from 99 to 136 molecules. The polarization energy ΔP of molecules outside the hemisphere was determined in macroscopic approximation according to (2.19). Calculated P_s and $\varphi_p = P_v-P_s$ values are presented in Table 5.4.

Table 5.4. Calculated electronic polarization energy P_s for charge carrier localized in the surface molecular layer of an anthracene crystal [1.62]

Planes, bounding off the crystal surface	P_s [eV]	$\varphi_p = P_v-P_s$ [eV]
ac	1.04	0.49
bc	1.10	0.43
ab	1.48	0.05

Note: $\varphi_p = P_v-P_s$ is the height of nearsurface "polarization barrier"; P_v is the electronic polarization energy in the bulk of the crystal: $P_v = 1.53$ eV (cf. Table 2.1).

Table 5.4 shows that for a surface bounded off by the cleavage face ab, with ca. 9 Å interlayer distance in the c' direction, the value of P_S differs only slighty from that of P_V, and the height of the "polarization barrier" is just hardly above kT level at room temperature. At the faces bounded off by ac- and bc planes of closest packing φ_p reaches values of 0.4-0.5 eV.

Figure 5.7 shows the dependence of φ_p on distance r in the b direction from the ac surface. The polarization barrier can be seen to penetrate to a depth of some 12 Å, i.e., down to the fourth molecular layer.

Charge carriers can "filter" through the "polarization barrier" along conducting "tubes" of dislocations emerging at the surface. Such "tubes" with a lower self-energy value of the charge carrier can form local "conducting" levels in some definite direction. The regions of emergence of dislocations on a free face, apparently, form suface levels of structural origin. If local pressure of typically average value of a few kilobar is formed in regions of compression of dislocation defects, the latter can, on emerging to the surface, create shallow local traps E_t^S of 0.1 eV depth (cf. Sect.5.2). Middle-angle boundaries emerging at the free surface and having a relative pressure $\delta p/p$ from 2 to 13 percent (cf. Table 3.8) can form surface levels E_t^S of depth between 0.05 and 0.2 eV.

More deeply situated surface centers of charge carrier trapping can be formed in the case of near-surface aggregation of edge dislocation (cf. Fig. 3.10a). Such dislocation aggregations at the surface of a single crystal or of polycrystalline grains can form local fields of stress close to the limit of mechanical strength of the crystal (cf. Sect.3.7.2). This means that in near-surface regions of such aggregations local pressures of 10-20 kbar are possible. In such compression regions surface traps for charge carriers E_t^S of 1.0-1.4 eV depth can be formed (cf. Table 5.3).

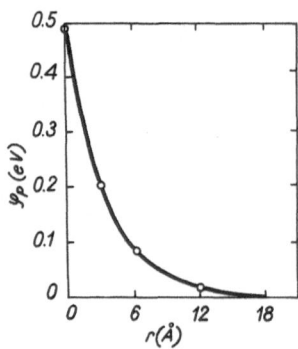

Fig. 5.7. Calculated near-surface potential barrier of polarization origin $\varphi_p(r)$ = P_V - $P_S(r)$, as dependent on distance r of the localized charge carrier in the b direction from the ac surface of an anthracene crystal, according to [3.11]

Structural surface traps can act equally as trapping states for electrons and for holes. However, owing to asymmetry of "self-trapping" (cf. Sect. 5.15), traps for holes are situated more deeply than those for electrons.

The downward bend of conduction levels at the surface of the crystal, characteristic of p-conductivity-type molecular crystals, is, most likely, due to prevalence of trapping of majority charge carriers, viz. holes in the surface traps of structural origin (cf. Sect.5.16).

Deep surface levels of bipolar trapping of charge carriers can also be formed by polar impurity molecules adsorbed at the surface or situated in the near-surface layer of the crystal and possessing a considerable permanent dipole moment [1.66].

5.5. Local States of Polarization Origin in the Vicinity of a Lattice Vacancy

If a charge carrier approaches the vicinity of a lattice vacancy (cf. Sect. 3.3), a situation arises, similar to that of a free surface, only on a smaller scale. We also observe a lowering of electronic polarization energy $(-\Delta P)$ of the charge carrier localized on molecules adjacent to the vacancy. The only difference is that in this case only one term, namely, that with the coordinates of the lattice vacancy, disappears from the polarization energy W_{i-d} sum in the expression (2.12) describing charge-induced dipole interaction. This means that a potential barrier φ_p is formed around the vacancy, causing scattering of charge carriers.

SCPF calculation (cf. Sect.2.4.2) of local trapping state energy spectrum in the vicinity of a vacancy in an anthracene crystal were performed by SILINSH and JURGIS [3.11].

For a vacancy with coordinates (0,0,0) the maximum value of the potential barrier $\varphi_V(r) = -\Delta P$ and of its radius r_V lies in the symmetrical directions [110], [$\bar{1}\bar{1}0$], [$\bar{1}10$], and [1$\bar{1}0$] in the ab plane of the crystal. Figure 5.8 presents, as an example, the calculated $\varphi_V(r)$ dependence in directions [110] and [$\bar{1}\bar{1}0$] of an anthracene crystal.

As can be seen from Fig.5.8, φ_V attains maximum value of 0.06 eV when the charge carrier is localized on the nearest neighboring molecules at the vacancy. The r_V value in these directions is ca. 16 Å.

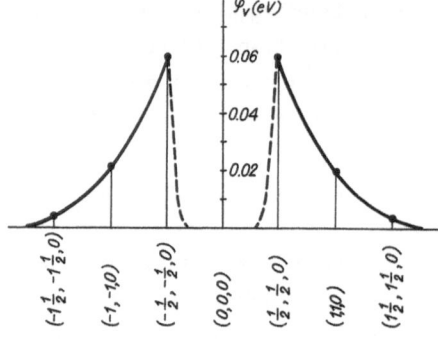

Fig. 5.8. Calculated potential barrier of polarization origin $\varphi_V(r) = -\Delta P$ in the [110] and [$\bar{1}\bar{1}$0] directions around a vacancy situated in the (0,0,0) lattice site in the ab plane of an anthracene crystal, according to [3.11]. The abscissa corresponds to the coordinates of the molecule on which the charge carrier is supposed to be localized

For a (0,1,0) molecule (and for corresponding symmetric molecules) the $\varphi_V = -\Delta P$ value equals 0.04 eV, and for a (1,0,0) molecule (and its symmetric equivalents) it is only 0.01 eV. Molecules outside the ab plane are practically insensitive to the presence of a vacancy. Thus, only some 14 molecules lying in the ab plane of the crystal around the vacancy take part in the formation of a polarization barrier. The cross section of such a vacancy barrier $q = \pi r_V^2$ is of the order of 8×10^{-14} cm^2.

Since lattice vacancy density at room temperature is 1.5×10^{14} cm^{-3} for anthracene, and 7×10^{15} cm^{-3} for naphthalene (cf. Table 3.2), the total number of molecules participating in creating local centers in the vicinity of a lattice vacancy is of the order of 2×10^{15} cm^{-3} for anthracene and 9×10^{16} cm^{-3} for naphthalene. It must also be borne in mind that with increase in temperature the concentration of vacancies and, correspondingly, the number of molecules participating in the creation of local centers in the vicinity of these vacancies grows exponentially (cf. Table 3.2). This circumstance might well be an additional factor determining the temperature dependence of carrier mobility.

In the case of vacancy discs (cf. Fig.3.26b) the problem of local states in the vicinity of vacancies is gradually reduced, with increase in disc diameter, to the problem of local surface states.

It has become rather obvious that in molecular crystals vacancies play only a minor role, taking part only in processes of antitrapping of charge carriers, very much unlike ionic crystals in which lattice vacancies create traps for charge carriers (cf. Sect.5.6). The only case when a lattice vacancy can become a trapping center for charge carriers in a molecular crystal is, if the molecules of the latter possess negative electron affinity. This is, for instance, the case in a benzene crystal.

5.6 Local Charge Carrier Trapping in Covalent, Ionic and Molecular Crystals

As already repeatedly emphasized (see, e.g., Sect.3.4), dislocation defects in molecular crystals, despite their apparent similarity in crystallographic topography and configuration with corresponding defects in covalent and ionic crystals, create local centers for charge carrier trapping of a fundamentally different physical nature. After having discussed in detail the nature of local trapping centers of excitons (cf. Sect.4.3) and of charge carriers (cf. Sects.5.2,3) of dislocation origin in molecular crystals we consider it instructive to return once more to the problem of these centers, in order to remove any misunderstandings. In particular, we wish to give a comparison of the energy structure and the nature of local states in covalent, ionic and molecular crystals.

Let us consider, as an example, an edge dislocation in a covalent crystal of the germanium type (Fig.5.9a), in an ionic crystal of the NaCl type (Fig.5.9b), and in a molecular crystal of the anthracene type (Fig.5.9c). Formally, from the point of view of crystallographic topography, the given dislocation defects are similar in all three cases. However, their energy structure and electronic properties differ considerably, as is evident from Fig.5.9d-f.

In a covalent crystal (Fig.5.9a) the last atom of the extra plane possesses a free valency which forms an electron trapping center below the dislocation line, with a corresponding discrete monoenergetic level in the forbidden energy gap (Fig.5.9d). The compression and dilation regions of the dislocation core, in their turn, modulate the width of the forbidden energy gap ΔE_G, as shown in Fig.5.10. The initial forbidden energy gap ΔE_G is widened in the region of compression (C) by a value of δE. Superimposition of closely situated dilation regions (D) lowers this value by $\delta E'$ or approximately by $\delta E/2$ for the whole structure [3.22].

Thus we get the following approximated expression for effective bandwidth $(\Delta E_G)_{eff}$ for the region of dislocation defects in a crystal of germanium type

$$(\Delta E_G)_{eff} \approx \Delta E_G + \delta E' \approx \Delta E_G + \frac{\delta E}{2} \quad . \tag{5.3}$$

This shows that in covalent crystals defect regions of lattice compression and dilation do not form local trapping centers, as in the case of molecular crystals. They only modulate the potential at the bottom of the conductivity

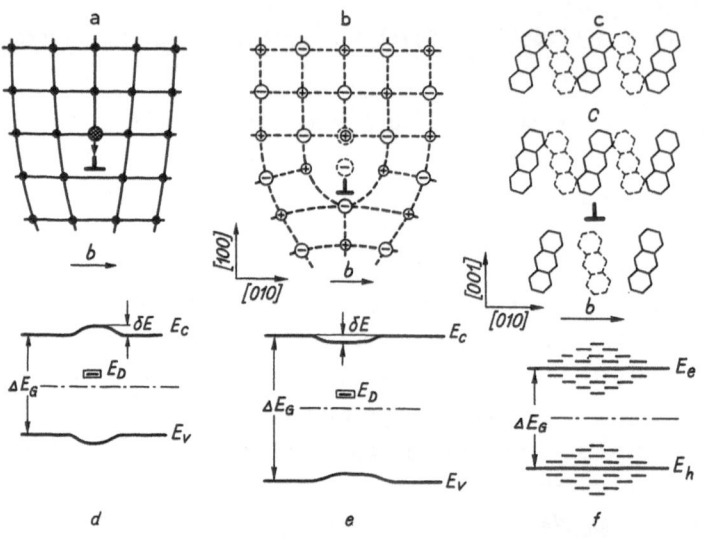

Fig. 5.9a-f. Schematic picture of an edge dislocation in a covalent (a), ionic (b), and a molecular (c) crystal, causing formation of corresponding local charge carrier trapping states in the energy diagram of the crystal (d-f)

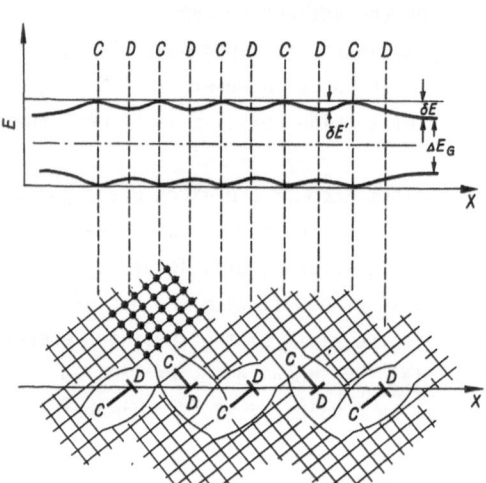

Fig. 5.10. Diagram of band structure for a covalent crystal, taking into account the effects of lattice compression (C) and dilation (D) at a grain boundary on forbidden energy gap width ΔE_G [3.22]

band E_C and at the top of the valency band E_V (Figs.5.9d and 5.10), thus only affecting transfer processes of delocalized charge carriers.

In an ionic crystal (Fig.5.9b) the extreme ion of the extra plane breaks the symmetry of the crystalline Coulomb field. If the extra plane ends with a positive ion, as is the case in Fig.5.9b, then an electron trapping center is created below the dislocation line with a corresponding discrete level inside the forbidden gap ΔE_G (Fig.5.9e). If, on the other hand, the extra plane terminates with a negative ion, a discrete trapping level for holes is created [3.22]. As regards compression regions of the dislocation defect, they only deform the band potential (Fig.5.9e), but do not create local centers, owing to low compressibility of the ionic crystals.

In molecular crystals in which molecular interaction takes place through weak Van der Waals dispersion forces (cf. Sect.1.1), the molecule at the end of the extra plane is completely identical with the other molecules of the crystal, as far as its electronic properties are concerned (cf. Sect. 3.4). In this case no free valencies are formed, as in covalent crystals, and no break in the crystalline field is caused, as in ionic crystals. In other words, there can be no formation of any kind of trapping centers in intermolecular interspace or vacancies in the dislocation core (cf. Sect. 3.3). All local states of a molecular crystal are due to localization of charge carriers directly on particular molecules of the crystal. Dislocation defects in a molecular crystal cannot therefore create monoenergetic discrete trapping states in the forbidden energy gap. Instead, local states of polarization origin are formed in compression and dilation regions of the lattice. Each molecule in the dislocation core has a corresponding energy level. The latter is situated below the level of electronic conductivity E_e

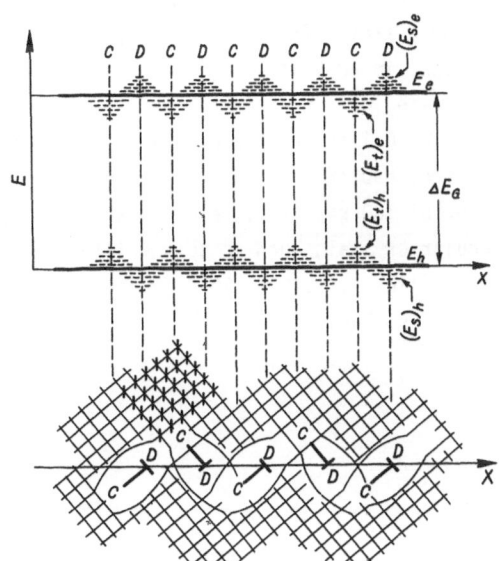

Fig. 5.11. Energy diagram of the formation of local states on grain boundary in compression (C) and dilation (D) regions of the lattice for a molecular crystal; E - self-energy of the charge carrier; $(E_t)_e$, $(E_t)_h$ - trapping states for an electron and a hole respectively; $(E_s)_e$, $(E_s)_h$ - corresponding "antitrapping" states for charge carriers

and above the level of hole conductivity E_h in compressed regions. The positions of these levels are reversed in regions of dilation (cf. Figs.5.9f and 5.11).

These local states are bipolar and act in an equivalent way as traps both for electrons and holes (for small asymmetry in electron and hole trapping in polarization traps see Sect.5.15).

The conductivity levels E_e and E_h in a molecular crystal are of polarization origin (cf. Sect.2.2) and characterize the self-energy of localized charge carriers in a perfect crystal. Hence, the formation of local states around these levels in defect regions of the crystal has, physically speaking, nothing in common with modulation of conduction bands for delocalized charge carriers in covalent or ionic crystals. Besides, the depth E_t of these local states, which are of a different physical nature, considerably exceeds the corresponding deformation values of covalent crystal bands at the same pressure in the compression region. Thus, for instance, the specific widening of the forbidden energy gap ΔE_G in germanium at hydrostatic pressure is only $[d(\Delta E_G)/dp]_T = 5 \times 10^{-6}$ eV/bar [3.22]. On the other hand, specific decrease in width of the energy gap ΔE_G for an anthracene crystal at hydrostatic pressure has a value of $[d(\Delta E_G)/dp]_T = -80 \times 10^{-6}$ eV/bar = -0.08 eV/kbar due to increase in electronic polarization energy. In other words, we have here a value exceeding by a factor of 16 the corresponding value for germanium, and of opposite sign, in addition. It is easy to understand therefore that in defect regions with local pressure of a few kilobar local trapping centers for carriers of ca. 0.1 eV depth are formed. At the limit, it is possible to reduce the energy gap ΔE_G even to zero value at high hydrostatic pressures of the order of 100 kbar (cf., e.g. [3.96]).

For some more closely packed molecular crystals such as tetrathiotetracene (TTT), the $(d(\Delta E_G)/dp)_T$ value is smaller, viz. -15×10^{-6} eV/bar, as reported by SHIROTANI et al [4.18] (cf. also Table 4.3).

*.... the plurality of the world signifies
neither absurdity nor chaos but
a challenge to look for new logical
relationships between phenomena.*

J.L.Locher

In Preface to "The World of M.C.Escher."

5.7 Randomizing Factors Determining Gaussian Distribution of Local States of Structural Origin

Since local states of polarization origin in molecular crystals are due to local lattice imperfections, one could, formally, give a description of the topography of local states by means of the elastic stress field causing molecular displacement from the equilibrium state of the ideal lattice. Each molecule which has been displaced from its initial equilibrium state actually forms a local state with respect to the regular lattice and may possess an altered self-energy value for a charge carrier (or exciton) localized on the given molecule.

If it had been possible to plot a microtopographic map of elastic stress fields and of corresponding deformation displacements of molecules in the crystal, it would produce, at the same time, a microtopograph of all corresponding local states. But if the coordinates of all displaced molecules are known, then the complete energy spectrum of these states can be easily calculated.

This problem can be solved without difficulties for the case of hydrostatic compression of the crystal, when the coordinates of molecules in the deformed lattice are known. In this case the corresponding change ΔP in electronic polarization energy of the new "local" states can be easily found (cf. Sect.5.1).

The problem becomes considerably more complex in the case of particular structural defects in a real crystal. A rough estimate of the coordinates of displaced molecules and of the energy spectrum of the local states in the given faulty structure can be obtained, applying a number of approximations and simplifications, only for such a simple structural defect as a planar stacking fault ribbon in an anthracene crystal (cf. Sect.5.3 and Figs.3.21, 22).

In real crystals with random distribution of structural defects stress fields seem to have such a complicated microtopography that the local states of deformation-caused molecular displacements and the corresponding energy spectrum of these states can be approached only by statistical methods.

Let us briefly consider the randomizing factors causing statistical distribution of the energy spectrum of local states for various types of dislocational and other defects in a real molecular crystal (cf. Chap.3).

Even for the very simplest type of dislocation defect — for a single straight-line edge dislocation — the distribution of the stress field around the dislocation core is far from simple (cf. Fig.3.4). In the quadrant above the dislocation line fields of compression, and below it fields of dilation are formed, but in the lateral quadrants complex fields of simultaneous compression and dilation appear, acting at right angles to each other (cf. also Fig.3.7). Accordingly, the local state energy, as created by this dislocation, depends in a very complex way on the distance from the dislocation core, as well as on the angle with respect to the Burgers vector, and it can assume both positive ($\Delta P > 0$) and negative ($\Delta P < 0$) values (cf. Fig.5.6).

The fields of elastic stress are still more complicated in case of curved-line dislocations forming closed loops (cf. Fig.3.26) or dislocation networks (cf. Figs.3.16,23,24). Here the energy spectrum of local states around such complex formations must obviously be of statistical nature. Thus, e.g., for a closed loop of edge dislocation forming a disc of excess molecules in the plane of the loop (cf. Fig.3.26a) the randomizing factor determining the increase ΔP for the molecules inside the disc is the diameter of the latter (or, more generally, its shape and size). The same applies to a vacancy disc that can be formed via condensation of vacancies (cf. Fig.3.26b). The shape of the potential barrier of local states around such a vacancy disc is also randomized by size and shape of the given disc.

Since dislocations interact through their own fields of elastic stress, the elastic fields between the dislocations are of highly complex configuration and change both in magnitude and in direction (cf. Fig.3.9a). In this case distance r_{12} and relative position of slip planes of both dislocations, as well as distances r_{1i} and r_{2i} between the given molecule i and these dislocations, and the angles between the radius vectors r_{1i} and r_{2i} and the Burgers vector of the dislocation act as randomizing factors for the local state energy spectrum. Dynamic randomization is effected through thermal activation of motion of dislocations. This can change considerably the positions of dislocations with respect to each other, causing corresponding changes in the microtopography of stress fields in their vicinity. This, in its turn, determines statistic distribution and concentration of local states of polarization origin. Such a view is supported by experimental evidence on thermal treatment of defective crystals (cf. Sects.4.3 and 5.14).

At considerable dislocation densities a statistical process of aggregation of dislocations takes place, and formation of assemblies sets in. The latter can be of completely random distribution (dislocation ensembles), of locally random distribution, or arranged in definite alignments with respect to some random distribution parameter of the aggregation (cf. Sect.3.7).

The highest degree of randomization of parameters can be found in dislocation ensembles with a completely stochastic distribution of dislocations. As already mentioned in Sect.3.7.3, such dislocation ensembles and their stress fields can be treated only by means of statistical methods (cf., e.g. [3.51]). For a general description of such ensembles statistically averaged values are used, such as mean density $\bar{\rho}$, mean distance $\bar{\lambda}$ between dislocations, orientational distribution function $\Phi_b(\ell)$, mean amplitude of stress $\bar{\sigma}_A$, etc. (cf. Sect.3.7.3).

For statistical analysis of the properties of a dislocation ensemble, particularly from the aspect of local state formation, it is useful to consider local dislocation densities in limited regions of the crystal. According to this approach the dislocations of the ensemble in the vicinity of some arbitrary chosen point A are subdivided into regions with different local dislocation density [3.51] (Fig.5.12). By counting the dislocations (according to the number of etch pits, or otherwise) it is possible to determine their number in each sector (cf. Fig.5.12) and thus describe the density in the given local region by means of some function $\rho(r,\theta)$ (r, θ denoting polar coordinates) [3.51]

$$\rho(r,\theta) \rightarrow \frac{n_{ij}}{S_{ij}} = \frac{n_{ij}j^*}{\pi\bar{\lambda}^2(2i + 1)} \quad , \tag{5.4}$$

where n_{ij} is the number of dislocations in the sector within the ring i and the azimuthal zone j (cf. Fig.5.12); S_{ij} is the area of this sector, j^* is the total number of azimuthal zones, $i = 1,2,3,\ldots, R_0/\bar{\lambda}$; R_0 is the size of the crystal.

By means of such a statistical approach to the description of a dislocation ensemble STRUNIN [3.50] obtained theoretically the distribution of stress values $\sigma_{k\ell}$ and the dispersion $D\sigma_{k\ell}$ of this distribution

$$D\sigma_{k\ell} = \frac{\mu^2 b^2}{K\pi(1-\nu)^2}\,\bar{\rho}\sum_{i=1}^{R_0/a}\frac{1}{i} \approx \frac{\mu^2 b^2}{K\pi(1-\nu)^2}\,\bar{\rho}\left(\ln\frac{R_0}{a}+0,557\right) \quad , \tag{5.5}$$

where K is a constant depending on k and ℓ; a is the lattice parameter.

The dispersion of internal stress $\sigma_{k\ell}$ can be seen to be proportional to the mean dislocation density $\bar{\rho}$ and to depend logarithmically on the size R_0 of the crystal.

For statistical analysis of the properties of a dislocation ensemble in some local region one just replaces the mean dislocation density $\bar{\rho}$ in (5,5) by the dislocation distribution in the rings around point A (cf. Fig.5.12). Such an approach leads to the following form of expression (5.5) [3.51]:

$$D\sigma_{k\ell}\left(\frac{\mu b}{\sqrt{K\pi}(1-\nu)\bar{\lambda}}\right)^{-2} = n_0 + \sum_{i=1}^{R_0/\bar{\lambda}}\frac{n_i}{i(2i+1)} \quad , \tag{5.6}$$

where n_0 and n_i are the numbers of dislocations in the central circle and in the ith ring, respectively (cf. Fig.5.12).

Expression (5.6) describes the statistical properties of the ensemble both from the local density aspect, as well as from the aspect of dislocation distribution in the ensemble as a whole.

Unlike a dislocation ensemble, aggregations of the type, presented in Fig.3.10a, reveal elements of some internal structure and even of symmetry. However, even such aggregations, particularly the one presented in Fig.3.10b, are randomized to a high degree, considering that the number of possible con-

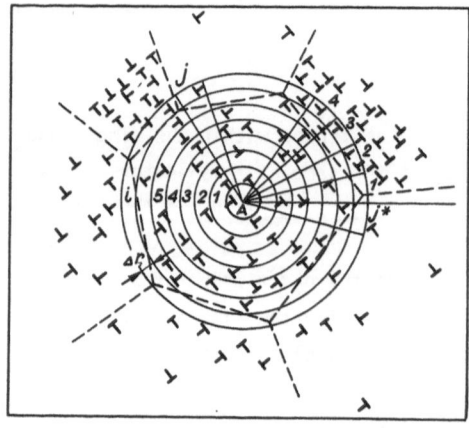

Fig. 5.12. Schematic picture of a dislocational ensemble near a certain fixed point A. Broken lines partition off regions with random distribution of dislocations but with different local density [3.51]

figurations in such a dislocation "tangle" is practically unlimited and can form an infinite variety of different configurations of complex stress fields.

The most regularly arranged aggregations of dislocations are vertical alignments of edge dislocations forming as the result of polygonization linearized, small- and middle-angle boundaries (cf. Figs.3.9b and 3.11). The basic randomizing factor in this case is the mean distance D between dislocations or, correspondingly, the angle between disoriented grain boundaries of the crystal (cf. Sect.3.8). The value of θ can change within a wide range — from a few seconds to tens of degrees. For this reason the relative compression of the lattice $\delta p/p$ at the boundary (cf. Table 3.8) and the corresponding depth ΔP of charge carrier traps are of randomized statistical nature. The statistical distribution of these local centers depends also in a random way on boundary length and on the existence of other boundaries, dislocation aggregations or vacancies in the vicinity.

Comparatively large local compression regions are formed at horizontal dislocation alignments (cf. Fig.3.10). In this case the mean distance D between dislocations and the length of the horizontal wall act as randomizing factors.

The energy spectrum of local states in stacking faults (cf. Sect.3.9) is randomized by the statistical distribution of width and length of faulty stacking ribbons, as well as by random formation of various collective configuations of planar stacking faults, such as a sequence or stack of faulty ribbons (cf. Sect.3.11).

The appearance of single fixing dislocations in nonbasal planes of a crystal is, generally, an essential randomizing factor. They can act as a "pin" for other dislocations, small-angle boundaries or stacking faults in the basal plane. Such "pins" create potential barriers for other defective structures in the vicinity of which local aggregations of dislocations can arise. The latter may form a complex topographical pattern of microfields of elastic stress (cf. Sect.3.11).

We see thus that the scope of randomizing factors which determine the formation of various structural defects in molecular crystals is considerably large. In turn, this circumstance appears to be decisive in determining the statistical nature of the energy spectrum of local states of polarization origin. It is natural therefore that for a phenomenological description of their energy spectrum Gaussian approximation would be the most adequate [cf. in Sect.3.1 (3.5) and Fig.3.2].

Gaussian distribution of local state energy spectrum was earlier proposed by us as a general working hypothesis [1.66] (cf. Sect.3.1). It may be suggested that the feasibility of such an approach has now been qualitatively supported by the preceding analysis of the statistical nature of the formation of local states of structural origin. This approach logically follows from general rules of statistics (cf., e.g. [5.7]). If any physical quantity is characterized by a distribution function and if statistical deviations are caused by an unlimited number of quite random, independent and uncontrollable events; if, furthermore, the deviations from the mean value are symmetrical with respect to the latter and the frequency of small deviations considerably exceeds that of large deviations, then the given value is, as a rule, most adequately, described by normal Gaussian distribution. It has the shape of a symmetrical bell-like curve (cf. Figs.3.1,2) which can be uniquely defined by means of two parameters: the mean value ξ of the respective statistical quantity and the dispersion of its distribution $D\xi$ [cf. (3.5), where we have $\bar{\xi} = E_t$ and $D\xi = \sigma^2$].

It must be borne in mind that the Gaussian approximation, by force of its very nature, is purely phenomenological. It might be worth recalling, in this connection, a witty remark by SQUIRES [5.8], saying that experimentators believe in the Gaussian distribution law relying on the proof by mathematicians, whilst mathematicians rely on its experimental evidence (cf. also [Ref.5.9, p.363]).

The discovery of this universal, although theoretically not sufficiently substantiated law of the most probable statistical distribution of random values is, most likely, due to Gauss' penetrating intuition. The law was initially postulated by Gauss as a basis for analysis of measurement errors of earth magnetism, and its universal character was appreciated only later. The postulation of this most important law for statistical analysis was, apparently, stimulated by the versatility of scientific interests and search for an organic link between the pure and applied aspects of mathematics, being a feature of creative thought most characteristic for both greatest mathematicians of all times — Friedrich Gauss and Norbert Wiener.

In a large number of statistical problems the Gaussian distribution is usually assumed a priori, and its parameters $\bar{\xi}$ and $D\xi$ are estimated from empirical data [5.7]. In our case the situation is more favorable. As will be shown further (cf. Sect.5.9), Gaussian approximation will be introduced through a phenomenological theory of space charge limited currents (cf. [1.68]). The latter will finally lead us to an experimental determination of distribution parameters E_t and σ of local charge carrier trapping states, at the same time establishing the validity of the Gaussian approximation itself (cf. Sect.5.10).

A more rigorous approach would eventually demand more elaborate argumentation. First, it would have to be shown that already the distribution of internal stress fields $\sigma_{k\ell}$ of a defective crystal might be adequately described by the Gaussian distribution law. Secondly, that the deformational displacements of the molecules in defect regions are unambiguously determined by the distribution of internal stress fields. And finally, it ought to be shown, and more strictly than in [1.66] (cf. Sect. 3.1), that the dispersion of molecular displacements uniquely determines the dispersion in the Gaussian distribution of the energy spectrum of local charge carrier trapping states in the defect regions of a deformed crystal.

Such successive proof of the applicability of Gaussian approximation is, unfortunately, outside our possibilities. There does not exist, at present, a theoretical method of exactly describing the statistical distribution of stress fields in a real crystal containing such a variety of structural defects as discussed above. Neither are there any direct experimental methods of studying the structure of the stress fields in such a crystal [3.51].

An encouraging approach has been proposed by STRUNIN [3.50] who succeeded in obtaining probability distribution for the internal stress tensor $\sigma_{k\ell}$ for dislocations depositioned at random. He shows that these components are situated symmetrically with respect to $\sigma_{k\ell} = 0$, and the dispersion of their distribution agrees with (5.5). Strunin's results reveal an interesting peculiarity in the given distribution, which may be of great importance for explaining the appearance of deep structural traps in molecular crystals. It turns out that the distribution of $\sigma_{k\ell}$ can be approximated by the normal distribution only in the vicinity of $\sigma_{k\ell} = 0$, i.e., for small stresses, whereas in the tail regions of the Gaussian bell there is a positive excess, with $\sigma_{k\ell}$ distribution considerably differing from the normal one. In other words, the probability of the occurrence of larger values for the components of the internal stress tensor $\sigma_{k\ell}$ is higher than at normal distribution. The positive excess in distribution is of the order of [3.50]

$$E_{\sigma_{k\ell}} \approx \frac{1}{\bar{\rho}a^2 \left(\ln \frac{R_0}{a}\right)^2} . \qquad (5.7)$$

This means that at random disposition of dislocations in an ensemble the excess $E_{\sigma_{k\ell}}$ in distribution of internal stresses can be sufficiently large, being inversely proportional to the mean dislocation density $\bar{\rho}$ and the square of the logarithm of crystal size R_0.

For ascertaining the cause of positive excess of large stress components it is necessary to take into account the logarithmic divergence of the series describing dispersion $D\sigma_{k\ell}$ of $\sigma_{k\ell}$ distribution in (5.5). This means that not only the most closely situated dislocations contribute towards the $\sigma_{k\ell}$ value, but also dislocations positioned at a considerable distance from the point under consideration. In other words, in dislocation ensembles of a defective crystal noncompensated distant-acting highly inhomogeneous stress fields can be formed. The dispersion of the latter can reach considerable values [3.51].

The field of internal stresses $\sigma_{k\ell}$ in a crystal can be studied experimentally by observing the distribution of starting stress values τ_c of dislocation motion and their path lengths till stopping point. STRUNIN [3.50], for instance, used experimental results obtained by PREDVODITELEV et al. [5.10] on starting stress values τ_c of the motion of fresh dislocations upon compression of a NaCl crystal. He found that out of 1600 observed dislocations 1426 satisfy normal distribution conditions, whilst 176 dislocations, i.e., 11 percent of the total amount, exhibit positive excess. This means that in the given case about one tenth of all the stress components $\sigma_{k\ell}$ determining the spread of starting stress τ_c have a value higher than 3 $D\sigma_{k\ell}$, thus producing the above-mentioned excess of normal distribution.

Hence, internal uncompensated distant-acting stress field components of a dislocation ensemble can create strong local fluctuations of stress at the tail end of the Gaussian distribution; they, consequently, can be the source of formation of deep structural traps for charge carriers.

Other forms of collective dislocation defects discussed above are subject to their own randomizing factors and appear to form their own kinds of distributions which need not necessarily be Gaussian. This applies to such defects as vertical and horizontal alignments of dislocations randomized by some particular parameter, dislocation aggregations and stacking faults. If, however, such distributions superimpose, then one of the most remarkable theorems of statistics — Lyapunov's central limiting theorem — comes into force [5.7].

Lyapunov showed that the distribution law for a normalized sum of independent random summands $\xi_1 + \xi_2 + \ldots + \xi_n$ is close to Gaussian for large values of n and small values of the ratio [5.11]

$$\frac{\sum_{k=1}^{n} M|\xi_k - M\xi_k|^3}{\left(\sum_{k=1}^{n} D\xi_k\right)^{3/2}} \, . \tag{5.8}$$

In this ration $M\xi_k = \bar{\xi}_k$ denotes the mean value (the center of distribution) of the random quantity ξ_k.

Thus, according to Lyapunov's central limiting theorem the sum of a large number of random summands is distributed according to the Gauss law, irrespective of the distribution of the summands themselves. (For a proof of Lyspunov's theorem cf., e.g. [5.7]).

Expression (5.8) does not work, if the dispersion of a small number of summands exceeds the dispersion of the other summands. It also does not apply in some special cases, for instance, in case of random quantities $\xi(t)$ in processes of low probability, described by Poisson distribution (cf. [Ref.5.11, p. 602]).

The central limiting theorem of statistics in the formulation of which, apart from Gauss, such great mathematicians as Moivre, Laplace, Tchebyshev and Lyapunov have taken part, demonstrates the great, almost universal importance of the Gaussian law in the description of random processes.

The Gaussian law is also the best approximation to the binomial distribution at large values of N (cf., e.g. [5.12]). No wonder therefore that in all statistical problems containing a large number of random quantities, the Gaussian distribution yields, as a rule, optimal approximation.

We see thus that the Lyapunov theorem provides a common Gaussian approximation for the description of local deformations of a molecular crystal, as well as for corresponding local changes of electronic polarization energy ΔP, created by various lattice defects possessing their own distribution functions. According to this theorem all these distributions can be enveloped by a common Gaussian curve.

Such a summary Gaussian distribution of stress fields $\sigma_{k\ell}$ in a crystal, symmetrical with respect to $\sigma_{k\ell} = 0$, comprises both local compression and local dilation regions of the lattice, created by structural defects. Corresponding local states of polarization origin form, according to the model discussed in Sect.3.1, two identical Gaussian distributions in the energy diagram of the crystal. These are the distributions of $G_e(E)$ type, one centered at electron conductivity level E_e, the other at hole conductivity level E_h (cf. Figs.3.1,2). The first distribution can be obtained from (3.5) at $E_t = 0$, the second one at $E_t = \Delta E_G$.

If the separate summands do not satisfy condition (5.8) and form an excess outside the bounds of the distribution $G_e(E)$, they can form a set of Gaussian distribution with the center inside the forbidden energy gap, thus exhibiting $G_g(E)$-type distribution (cf. Fig.3.2).

The summary envelope of the $G_e(E)$ and $G_g(E)$ distributions [cf. (3.5) and Fig.3.2] inside the forbidden energy gap can, in this case, be described by the expression

$$N(E) = \sum_i \frac{(N_t)_i}{\sqrt{2\pi}\sigma_i} \exp\left[-\frac{(E-E_t)_i^2}{2\sigma_i^2}\right] \quad . \tag{5.9}$$

The possibility of determining experimentally several Gaussian distribution sets of local states inside the forbidden energy gap may be limited by the resolving power of the experimental method employed. It will be dealt with in greater detail in Sect.5.10.

5.8 Investigation of Local Trapping States by Method of Space Charge Limited Currents

5.8.1. General Considerations

The method of space charge limited currents (SCLC) is the most widely used and reliable one for studying the energy spectra and density of local charge carrier trapping states in insulators and high-resistance semiconductors. The method is based on monopolar injection of charge carriers from an ohmic contact into the bulk of the insulator under study.

The problem of electron injection from an ohmic contact into an insulator was first considered in the classical work by MOTT and GURNEY [2.45]. In principle it is completely analogous to injection of electrons from a thermocathode into vacuum.

The ideas of Mott and Gurney were subsequently developed into an accomplished and consistent theory of space charge limited currents (SCLC) in the works by ROSE [5.13,14] and LAMPERT [1.76,5.15,16]. This work formed the basis for a whole field of electrophysical research connected with the study of injection currents in insulators. The results of a number of later studies demonstrated that injection currents can serve as an excellent "probe" for obtaining information on local charge carrier trapping states in various high-resistance semiconductors and insulators, including molecular crystals (cf. e.g. [1.75,76,2.134,3.1,5.17]).

A characteristic feature of SCLC or monopolar injection currents (these terms are used synonymously in literature) is the superlinear dark current density j dependence on voltage U [1.76,5.14]

$$j \sim U^n \quad , \qquad\qquad\qquad\qquad\qquad (5.10)$$

where $n \geq 2$, as well as superlinear dependence of j on thickness of sample L

$$j \sim \frac{1}{L^\ell} \quad , \qquad\qquad\qquad\qquad\qquad (5.11)$$

where $\ell \geq 3$.

The structure and shape of the $j = f(U)$ and $j = f(L)$ curves contain useful information on local states, their energy spectrum and average density. The SCLC method permits detection of trapping states with densities as low as 10^{12} cm^{-3} [1.76] (cf. Table 5.8).

A profound understanding of the properties of monopolar injection currents may be gained through comparatively simple phenomenological analysis [1.76]. This kind of analysis, introduced by ROSE [5.13,14], attempts to describe only the most general features of the mechanisms of SCL current flow under conditions of monopolar injection. Such approach considerably simplifies SCLC theory, providing at the same time a sufficiently complete picture of the physics of monopolar injection, as well as information on the distribution and density of local states [1.76]. It is significant that LAMPERT and MARK dedicated their excellent monograph on current injection in solids [1.76] to Albert Rose, the founder of SCLC theory who "chose simplicity over precision and thereby gave to us all insight".

The quantities that the SCLC theory must deal with are the density of free and trapped charge carriers, free carrier drift mobility, and electric field intensity. The phenomenological analysis ignores the spatial variation of these quantities and, instead, concerns itself only with their respective averaged values and relations connecting them. Such an approach is justified, since for planar current flow all of these quantities change relatively little along the thickness L of the sample, almost throughout the whole distance from cathode to anode. Hence all quantities considered are sufficiently well represented by their average values [1.76].

The final result is that the relations derived by phenomenological analysis, e.g., current-voltage (CV) characteristics, their temperature dependences, etc., are correct in their functional dependences on all physical quantities and, at worst, contain only an incorrect numerical coefficient, the inaccuracy of the latter rarely exceeding a factor of 2 [1.76]. This fully justifies the application of such a simplified phenomenological analysis.

5.8.2 Injecting and Blocking Contacts

Before we pass over to a discussion of SCLC theory and its applications, a few words ought to be said on the problem of contacts.

In order to provide monopolar injection it is obviously necessary to obtain an injecting contact on one surface and a blocking one on the opposite surface of the crystal.

It ought to be stressed from the very outset that the properties of an injecting contact, as well as the mechanism of injection are considerably different in the case of traditional low-resistance semiconductors and of insulators of the molecular crystal type. It can lead to serious misunderstandings if we call every contact that provides good injection an "ohmic" one, without stipulating the concrete mechanism of injection. This is particularly essential in the comparison of traditional semiconductors and insulators.

An ohmic contact is, generally speaking, a contact constituting a reservoir of charge carriers which can, if necessary, pass over into the bulk of a semiconductor or an insulator [5.14].

An ideal ohmic contact satisfies the condition [1.75]

$$n(0) = \infty \quad , \tag{5.12}$$

where $n(0)$ is the charge carrier concentration at the boundary between contact and sample.

However, practically any contact is an ohmic one if the following inequality holds [1.75]

$$n(0) \gg \bar{n} \quad , \tag{5.13}$$

where $\bar{n}$ is the mean density of free charge carriers in the bulk of the sample.

For better comparison of the concept "ohmic", as used for traditional semiconductors and insulators, let us briefly consider the process of the formation and properties of a contact, typical for both cases.

When an electric contact is formed between a metal electrode and a semiconductor of germanium type, electron transfer by diffusion takes place, as a rule. As a result the difference between the Fermi level of the metal $(E_F)_{Me}$ and that of the semiconductor $(E_F)_{SC}$ become equalized, and a contact potential barrier $\Delta\varphi_C$ is created

$$\Delta\varphi_C = (E_F)_{Me} - (E_F)_{SC} = \varphi_{Me} - \varphi_{SC} \quad ,$$

where φ_{Me} and φ_{SC} are the work functions of metal and semiconductor, respectively. Depending on the relations between the values of φ_{Me} and φ_{SC} and on the type of conductivity, either an ohmic or a blocking contact is created. Thus, for instance, if we have an n-type semiconductor, with $\varphi_{Me} < \varphi_{SC}$, an ohmic contact is formed (Fig.5.13a), but with $\varphi_{Me} > \varphi_{SC}$ a blocking one.

A characteristic feature of an ohmic contact presented in Fig.5.13a is that it remains an ohmic one at arbitrarily low electric field intensity and retains its "ohmicity" at any value of the latter. As to a blocking contact (Fig.5.13b) its potential barrier value depends on the bias of the external electric field. The latter cannot, however, convert such a contact into an ohmic one.

Of quite different nature is the contact of a metal electrode with an insulator of molecular crystal type. First of all, a metal usually forms a neutral contact with an insulator [5.14]. In the absence of an external field such a contact is, as a rule, a blocking one. Its formation is not accompanied by equalization of Fermi levels and creation of a contact poten-

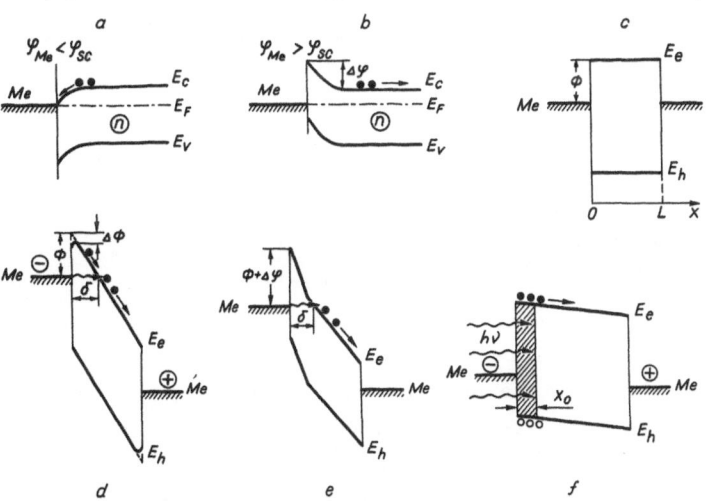

Fig. 5.13a-f. Schematic energy diagrams of a metal-semiconductor (a,b) and of a metal-insulator (molecular crystal) (c-f) contact; (a) ohmic injecting contact of n-type semiconductor; (b) blocking contact of n-type semiconductor; (c) neutral blocking contact of a molecular crystal; (d) injecting quasi-ohmic contact of a molecular crystal; (e) injecting contact of a molecular crystal in the presence of a near-surface potential barrier $\Delta\varphi$; (f) photogenerated reservoir contact in a molecular crystal; x_0 denotes depth of penetration of light

tial barrier (cf. Fig.5.13c). Application of an external electric field $\mathscr{E}$, which can reach rather high values in the case of an insulator, causes considerable incline of conductivity levels of the crystal (Fig.5.13d) and, as field intensity reaches a value of $\mathscr{E} \gtrsim \mathscr{E}_{inj}$, noticeable tunneling of carriers through the barrier layer sets in. Accordingly, at electric field intensity values $\mathscr{E} \gtrsim \mathscr{E}_{inj}$ such a blocking contact becomes an ohmic one and statisfies condition (5.13) which is still less rigorous for insulators, owing to the small value of $\bar{n}$. With tunneling of charge carriers such contact can fully provide the injection level necessary for SCLC conditions. It can therefore, according to inequality (5.13), be considered as a full-fledged "ohmic" contact. However, as can be seen from comparison of diagrams a and d in Fig.5.13, the contact properties and the mechanism of injection are essentially different in both cases. It seems therefore appropriate to use a more flexible term — "quasi-ohmic" contact, for desribing the injecting properties of a neutral injecting contact (cf. Fig.5.13d).

In addition to tunneling, "quasi-ohmicity" of a neutral contact (Fig. 5.13d) can be provided by means of thermoelectric emission of electrons from the metal into the insulator. This can take place if the height Φ of the barrier is not too large. Thermoelectric emission current is known to have a tendency towards saturation with increase in field intensity $\mathscr{E}$. Saturation sets in at field intensities (in V/cm) according to the condition [5.14]

$$\mathscr{E}_S = \frac{v}{4\mu} \quad , \qquad\qquad (5.14)$$

where v is the thermal velocity of electrons; μ is their mobility.

At values of $v \approx 10^6$ cm $\cdot$ s^{-1} and $\mu = (10^{-2}-1)$ cm^2/(V $\cdot$ s) we get limiting values of $\mathscr{E}_S$ of the order of 2.5×10^5-10^7 V/cm. Consequently, within the range of electric field intensities at which SCLC are usually studied (10^4-10^6 V/cm) "quasi-ohmicity" of the injection current can be maintained both by means of tunneling, as well as by thermoelectron emission.

It must be borne in mind that effectivity of thermoelectron emisson increases with growing field intensity $\mathscr{E}$, due to the Schottky effect (cf., e.g. [5.17]) which lowers the height of the potential barrier by a value of $\Delta\Phi$ (cf. Fig.5.13d), equal to

$$\Delta\Phi = e^{3/2} \mathscr{E}^{1/2} \quad . \qquad\qquad (5.15)$$

In some cases the "neutrality" of a metal-organic crystal contact is distorted by the near-contact field causing a certain bend of energy levels at the surface. Such a bend is, as a rule, connected with the existence of surface levels for carrier trapping in the crystal (cf. Sect.5.4). Such a near-surface barrier can, depending on its sign and thickness, enhance or inhibit corresponding injection processes of tunneling or of thermoelectron emission (cf. Fig.5.13e).

From metal electrodes, in which the Fermi level is lower than the middle of the forbidden energy gap of the molecular crystal, hole injection is usually the dominating one. In anthracene-type crystals this condition holds for most high-work-function metals, such as Au, Ag, Al, Pb, etc., which all are typical injectors of holes. Electron injection can be achieved from metal electrodes with a lower work function φ_{Me} having $(E_F)_{Me}$ above the middle of the energy gap. Alkali metal (K, Na) amalgams are frequently used as effective electron injectors [2.133,5.3,5]. A number of various electrolytic contacts with monopolar injection capacity have also been suggested (a review of electrolytic contacts for monopolar injection in an anthracene crystal can be found in [Ref. 1.75, p.6]).

It has been frequently observed, especially in thin-layer sandwiched samples of Me/molecular crystal/Me type, that the injection efficiency is not actually determined by the work function of the metal, but by effective barriers to bulk injection formed by charge carriers trapped near the injecting contact. This is due to the widely recognized circumstance that the trap density is likely to be much higher near the insulator surface than in the bulk (cf. Sect.5.11). It has been observed by a number of authors (see, e.g. [1.104,2.134,3.1,5.18,19]) that the higher barrier for bulk injection usually occurs on the nonsubstrate side of the sample, namely, at the upper electrode in vacuum-evaporated thin-layer cells. The blocking effect in this case is attributed to very high trap density near the upper electrode causing effective trapping at low voltages of practically all injected charge carriers [5.18,19]. Thus an actually injecting contact may be converted into a blocking one owing to asymmetric spatial distribution of trapping centers. The region of high trap density near a metal-insulator interface provides effective blocking for the bulk at low voltages [5.19]. Only at higher field intensities, when these near-electrode trapping states are filled, a bulk-controlled space charge limited current flow occurs. This causes rectification at low voltages [1.104,2.134,3.1,5.18,19], as well as considerable photovoltaic effect [5.18,19]. Such a quasi-blocking contact is, as a rule, very effectively formed if the upper electrode is aluminium in evaporated thin-layer cells [1.104,2.134].

An extended SCLC theory for spatially nonuniform trap distribution in insulators has been recently proposed by BONHAM [5.20] (cf. Sect.5.11). He discussed also in some detail the problem of asymmetric trap distribution in the near-electrode region, giving rise to blocking and rectification effects at low voltages.

It is also easy to create a photo-generated reservoir contact in molecular single crystals. In this case charge carriers (electrons and holes) are generated simultaneously by strongly absorbed light in a thin layer under a semi-transparent metal electrode within the spectral range of $h\nu \gtrsim \Delta E_G$ and at absorption coefficient values $\alpha = 10^4$-10^5 cm^{-1}, which corresponds to penetration depth of light $x_0 \approx 10^{-4}$-10^{-5} cm (cf. Fig.5.13f). Depending on the field direction it is possible in this case to inject either electrons or holes into the bulk of the crystal [1.76].

The usual sketching of schematic energy diagrams, like the ones presented in Fig.5.13, might suggest that the contact problem is relatively simple and straightforward. Actually, in experimental practice the problem of fabricating a suitable pair of injecting and blocking contacts for monopolar injection in molecular crystal samples is a rather difficult one. Frequently a significant portion of the research program of the electronic properties of high-resistance semiconductors and insulators is spent solving, or trying to solve, the contact problems [1.75,76,2.134,3.1]. Even updated technology of contacts is in a primitive state in the molecular crystal field and may be regarded more as art than science.

Although the problem of the injecting contact can hardly be avoided by the experimentalist, it generally need not be faced in theoretical approach, at least in the framework of phenomenological SCLC theory of constant-field approximation. As a matter of fact, the behavior of monopolar SCL injection currents is, generally, completely dominated by the bulk properties of the sample. Thus it is possible to construct an adequate SCLC theory of monopolar injection neglecting the diffusion current contributions to the total injection current. Concomitantly the problem of the detailed structure of the injecting contact is entirely circumvented through the use of an idealized boundary condition, namely, vanishing of the electric field intensity at the injecting electrode interface. Although this boundary condition yields unphysical properties in the immediate vicinity of the injecting contact [the injected free-carrier concentration going to infinity at the electrode interface, cf. (5.12)] it nevertheless gives a correct description of the current-flow properties over the bulk of the sample and thus makes SCLC theory applicable for the study of local states in the bulk of the crystal [1.75,76,68]. In this case an enormous simplification of the theoretical problem is attained at a very small price indeed, namely, an unphysical description of the current flow properties in the immediate vicinity of the injecting electrode [1.76].

Except under very special circumstances, e.g., at low applied voltages and very small cathode-anode spacings, contact properties and diffusion currents play a negligible role in the determination of CV characteristics (especially in the superlinear region) and can be accordingly ignored in phenomenological SCLC theory (cf. Sects.5.8.3 and 9.2).

5.8.3 Conventional SCLC Theories of Discrete and Exponential Approximation of Trap Distribution

Traditionally charge carrier trapping states in real insulators are treated in the framework of two extreme SCLC theories, one assuming discrete, mono-energetic, the other quasi-continuous exponential trap distribution.

The approximation of discrete trap distribution provides a satisfactory working description in SCLC terms for the investigation of quasi-discrete trapping states occurring mainly in single-crystal materials.

On the other hand, SCLC theory of exponential approximation has been wide-ly used for the investigation of trapping states with quasi-continuous energy spectrum of wide dispersion dominating, as a rule, in amorphous and polycrys-talline samples, e.g. evaporated thin layers of poorly defined crystallinity. These extreme approaches are *ad hoc* assumptions mainly to simplify theoretical treatment of corresponding SCLC theories. As will be shown later, they provide a rather poor working description for the vast variety of trapping states in real molecular crystals having presumably Gaussian character of energy spec-trum with different values of dispersion parameter (cf. Sect.5.8.5).

A simple phenomenological treatment of SCLC theories for both discrete and exponential approximations is given by ROSE [5.14]. In the monograph by LAMPERT and MARK [1.76] the subject is covered at several levels, ranging from purely phenomenological to the purely analytical. In the latter ap-proach, the characteristic differential equations are treated approximately and, where possible, more rigorously. Earlier applications of both tradi-tional SCLC theories for studies of organic solids are discussed in the review article by HELFRICH [1.75] (see also [2.134,3.1,3]).

The reader may find detailed information in the above-mentioned sources. In this section we shall give only illustrative energy diagrams and final expressions of the phenomenological SCLC theories of discrete and exponen-tial trap distributions, necessary for further analysis of applicability limits for both approaches.

SCLC Theory for an Insulator With Discrete Trap Distribution [1.75,76,5.14]

A typical energy diagram of an insulator with a single set of *discrete* trap-ping states for electrons, located at energy E_t, is given in Fig.5.14.

At thermal equilibrium (cf. Fig.5.14a) free electron concentration n_0 is given by the familiar expression

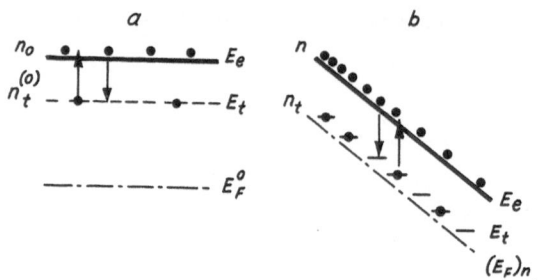

Fig. 5.14a,b. Energy diagram of an insulator with a single set of discrete traps at E_t for electrons: (a) at thermal equilibrium, characterized by the thermodynamic Fermi level E_F^0; (b) quasi-thermal equilibrium at steady-state electron injection, characterized by the quasi-Fermi level $(E_F)_n$

$$n_0 = N_e \, \hat{\exp}\left(-\frac{|E_F^0, E_e|}{kT}\right) \quad , \tag{5.16}$$

where N_e is the effective density of states on conductivity level E_e; E_F^0 is the thermodynamic Fermi level; $|E_F, E_e|$ is the energy interval (absolute value) between the corresponding levels.

The equilibrium concentration $n_t^{(0)}$ of filled electron traps at level E_t (cf. Fig.5.14a) is then given by the familiar Fermi-Dirac expression

$$n_t^{(0)} = \frac{N_t}{1 + \frac{1}{g}\exp\left(\frac{E_t - E_F^0}{kT}\right)} = \frac{N_t}{1 + \frac{N_e}{gn_0}\exp\left(\frac{E_t - E_e}{kT}\right)} \quad , \tag{5.17}$$

where N_t is the total concentration of traps and g the degeneracy factor (statistical weight) for traps.

Now, as clearly seen from Fig.5.14a, this equilibrium trap occupancy results from a balance between capture of electrons (downward arrow), and their thermal reemission into conductivity level (upward arrow). The most crucial point for the development of injection theory is the fact that the presence of a moderate electric field will not affect these elementary microscopic processes of electron capture and thermal reemission. Thus, in the presence of the applied field, if it is not too strong, the balance between free and trapped electrons is altered *only* through the change in free-electron concentration accompanying injection [1.76]. In other words, the balance between free and trapped electrons is reached at steady-state injection as if the crystal were in thermal equilibrium (cf. Fig.5.14b) only with the free- electron concentrationn, achieved under injection, instead of the true thermal equilibrium concentration n_0. The corresponding electron quasi-Fermi level $(E_F)_n$ is related to n, by definition, exactly as n_0 is related to E_F^0 in (5.16), viz.

$$n = n_i + n_0 = N_e \exp\left(- \frac{|(E_F)_n, E_e|}{kT}\right) \quad, \tag{5.18}$$

where n_i is the average injected excess free-electron concentration.

The trapped-electron concentration n_t is then given by an expression analogous to (5.17)

$$n_t = n_t^{(i)} + n_t^{(0)} = \frac{N_t}{1 + \frac{1}{g} \exp\left(\frac{E_t - (E_F)_n}{kT}\right)}$$

$$= \frac{N_t}{1 + \frac{N_e}{gn} \exp\left(\frac{E_t - E_e}{kT}\right)} \quad, \tag{5.19}$$

where $n_t^{(i)}$ is the average injected excess trapped-electron concentration.

An electron trap at E_t is said to be shallow if E_F lies below E_t (cf. Fig.5.14), viz.

$$\frac{E_t - E_F}{kT} > 1. \tag{5.20}$$

In this case the ratio n/n_t equals

$$\frac{n}{n_t} = \frac{N_e}{gN_t} \exp \frac{E_t - E_e}{kT} = \theta, \tag{5.21}$$

where θ is a constant, independent of injection level, as long as the trap remains shallow, according to (5.20). Phenomenological SCLC theory gives the following final expression relating the current density j, voltage U, and sample thickness L, for a single set of shallow discrete trapping states:

$$j \approx \theta \varepsilon \mu \frac{U^2}{L^3} \quad, \tag{5.22}$$

where μ is the free-electron drift mobility and ε the dielectric constant. (The correct, analytical derived result differs only by a numerical factor 9/8 [1.76]).

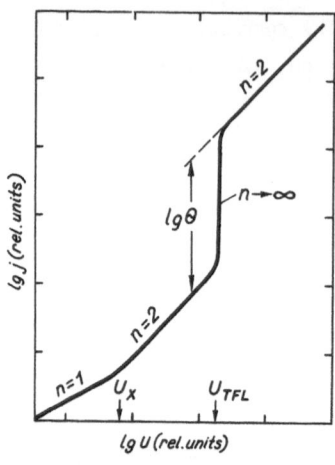

Fig. 5.15. Prototype current-voltage characteristic on log-log plot for monopolar injection of charge carriers into insulator
possessing a single set of discrete trapping states (cf.Fig.5.14)

Prototype current-voltage (CV) characteristic, according to (5.22), is
presented in Fig.5.15.

As may be seen from Fig.5.15, the given CV characteristic falls into
four subregions with different slope values $n = d(\log j)/d(\log U)$.

In the first region, at sufficiently low voltages when $n_i < n_0$, Ohm's
law is observed. The second region begins at the voltage U_x when $n_i \gtrsim n_0$,
and a crossover from Ohm's law to the SCLC shallow-trap square-law (5.22)
takes place. The approximate U_x value is given by

$$U_x \approx \frac{en_0L^2}{\theta\varepsilon} \quad . \tag{5.23}$$

In the third extremely narrow region a change from square-law to trap-filled
limit takes place, as the quasi-Fermi level $(E_F)_n$ crosses the discrete trapping level E_t, causing a step-like increase in current by a value of θ with
$n \to \infty$. The corresponding trap-filled limit voltage U_{TFL} is given by

$$U_{TFL} = \frac{eN_tL^2}{\varepsilon} \quad . \tag{5.24}$$

Expression (5.24) is often used for determination of total trap concentration N_t from experimental CV characteristics.

Finally, in the fourth region at voltages $U \gtrsim 2U_{TFL}$ the CV characteristic
follows the square law of a trap-free insulator

$$j = \frac{9}{8} \varepsilon\mu \frac{U^2}{L^3} \quad . \tag{5.25}$$

This law is often called Child's law for solid-state injection. In the case of two or more sets of discrete traps one should observe a step-like CV characteristic with the prototype characteristic (Fig.5.15) as a basic building block of such complex $j = f(U)$ dependence [1.76].

For an insulator with shallow traps it is convenient to define the concept of effective drift mobility μ_{ef}.

At steady-state monopolar injection the population n_t of the electron traps will be in quasi-thermal equilibrium with n, as discussed above. Then the entire body of injected electrons $n + n_t \approx n_t$, will have the following drift mobility

$$\mu_{ef} = \frac{n}{n_t} \mu = \theta\mu \quad . \tag{5.26}$$

In terms of μ_{ef} we have

$$(t_x)_{ef} = \tau_\Omega \quad , \tag{5.27}$$

where $(t_x)_{ef}$ is the effective transit time of a charge carrier between cathode and anode of the entire body of injected charge at voltage U_x

$$(t_x)_{ef} = \frac{L^2}{\mu_{ef} U_x} \quad ; \tag{5.28}$$

and τ_Ω is the ohmic relaxation time given by

$$\tau_\Omega = \frac{\varepsilon}{en_0\mu} = \frac{\varepsilon}{\sigma_0} \quad . \tag{5.29}$$

SCLC Theory for an Insulator With Exponential Trap Distribution [1.75,76,5.14]

According to the model of exponential approximation (cf. Fig.5.16), the energy dependence of trap distribution N(E) can be described by the following expression:

$$N(E) = \frac{N_t}{kT_c} \exp\left(- \frac{|E,E_e|}{kT_c} \right) \quad , \tag{5.30}$$

where N_t is the total concentration of traps and T_c is the so-called characteristic temperature of trap distribution.

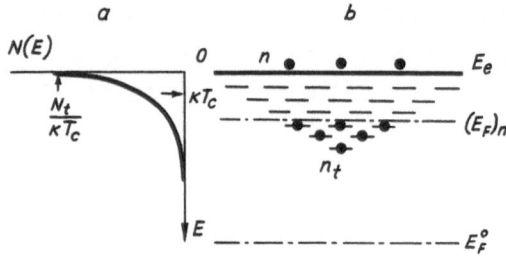

Fig. 5.16. Energy diagram of an insulator with exponential trap distribution: (a) energy-dependence of trap distribution N(E), according to the exponential model [cf.(5.30)]; (b) quasi-thermal equilibrium at steady-state electron injection, characterized by the quasi-Fermi level $(E_F)_n$

The phenomenological parameter T_c has no direct physical meaning, it is simply used in the exponential model to characterize the rate of decrease of trap concentration with increasing energy E (cf. Fig.5.16a). For example, $T_c \to \infty$ corresponds to uniform distribution of traps. It is usually assumed that T_c is equal or greater than the surrounding temperature T, viz. $T_c \gtrsim T$. If $T_c < T$ the situation is reduced to that of shallow trapping described by (5.22).

For exponential trap distribution (cf. Fig.5.16) concentration of free (n) and trapped (n_t) injected carriers in the SCLC regime are given, respectively, by

$$n = N_e \exp\left(- \frac{|(E_F)_n, E_e|}{kT}\right) \tag{5.31}$$

and

$$n_t = kT_c N_t \exp\left(- \frac{|(E_F)_n, E_e|}{kT_c}\right) . \tag{5.32}$$

Finally, the j = f(U,L) dependence in the SCLC regime of injection is given by

$$j = e\mu N_e \left(\frac{\varepsilon}{ekT_c N_t}\right)^m \frac{U^{m+1}}{L^{2m+1}} , \tag{5.33}$$

with $m = \frac{T_c}{T}$.

Since for exponential trap distribution $m = T_c/T > 1$, it follows that current j increases superlinearly with growing voltage U

$$j \sim U^n , \tag{5.34}$$

where n = m+1 > 2, and decreases still more steeply with increasing sample thickness L

$$j \sim \frac{1}{L^{\ell}} \quad , \tag{5.35}$$

where $\ell = 2m+1 > 3$.

Prototype CV characteristics for exponential trap distribution are presented in Fig.5.17. The crossover voltage U_x from Ohm's law to space charge controlled current regime is given in this case by [3.3,5.21]

$$U_x = U_m \exp\left(- \frac{|E_F^0, E_e|}{kT_c}\right) \exp\left(- \frac{T}{T_c}\right) \quad , \tag{5.36a}$$

where U_m is the trap-filled voltage

$$U_m = \frac{N_t e L^2}{\varepsilon} \quad . \tag{5.36b}$$

(It should be mentioned that although the trap-filled limit voltages for exponential U_m and discrete trap distribution U_{TFL} seem formally to be

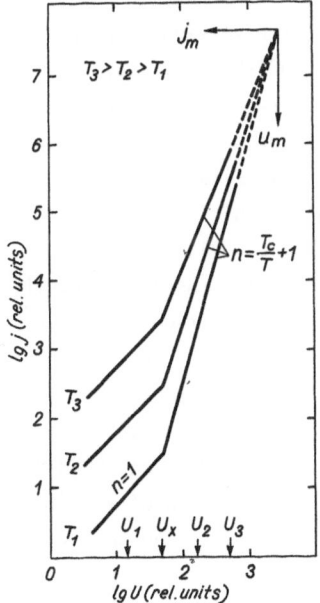

Fig. 5.17. Prototype current-voltage characteristics on log-log plot for exponential trap distribution (cf. Fig.5.16), with temperature T as parameter

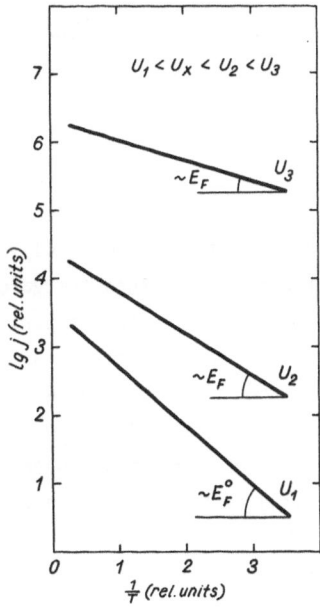

Fig. 5.18. lg j = f(1/T) curves, with voltage U as parameter, as obtained from CV characteristics in Fig.5.17 at voltages U_1, U_2 and U_3, respectively

equivalent [cf. (5.24,36b)], their physical meaning and experimental methods of determination differ considerably, as may be easily seen by comparing Figs.5.15,17).

As follows from Fig.5.17, the CV characteristics in the superlinear region at $U > U_x$ exhibit constant slope n value, depending only on temperature T. Still more important is the fact that this constant slope value n is directly determined by the trap distribution parameter T_c value. It should be emphasized that such direct correlation between CV curve slope and trap distribution is the most characteristic feature of exponential approximation, and we will often refer to it in further discussions analyzing the applicability limits of this approximation.

Finally, as can be seen from Fig.5.18, the temperature dependences of CV characteristics allows one to determine from the slopes of $lg\ j = f(1/T)$ curves the activation energy of intrinsic dark conductivity in the ohmic region, or, correspondingly, the position of the Fermi level E_F^0, as well as the shift of the quasi-Fermi level E_F at different injecting voltages in the superlinear region of CV characteristics.

Applicability Limits of Diffusion-Free SCLC Theory Approximation

In Sect.5.8.2 the conceptual basis of a diffusion-free version of phenomenological SCLC theory was discussed. It was concluded that, for sufficiently high applied voltages diffusion effects ought to be negligible. The question, however, arises about the limiting voltage value determining the applicability of diffusion-free SCLC theory in realistic situations.

General considerations show (see, e.g, [1.75]) that diffusion currents can be regarded as completely negligible if the applied voltage U is considerably greater than the value of kT/e, i.e., U >> kT/e. As may be easily shown, already at room temperatures (kT = 0.025 eV) the voltage of 2 V is by two orders of magnitude greater than the value of kT/e and, consequently, even at such low voltages diffusion-free SCLC theory may be regarded as valid.

Recently a series of papers by BONHAM and JARVIS [5.22,23], and BONHAM [5.20,24] have been devoted to analytical studies of diffusion effects in the SCLC theory for insulators. The authors performed a detailed numerical analysis of CV characteristics and potential distribution for an insulator with one ohmic and one blocking electrode, including diffusion effects [5.23]. They have demonstrated that the electric field across most of the

field. A constant-field approximation is shown to give a good description of CV curves when the injected charge density is small at the blocking electrode (good blocking conditions), and the current lies well below the purely space charge limited current which would flow if both electrodes were ohmic [5.22,23]. According to BONHAM [5.20] the diffusion-free SCLC theory may safely be used at applied voltages $U \gtrsim 5$ V.

5.8.4 Criteria for Validity of SCLC Conditions

For adequate use of SCLC theory dependences for the determination of local trapping state parameters it is always necessary to ascertain to what extent the conditions for SCLC are valid in the given interval of electric field intensity.

Let us briefly discuss some basic criteria for validity of SCLC conditions.

1) A phenomenological criterion for experimental ascertainment of fulfillment of SCLC conditions was proposed by SILINSH and TAURE [5.25,26]. The $j = f(U)$ and $j = f(L)$ dependences under SCLC conditions can be expressed in the following general form:

$$j \sim \frac{U^n}{L^{\ell}} = \frac{U^{m+1}}{L^{2m+1}} \quad . \tag{5.37}$$

The left-hand side of (5.37) represents the experimentally determined $j = f(U)$ and $j = f(L)$ dependences, with the corresponding superlinear exponents n and ℓ; the right-hand side of (5.37) gives the respective SCLC theory dependences (with $m = 1$ for square law [cf. (5.22) and $m > 1$ for supersquare law, cf. (5.33)]).

The validity of (5.37) serves as an experimental confirmation of SCLC conditions in the specimen within the given range of electric field intensities. In the case of SCLC conditions the exponent m must have one and the same value $m = m_U$ (at $T = $ const) for j dependence on voltage U and $m = m_L$ for dependence of j on thickness L of the sample, i.e., $m_U = m_L$.

The phenomenological criterion for validity of SCLC conditions is, consequently [5.25,26]

$$\frac{m_U}{m_L} = \frac{2(n - 1)}{\ell - 1} = 1 \quad . \tag{5.38}$$

The problem of validity of SCLC conditions, as determined by (5.38), has been studied in some detail in the investigations of SILINSH et al. [2.134, 5.26], BALODE et al. [3.1], and TAURE [5.27] for thin-layer specimens of a large number of aromatic and heterocyclic organic molecular crystals. They have found that the criterion $m_U/m_L \approx 1$ is valid within the field intensity range between $\mathscr{E} = \mathscr{E}_x$ (crossover field to superlinear CV characteristic; typical value of $\mathscr{E}_x$ being from 8×10^3 to 2×10^4 V/cm) and $\mathscr{E} = 10^5$-10^6 V/cm.

At field intensities $\mathscr{E} > 10^5$-10^6 V/cm the ratio m_U/m_L usually becomes considerably higher than unity ($m_U/m_L \gg 1$). This indicates a steeper rise of the $j = f(U)$ curve, as compared to that of $j = f(L)$.

Figures 5.19,20 present typical examples of applying the m_U/m_L criterion according to (5.38). As can be seen from these figures, the field intensity interval of SCLC validity is not the same for different substances. In some cases it is relatively narrow (cf. Figs.5.19,20a), covering only 0.5-1 de-cades of change in field intensity $\mathscr{E}$. Only in exceptional cases, such as that of indandione-1,3 pyridinium betaine (IPB) (cf. Fig.5.20b) the condi-tion $m_U/m_L \approx 1$ remains valid over two orders of $\mathscr{E}$ values.

It stands to reason that a $\mathscr{E} > 10^5$-10^6 V/cm, when a rise in the $j = f(U)$ curve sets in and the ratio m_U/m_L becomes considerably higher than unity (cf. Fig.5.19), we, apparently, observe some form of field-enhanced space charge limited current for which traditional SCLC theory cannot be applied any more.

A theoretical analysis of field-enhanced SCLC for Coulomb type trapping centers is given, e.g., by MURGETROYD [5.28] and RAYKERUS [5.29]. In this case the additional increase in current at high field intensities under SCLC conditions is attributed to decrease in effective trap depth, due to the well-known Poole-Frenkel effect. Such an approach is not, however, ap-plicable for trapping states of polarization origin in molecular crystals. Some other possible mechanisms of field-enhanced SCLC should be suggested in this case. For example, the elementary microscopic electron capture pro-cess (cf. Sect.5.8.3 and Fig.5.14) can be affected if the charge carriers are substantially "heated up" by applying a high electric field, since the electron-capture cross section is strongly dependent on electron energy (cf., e.g. [1.76]). Another possible high-field effect, causing the enhance-ment of current flow, could be drift mobility dependence on applied field. For example, the drift velocity v above some critical field strength $\mathscr{E}_1$, i.e., at $\mathscr{E} > \mathscr{E}_1$ may increase as the square root of the applied field, name-ly, $v = \mu \, (\mathscr{E}_1 \mathscr{E})^{1/2}$, where μ is the low-field field-independent mobility. Such an effect is characteristic for "warm" electrons when acoustic phonon scat-tering is dominant [1.76]. We shall not, however, discuss, further the pos-sible field-enhanced SCLC mechanisms in molecular crystals since it lies outside the scope of our interests.

It is interesting to mention that, according to [5.26], in thin-layer specimens of some heterocyclic substances, such as certain molecular asso-ciates — derivatives of indandione-1,3, the injection range of SCLC is alto-gether absent. In these cases the ohmic region of the CV characteristic is superseded by a superlinear one at $U > U_x$. The ratio m_U/m_L, however, considerably exceeds unity ($m_U/m_L \gg 1$), and the superlinear $j = f(L)$ dependence, characteristic of SCLC, is not observed at all.

This example shows that use of SCLC theory for determining the energy spectrum of traps requires, apart from $j = f(U)$ measurements, also an investigation of $j = f(L)$ dependence, in order to find the field intensity range after (5.38), within which pure SCLC conditions are valid. Unfortunately, there exists a large number of reported dark characteristic measurement data for anthracene, tetracene, phthalocyanine and other single crystals, presumably performed under SCLC conditions (cf. [1.75,76]), however, without any reference to $j = f(L)$ dependence and to verification of the

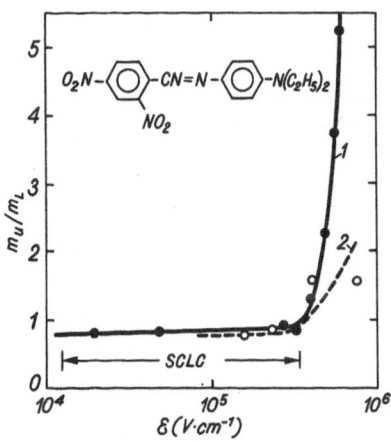

Fig. 5.19. Experimental determination of the field intensity range within which condition (5.38) for SCLC holds in the case of a thin-layer Schiff base (SB) cell: Al/SB/Au: 1 - for sample thickness L = 0.4 μm; 2 - for L = 0.6 μm [3.1]

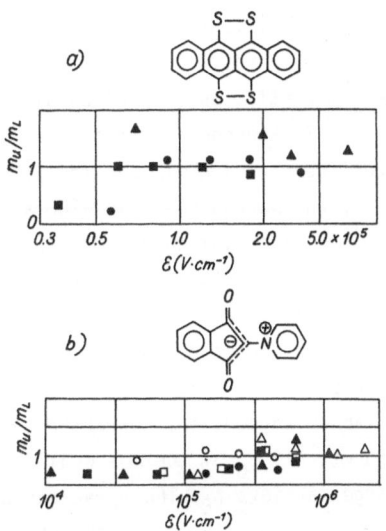

Fig. 5.20. Experimental determination of the field intensity & range within which condition (5.38) for SCLC holds, according to [5.27]: (a) for a thin-layer tetrathiotetracene (TTT) cell Al/TTT/Au [at thickness L: (▲) 0.7, (●) 1.1, (■) 1.7 μm]; (b) for a thin-layer indandione-1,3-pyridinium betaine cell (IPB) Au/IPB/Au [of thickness L: (▲) 0.9, (■) 1.6, (●) 2.3 μm], with positive (hole injecting) polarity on the lower Au electrode; (Δ, □, ○) for the same thickness and with positive (hole injecting) polarity on the upper Al electrode

existence of SCLC conditions. This is, of course, understandable, since work with single crystal samples of various thickness practically excludes the possibility of ensuring identical injection parameters of contacts and, accordingly, of obtaining reliable $j = f(L)$ dependences. Such data can be safely secured working with vacuum-evaporated thin layers of organic crystals, maintaining identical technological conditions in obtaining injecting metal contacts [2.134,3.1,3,5.26].

In our studies we used, as a rule, a method described in detail in [3.1] for preparing thin-layer samples of aromatic and heterocyclic organic crystals. This method makes it possible to obtain, under identical evaporation conditions, thin-layer cells of different sample thickness on a common substrate, thus providing the necessary requirements for studying all the basic SCLC dependences, including $j = f(L)$ dependence and criterion (5.38).

2) An additional criterion for onset of SCLC conditions is the behavior of crossover voltage U_x from ohmic to superlinear SCLC region of the CV characteristic. Under SCLC conditions U_x is well reproducible within 10-20 % [5.26]. If the ohmic part of the CV characteristic passes directly into the superlinear region without the SCLC range the crossover voltage reproduces poorly and spreads over 100-200 and more % [5.26].

3) An essential condition for SCLC is $t_{tr} \leq \tau_\Omega$ where t_{tr} is the effective transit time of the charge carrier and τ_Ω the ohmic relaxation time [cf. (5.27-29)]. It determines the upper limit of equilibrium specific conductivity σ_0 of the sample, at which it is still possible to achieve SCLC conditions.

Table 5.5 contains characteristic τ_Ω values, as dependent on specific conductivity σ_0 at thermal equilibrium and on equilibrium concentration of charge carriers n_0. The τ_Ω values are given for two typical carrier mobilities: $\mu = 1$ cm^2/(V·s) (anthracene-type single crystals) and $\mu = 10^{-3}$ cm^2/ (V·s) (vacuum-evaporated layers of aromatic and heterocyclic organic crystals).

Transit time t_{tr} of a charge carrier in the voltage range of $U < U_x$ is, typically, of the order of $t_{tr} \sim 10^{-5}$ s, both for thin single crystals of anthracene [1.75], as well as for vacuum-evaporated thin layers of other organic crystals [2.141,5.30]. The upper limit of conductivity σ_0 of a sample in which SCLC conditions can still be observed is, according to Table 5.5, of the order of 10^{-8} Ω^{-1} cm^{-1}. This means that SCLC conditions can be obtained both in organic insulators ($\rho_0 \geq 10^{12}$ Ωcm), as well as in high-resistance organic semiconductors ($\rho_0 \gtrsim 10^8$ Ωcm).

Table 5.5. Values of ohmic relaxation time τ_Ω at thermal equilibrium, as dependent on specific conductivity σ_0 and charge carrier concentration n_0

σ_0 [Ω^{-1} cm^{-1}]	n_0 [cm^{-3}]		τ_Ω
	for $\mu = 1$ cm^2/V·s	for $\mu = 10^{-3}$ cm^2/V·s	[s]
10^{-12}	6.3×10^6	6.3×10^9	0.27
10^{-10}	6.3×10^8	6.3×10^{11}	$\sim 3 \times 10^{-3}$
10^{-8}	6.3×10^{10}	6.3×10^{13}	3×10^{-5}
10^{-6}	6.3×10^{12}	6.3×10^{15}	3×10^{-7}

In some cases it is necessary to use special artifices for obtaining SCLC conditons. Thus, for instance, superlinear CV characteristics are observed in comparatively low-resistance tetrathiotetracene (TTT) layers possessing ρ_n values of the order of 10^8 Ωcm at room temperature. Superlinearity of $j = f(L)$ is, however, absent, and the ratio m_μ/m_L considerably exceeds unity ($m_\mu/m_L \gg 1$) [2.142,5.27]. Superlinearity of $j = f(U)$ is in this case most likely caused by Schottky effect. Such an interpretation is supported by the dependence of CV characteristics on polarity and the linear current j dependence on field strength $\mathscr{E}$ in Schottky coordinates ($\lg j \sim \mathscr{E}^{1/2}$) [2.142,5.27]. SCLC conditions can be obtained in such TTT layers only at temperatures T < 250 K when σ_0 is sufficiently low for relation $t_{tr} < \tau_\Omega$ to be valid. Criterion (5.38) is also satisfied in this case (cf. Fig.5.20a), but the range of field intensities $\mathscr{E}$ at which SCLC conditions hold is comparatively narrow—from 0.5×10^5 to 5×10^5 V/cm.

5.8.5 Difficulties in Interpreting Experimental CV Characteristics in Terms of Discrete and Exponential Trap Distribution Models

The phenomenological SCLC theory proposed and developed by ROSE and LAMPERT [1.76,5.13-15] was further extended in application to molecular crystals by a number of authors [1.75,76,2.134,3.1,3,5.25,26,31-34].

MARK and HELFRICH [5.31] were first to apply successfully the SCLC method to the study of trapping states in organic molecular crystals (anthracene and a number of lower polyphenyls). They showed that this method can be useful for investigating the energy spectrum of traps, as well as transfer and trapping processes in thin samples of organic single crystals.

Thus MARK and HELFRICH [5.31,32] found by SCLC method that, for instance, in vapor-grown anthracene single crystals there exist in considerable densities charge carrier traps with a quasi-continuous energy spectrum, which the authors approximated in terms of an exponential model with $kT_c = 0.05$-0.11 eV and $N_t = 4 \times 10^{15}$-4×10^{17} cm^{-3}. At the same time, they observed the presence of separate discrete traps situated ca. 0.9 eV above hole conductivity level E_h and having a concentration of the order of $N_t = 10^{11}$ cm^{-3}.

After these first papers a large number of studies were carried out using SCLC methods on single crystals (of naphthalene, anthracene, phthalocyanine, p-chloranyl, etc.), as well as on thin-layer systems of a great variety of low-molecular weight organic substances. It is not our task to go into this work in detail since there exists the above-mentioned excellent monograph by LAMPERT and MARK [1.76] on this subject, as well as a number of review articles (cf., e.g. [1.66,75,2.134,3.1]).

All this work shows that the assertion of LAMPERT and MARK (cf. [1.76]) that, at the present stage of solid-state techniques, investigation of injection currents has become if not the only, yet at least the basic method of studying local trapping states in insulators, including molecular crystals, is by no means exaggerated.

We consider it of interest to discuss here some general problems and difficulties in interpreting experimental CV characteristics in terms of both traditional SCLC theories (cf. Sect.5.8.3).

An analysis of a large number of CV characteristics of molecular single crystals and thin layers shows that the structure of CV curves is more complex than might have been theoretically expected from the discrete (cf. Fig. 5.15) or exponential (cf. Fig.5.17) trap distribution models. Therefore frequently difficulties arise in interpreting experimental CV curves in terms of one or the other model. Real CV characteristics do not, as a rule, reproduce the idealized curves, as shown in Figs.5.15,17. They lie somewhere inbetween or, else, outside the limits of these approximations (cf., e.g., experimental CV characteristic in Figs.5.21,23,36-40).

We shall dwell now in some detail on the interpretation difficulties illustrated by some typical experimental CV characteristics of molecular single crystals and thin layers. We shall also analyze a generalized prototype CV characteristic in Fig.5.22 in terms of exponential approximation. This shape of the curve corresponds to a large number of similar dark CV characteristics obtained by SCLC method for studying local trapping states in molecular crystals.

In the latter case we shall use (log j, log $\mathscr{E}$) coordinates instead of the usual (log j, log U) ones. Such a procedure lends itself to more convenient comparison of CV characteristics obtained from single crystal samples with thickness L of the order of 100 μm and more, at which the corresponding U values are of the order of 10^2-10^3 V—with the corresponding curves obtained from thin-layer cells with L of the order of 0.5-3.0 μm and with corresponding U values of only 1-100V.

We shall begin our illustrative analysis with data reported by HEILMEIER and WARFIELD [5.35] who observed two kinds of typical CV curves of dark conduction in phthalocyanine crystals under SCLC conditions (cf. Fig.5.21, curves 1 and 2). Curve 1 of Fig.5.21 was interpreted by the authors of [5.35] as a clear manifestation of the square law (5.22) presumably evidencing the presence of a discrete set of charge carrier trapping states. There is, however, one suspicious circumstance, namely, the range of square-law dependence $j \sim \mathscr{E}^2$ is extremely narrow, and the rise of the slope n of the CV curve is not steep enough at the field intensity of supposed trap-filled limit, i.e., at $\mathscr{E} = \mathscr{E}_{TFL} = U_{TFL}/L$ (cf. Fig.5.21, Curve 1, and Fig.5.15).

Assuming that the above interpretation is nevertheless valid, if with reservations, for Curve 1 in Fig.5.21, its inconsistency becomes obvious for the CV characteristic, as presented by Curve 2 in Fig.5.21. In this case the short square-law part of the $j = f(\mathscr{E})$ Curve with n = 2 is followed by a part with a slope value of n = 3. Such a monotonous increase in the slope of the CV characteristic indicates a quasi-continuous trap distribution and cannot be interpreted in terms of a discrete trap distribution model (cf. Fig.5.21, Curve 2, and Fig.5.15). This naturally leads to the idea of replacing the discrete distribution model proposed in [5.35] by a model of exponential distribution of traps (in such approximation we obtain a distribution parameter $kT_c \approx 0.05$ eV). One might, however, suggest that it would be more appropriate in this case to use a Gaussian distribution model which would describe from a single standpoint both types of CV characteristics, as represented by Curves 1 and 2 in Fig.5.21 (cf. Sect.5.9).

In some cases the choice of an adequate distribution model is impeded by purely methodological circumstances. If the CV characteristic is measured, e.g., for a relatively thick single crystal sample, then one may observe only the square-law part of the $j = f(\mathscr{E})$ curve within the range of accessible field intensities. It would not be safe to attempt to interpret such a CV characteristic as evidence in favor of a presence of discrete traps. An increase in field intensity might well reveal a region of the CV characteristic with slope n > 2, evidencing quasi-continuous trap distribution. Such a treatment of a "square-law" region in experimental CV characteristics of single crystals appears to be the source of assertions occasionally encountered in the literature and contending that we usually observe discrete charge carrier traps in single crystals, and quasi-continuous ones in thin-layer samples. One is entitled to speak safely of discrete trapping states only in cases of CV characteristics showing a pronounced threshold of trap-filled limit at $\mathscr{E} = \mathscr{E}_{TFL}$ (cf. Fig.5.15).

The most complex problems of interpreting experimental CV characteristics arise in cases when the superlinear slope of the $j = f(\mathscr{E})$ curve with n > 2 rises monotonously with field intensity, i.e., when we have $n = n(\mathcal{E})$. As already mentioned, a generalized CV characteristic of such type of dependence is presented in Fig.5.22. It is a prototype of most experimentally observed CV characteristics obtained both from thin-layer samples, as well as from single crystal specimens (cf., e.g. Figs.5.21,23,36-38). The exponential trap distribution model obviously does not fit such curves, since the model requires constant slope value, viz. $n = m + 1 = T_c/T+1$, at given temperature T [cf. (5.33) and Fig.5.17].

If the investigator has only the model of exponential approximation at his disposal, it is frequently erroneously assumed that condition $n = const$ applies only to the largest slope obtained in the given interval of field intensity change. This slope is then endowed with a quality which it does not possess, namely, that of indicating characteristic parameter kT_c of supposed exponential trap distribution, assuming that the condition $n = m + 1 = T_c/T+1$ is valid. Such an assertion is completely unfounded. Figure 5.22 clearly shows that it is possible to obtain a large variety of n values and

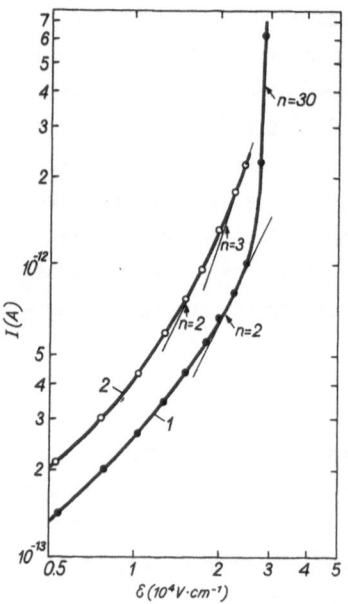

Fig. 5.21. Typical CV characteristics under SCLC conditions for phthalocyanine single crystals, according to [5.35]. Thickness of crystal $L \approx 2 \times 10^{-2}$ cm (cf. also [1.76])

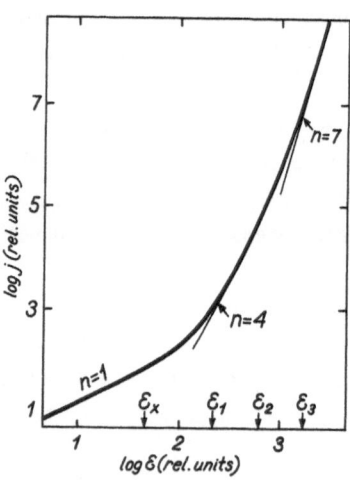

Fig. 5.22. A generalized prototype CV characteristic on log-log plot with monotonously increasing slope value n. In this case neither the discrete (cf. Fig.5.15), nor the exponential (cf. Fig.5.17) model is valid

thus deduce all sorts of corresponding "distribution parameters" kT_c, depending on the maximum value of $\mathscr{E}$ for which the $j = f(\mathscr{E})$ dependence is being studied. Thus we have, for instance, a value of $n = 4$ at $\mathscr{E}_{max} = \mathscr{E}_2$ (Fig.5.22) which yields $kT_c = 0.075$ eV. If we now take $\mathscr{E}_{max} = \mathscr{E}_3$, we get $n = 7$, obtaining $kT_c = 0.15$ eV, and so on. This leads us by pure intuition to the conclusion that for the given type of CV characteristics, as shown in Fig.5.22, the slope n cannot be indicative of trap distribution. Such CV characteristics can only be approximated, as will be shown in Sect.5.9, by means of a Gaussian model in which the distribution parameter σ does not directly depend on varying slope $n = f(\mathscr{E})$ of the CV characteristic shown in Fig.5.22.

Unfortunately, experimental CV characteristics of similar type are frequently subjected to exponential approximation. As a result kT_c values lying between 0.03 and 0.11 eV have been obtained for anthracene single crystals, which corresponds to maximum slope values of the $j = f(\mathscr{E})$ curve from $n = 2$ up to $n = 5.5$. It is not surprising therefore that kT_c values reported by various authors (cf. references in [1.66]) do not depend on the method of growing the anthracene single crystal samples (from gaseous phase, solution, or melt). Actually, the kT_c values obtained by such an approach are rather a characteristic of the CV-curve measurement technique than an indicator of real trap distribution in the sample under investigation, and hence devoid of any physical meaning. Therefore constancy of slope n of $j = f(\mathscr{E})$ curve in logarithmic coordinates [or, correspondingly, of the $j = f(U)$ curve] in the superlinear region of the CV characteristic with $n \geq 2$ within a wide range of field intensity $\mathscr{E}$ values is an absolutely indispensable condition for the validity of the exponential model of trap distribution, i.e.,

$$n = \frac{d(\lg j)}{d(\lg \mathscr{E})} = \frac{d(\lg j)}{d(\lg U)} = const \tag{5.39}$$

for $n \geq 2$.

Only under strict compliance with condition (5.39) can the slope value n be used as a quantitative indicator for determining distribution parameter kT_c of exponential approximation according to the expression [cf. (5.33)]

$$n = m + 1 = T_c'/T + 1 \tag{5.40}$$

at $T = const$.

An additional necessary condition for the validity of the exponential approximation model is the temperature dependence of CV characteristics which ought to be in agreement with (5.40), if T is taken as parameter. If this condition is observed, a family of CV characteristics of the type as in Fig.5.17 should be obtained, with constant slope n for all CV curves measured at different temperature T, all of them converging at the point with coordinates U_m, j_m.

Conditions (5.39,40) are necessary, but not sufficient criteria for the validity of exponential approximation. Constancy of slope ℓ in the $j = f(L)$ dependence must also take place

$$\ell = \frac{d(\lg j)}{d(\lg L)} = \text{const} \tag{5.41}$$

for $\ell \geq 3$.

This could be illustrated by an example. Figure 5.23 shows temperature dependences of CV characteristics for thin-layer samples of transbis-bindonilene (TBB) under conditions of monopolar injection of holes from a metallic Au electrode. The family of CV curves in Fig.5.23 reminds us by their shape of a family of idealized CV characteristics corresponding to exponential approximation (cf. Fig.5.17). True, the CV characteristics in Fig.5.23 do not exhibit a sharp crossover from ohmic to superlinear dependence. However, such characteristics are usually approximated exponentially. Yet, a more scrupulous investigation of $j = f(L)$ dependence of given TBB samples showed that condition (5.41) is obviously not valid (cf. Fig.5.31b), and therefore exponential approximation is not feasible. This example demonstrates the necessity of careful examination of all criteria of adequacy of the chosen model before using it for evaluation of distribution parameters of local trapping states.

Another type of incorrect interpretation of experimental CV characteristics, rather often found in the literature, is the approximation of a CV curve of the type as in Fig.5.22, by a "square law" [cf. (5.22)], especially in case of single crystal samples. Thus, for example, MANY et al. [5.36] have approximated CV characteristics of the given type for an anthracene single crystal (cf. Fig.5.37) by "square-law" dependence [cf. (5.22)], although the comparatively narrow range of square-law slope (n = 2) of the CV curve monotonously changes to higher slope values, reaching that of n = 8 (cf. Fig.5.37, Curve 2). The voltage corresponding to transition from square law to supersquare law dependence at U = 70-80 V is treated in this case without sufficient reason as trap-filled voltage U_{TFL}. Such incorrect approach led the authors of [5.36] to the conclusion that in the anthracene single crystals studied by them there exist discrete electron trapping centers with a concentration of $N_t = 10^{14}$ cm^{-3}. In reality the experimental CV characteristic in Fig.5.37, as obtained in [5.36] indicates quasi-continuous trap distribution, since we have here n = n(U) which can be more adequately approximated by Gaussian distribution of trapping states having distribution parameter $\sigma = 0.13 \pm 0.03$ eV and $N_t = (7-8) \times 10^{13}$ cm^{-3}. Similar incorrect interpretation of CV characteristics of the type shown in Fig.5.22, can be found also in a later paper by MANY et al. (cf. e.g. [5.37]).

An interesting point arising in this discussion of the limits of applicability of the exponential approximation model concerns the interdependence between total trap concentration N_t and parameter kT_c of the exponential approximation. This correlation has been studied experimentally by OWEN et al. [5.38], and REUCROFT and MULLINS [5.39]. It can be expressed according to NEŠPUREK et al. [5.40] in the following general way (cf. Fig.5.24):

$$\ln N_t = A + \frac{B}{kT_c} \qquad . \qquad (5.42)$$

OWEN and CHARLESBY [5.41] considered the dependence on Fig.5.24 as an artefact. The authors of [5.40], however, attempted to find a formal interpretation of (5.42). They considered the correlation effect (5.42), as manifest in Fig.5.24, to be a "compensation phenomenon" (similar to the effect of "compensation" in the expression $\sigma = \sigma_0 \exp(-E_a/kT)$ for dark conductivity). They showed, further, that formally this effect can be observed, if the family of CV characteristics from which N_t and kT_c values have been determined, shows a common point of convergence. But a family of CV characteristics will always have such a point of convergence in case of Gaussian distribution of traps. It corresponds to the passage of the Fermi quasi-level through the peak of the Gaussian bell (cf. Fig.5.26). It stands to reason therefore that this "compensation phenomenon", as in Fig.5.24, can be considered as an indirect support in favor of replacing exponential approximation of trap distribution by the more adequate Gaussian one (cf. Sect.5.9). It might be noted in this connection that if, according to [5.40], the slightly stepwise CV curve of two quasi-discrete Gaussian distributions is approximated by a straight line (cf. Fig.5.33), a "compensation phenomenon" will also be observed which indirectly proves the inadequacy of the exponential approximation.

Summing up the aforesaid one can state that a large number of experimental CV characteristics in molecular crystals, being of the type as in Fig. 5.22, cannot be satisfactorily described in terms of the exponential model. We shall therefore discuss a phenomenological SCLC theory for Gauss distribution of local charge carrier trapping states. We shall also demonstrate in the next sections that the Gaussian model (cf. Sect.3.1) is much more appropriate and feasible for the interpretation of CV characteristics of the above type than the exponential one. One might, perhaps, physically speaking, consider exponential distribution of traps at the very edge of the conductivity level (cf. Fig.5.16). In the middle of the energy gap, however, in the

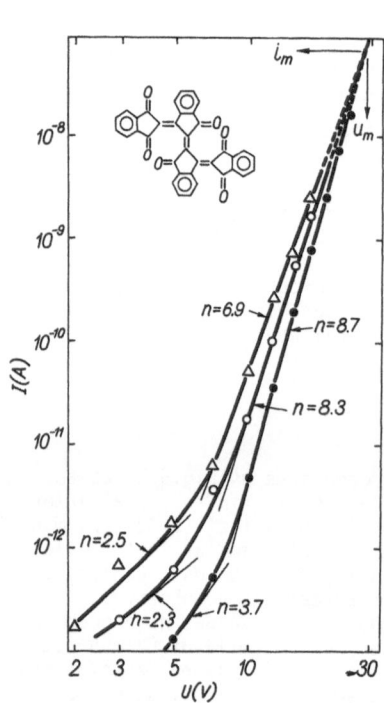

Fig. 5.23. Experimental CV characteristics for a thin-layer cell of trans-
bisbindonilene (TBB) Al/TBB/Au with
temperature T as parameter: (•) 242 K,
(o) 279 K, (Δ) 315 K

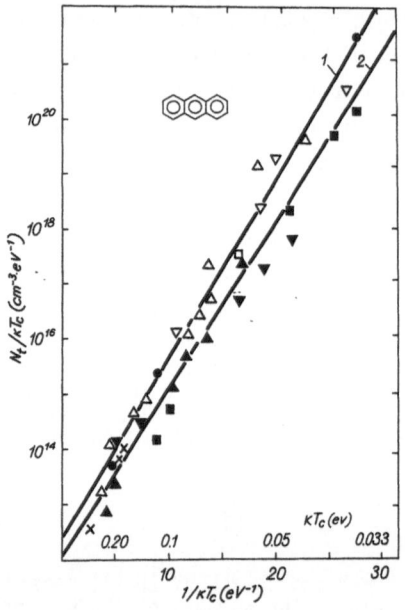

Fig. 5.24. Dependence of total concentration N_t/kT_C of charge carrier
traps on the value of $1/kT_C$ in the
case of exponential approximation
of CV characteristics for anthracene single crystals grown in various gaseous media; filled symbols
indicate injection of holes (h),
empty ones - injection of electrons (e): (•o) crystal grown in
nitrogen; (▲Δ) in helium; (■□) in
xenon; (×) in argon; (▼▽) *in vacuo*,
according to [5.38]

vicinity of some fixed level E_t such a distribution would be much too artificial and lacking any physical justification.

On the other hand, the concept of discrete monoenergetic levels is an
oversimplified and idealized model. If such discrete levels are created by
impurity centers or by specific structural defects, then one naturally
should take into account a real statistical distribution of the energy of
such levels. One can therefore, at best, only speak of quasi-discrete levels
with a dispersion of distribution.

The Gaussian distribution model of trapping states has the advantage of
describing various kinds of distributions in terms of the same approxima-

tion. It can be successfully applied to trapping states of the quasi-continuous $G_e(E)$ type of distribution, as well as to quasi-discrete distribution of the $G_g(E)$ type (cf. Fig.3.2), and also to a set of several Gaussian distributions inside the energy gap (cf. Sect.5.9.8). An essential point is that the model of discrete distribution of trapping states can be postulated as a special case of the Gaussian model with distribution parameter $\sigma \to 0$.

5.9 Phenomenological SCLC Theory for Molecular Crystals with Gaussian Distribution of Local Trapping States

5.9.1 Conceptual Basis

The idea of possible Gaussian distribution of local charge carrier trapping states in insulators was previously discussed by LANYON [5.21], SUSSMAN [3.3], and LAMPERT and MARK [Ref.1.76, p.26]. But these authors, as well as many others, have preferred, in practice, the conventional model of an exponential distribution of local states (see, e.g. [1.75,76,2.134,3.1,3] and references therein).

However, as already demonstrated in the previous section, the exponential model is an *ad hoc* assumption providing a mathematically comparatively simple way of obtaining the basic dependences of phenomenological SCLC theory (cf. Sect.5.8.3). In its final consequences the exponential model has proved to be incapable of describing the full variety of real current-voltage characteristics in materials with statistical quasi-continuous distribution of local trapping states. Thus insoluble difficulties arose in attempting to interpret many experimental CV characteristics in terms of the given approximation (cf. Sect.5.8.5).

The advantages of the Gaussian model in describing the energy spectra of local states of structure origin, both from a physical and statistical standpoint, have been discussed in detail in a number of papers by SILINSH et al. [1.62,66,68,5.6,42].

Although the SCLC theory of Gaussian trap distribution is more sophisticated than the exponential approach, it seemed worthwhile to overcome these formal difficulties [1.68]. Firstly, the Gaussian approach is physically self-consistent in the framework of a generalized phenomenological ionized state energy model for real molecular crystals, according to which the energy of conductivity levels as well as that of the trapping states is determined by electronic polarization of the crystal by charge carriers (cf. Sect.3.1). Secondly, it is statistically more appropriate for the description of local polarization energy deviations caused by a random distribution of trapping centers of structural origin in the crystal, as compared to the exponential approximation (cf. Sects.3.1 and 5.7).

For practical application of the Gaussian model in the study of local trapping states in molecular crystals a phenomenological SCLC theory in Gaussian approximation has been developed by NEŠPUREK and SILINSH [1.68]. They have also determined the validity range and main criteria for such approximation. We are now going to present in the next sections the basic dependences of SCLC theory of Gaussian approach and examples of its application for studying local trapping states in a number of aromatic and heterocyclic molecular crystals.

5.9.2 Basic SCLC Theory Equations

In order to derive analytical expressions for SCLC conditions, relating the current density j, voltage U, and sample thickness L for a single Gaussian distribution located at E_t (cf. Fig.3.2), the following set of equations has to be solved:

$$j = e\mu \frac{U}{L} N_e \exp\left(-\frac{E_F}{kT}\right);$$

(5.43)

$$\rho_t = \frac{\varepsilon U}{eL^2} \; ;$$

(5.44)

$$\rho_t = \int_{-\infty}^{\infty} G(E)F(E)dE = \frac{N_t}{\sqrt{2\pi}\sigma} \int_{-\infty}^{\infty} \frac{\exp\left(-\frac{(E - E_t)^2}{2\sigma^2}\right)dE}{1 + \exp\left(\frac{E - E_F}{kT}\right)} \; ,$$

(5.45)

where N_e is the density of states at the conductivity level; ε is the dielectrical permeability; ρ_t the trapped carrier density; G(E) the Gaussian distribution of local states; F(E) the Fermi-Dirac function, N_t the total density of local states; σ the dispersion parameter of Gaussian distribution, and E_F the quasi-Fermi level. The solution of this problem is connected with certain formal difficulties since the trapped carrier distribution integral (5.45) cannot be solved analytically, and various approximations or numerical integration procedures have to be used.

CROITORU and GRIGORESCU [5.43] and later NEŠPUREK and SMEJTEK [5.44,45], and BONHAM [5.46] have proposed different analytical approximations of current-voltage (CV) characteristics for a Gaussian distribution of local states.

In the work by NEŠPUREK and SILINSH [1.68] the validity range for various approximations has been determined by comparison of numerical integration and analytical approximation data. Calculations were performed with the following assumptions typical for phenomenological SCLC theory and inherent to the set of equations (5.43-45): a) the electric field is constant throughout the sample; b) the effects of carrier diffusion are negligible. These assumptions of diffusion-free approximation are obviously valid only for higher voltages in the SCLC region at $U \gtrsim U_x$ (cf. Sect.5.8.3).

Further, the authors considered monopolar electron injection from the cathode into the insulator crystal. Recombination processes and the effect of thermal equilibrium charge carriers may thus be neglected. Under these conditions, integral (5.45) for the distribution of trapped injected electrons near the anode n_{td} may be written as

$$n_{td} = \frac{N_t}{\sqrt{2\pi}\sigma} \int_{-\infty}^{\infty} \frac{\exp\left(-\frac{(E - E_t)^2}{2\sigma^2}\right)dE}{1 + \frac{N_e}{n_d}\exp\left(\frac{E}{kT}\right)} \quad , \tag{5.46}$$

where n_d is the density of free electrons at the anode.

5.9.3 Validity Range for Different Analytical SCLC Approximations

Numerical integration data show that there is no single analytical SCLC approximation available that would be adequate throughout the whole voltage range of the CV characteristic. It falls into four subregions, according to the voltage range and slope value $n = d(\lg j)/d(\lg U)$, within which each approximation is valid [1.68]. The four analytical approximations and their validity range are summarized in Table 5.6 and illustrated in Fig.5.25.

For the first (I) SCLC region, at low voltages with $n = 2$, the best approximation is provided by the quadratic-range formula (I) proposed by BONHAM [5.46]. This corresponds to filling traps at the "tail" part of the Gaussian curve.

For the second (II) superquadratic region with $n > 2$ an appropriate approximation may be reached by formula (II), also derived by BONHAM [5.46],

Table 5.6. Analytical SCLC approximations of Gaussian trap distribution

j_{SCLC} region	Approximation formulae	Validity range
I quadratic range [5.46]	$$j_{SCL} = e\mu\varepsilon\,\frac{U^2}{L^3}\,\frac{N_e}{N_t}\,\exp\left(-\frac{E_t + (\sigma^2/2kT)}{kT}\right) \qquad (I)$$	$n = 2$
II superquadratic range [5.46]	$$j_{SCL} = e\mu N_e \exp\left(-\frac{E_t}{kT} + 2\right)\frac{U}{L}$$ $$\times \exp\left\{-\frac{1}{kT}\sqrt{2\sigma^2}\left[1 + \left(\frac{2kT}{\sigma}\right)^2 \ln\left(\frac{eL^2 A}{\varepsilon U}\right)^{*)}\right]\right\} \qquad (II)$$	from $n > 2$ to $n \approx 4$ $E_F < E_t$ $U < 1/2\,U_{TFL}$
III medium-high voltage range [5.44]	$$j_{SCL} = \frac{e\mu U}{L}\,\frac{(\varepsilon U)^\alpha\, N_e\,\exp[-(E_t/kT)]}{(eL^2 N_t - \varepsilon U)^\alpha},$$ $$\text{where } \alpha = \sqrt{\frac{2\pi\sigma^2}{16k^2T^2} + 1} \qquad (III)$$	from $n > 3$ to $n \approx 8$ $E_F > E_t;\ U < U_{TFL}$ $n_d \lesssim 0.1n_{td}$
IV high-voltage TFL range [5.44]	$$j_{SCL} = e\mu n_d \frac{U}{L} \qquad (a)$$ $$U = \frac{eL^2}{\varepsilon}\left[n_d + \frac{N_t}{1 + [(N_e/n_d)\exp(-E_t/kT)]^{1/\alpha}}\right] \qquad (b)$$ (IV)	from $n > 3$ to $n \approx n_{TFL}$ from $U < U_{TFL}$ to $U \approx U_{TFL}$

*) $A = \dfrac{N_t}{2(2\pi)^{1/2}} \cdot \dfrac{kT}{\sigma}\, e^{1/2} \displaystyle\int_{-\infty}^{\infty} \dfrac{\exp[-(y^2/2)(kT/\sigma)^2]}{\cosh(y/2)}\, dy$, where $y = \dfrac{E - E_F}{kT}$

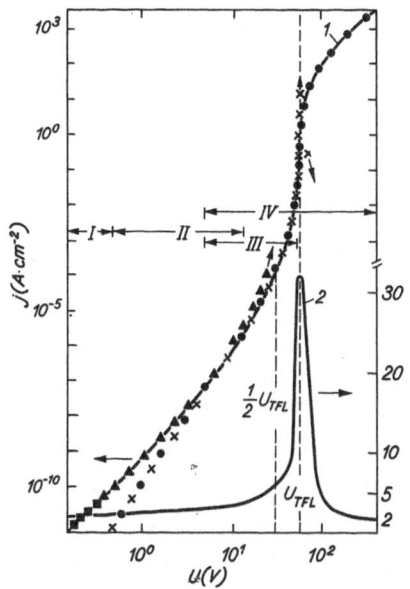

Fig. 5.25. (1) Calculated CV character-istics $\overline{J_{SCL}} = f(U)$ for molecular crystal with Gaussian $G_g(E)$ trap distribution; solid line: numerical integration data; (■) approximation by formula (I); (▲) by formula (II); (×) by formula (III); (●) by formula (IV) from Table 5.6. Parameters: $E_t = 0.5$ eV, $\sigma = 0.1$ eV; $T = 300$ K; $L = 10^{-4}$ cm; $N_t = 10^{16}$ cm^{-3}; $N_e = 10^{19}$ cm^{-3}; $\mu = 1$ cm^2/V·s. (2) Slope $n = f(U)$ of CV characteristic (1). U_{TFL} is trap-filled-limit voltage [1.68]

valid for the voltage range at which the quasi-Fermi level E_F passes over the steep portion of the Gaussian "bell" below the maximum E_t, i.e., in the energy interval $E_t > E_F > E_t - \sigma^2/kT$. This corresponds to the slope range from $n > 2$ up to $n \approx 4$ (see Fig.5.25).

The medium high-voltage region may be adequately described by approximation (III) proposed by NEŠPUREK and SMEJTEK [5.44]. This region includes the transition of E_F through the maximum E_t of the Gaussian curve and the post-maximum region ($U < U_{TFL}$). Approximation (III) is valid up to free and trapped carrier concentration ratio at the anode: $n_d/n_{td} \lesssim 0.1$. It does not, however, include the trap-filled-limit voltage U_{TFL} since at the $U \rightarrow U_{TFL}$ the denominator of approximation (III) exhibits singularity, viz. $(eL^2 N_t - \varepsilon U)^2 \rightarrow 0$, according to the expression for U_{TFL}

$$U_{TFL} = \frac{eL^2 N_t}{\varepsilon} \; . \tag{5.47}$$

The CV characteristic in the neighborhood of U_{TFL} is more adequately approximated by the set of equations (IV) [5.44].

As may be seen from Fig.5.25, the validity regions of approximations (II), (III) and (IV) partially overlap. Many reported experimental CV characteristics cover mainly regions (II) and (III) and may be adequately approximated by the corresponding formulae (cf. Table 5.6). If the total trap density N_t is not too high, the trap-filled-limit voltage U_{TFL} may be experimentally reached under SCLC conditions. In such a case the approximation set (IV) should be used.

The j_{SCL} in the trap-free SCLC region ($U > U_{TFL}$) with $n = 2$ (see Fig. 5.25) can be described by the conventional square-law fromula (cf. Sect. 5.8.3)

$$j_{SCL} = \frac{e\varepsilon\mu U^2}{L^3}$$ (5.48)

5.9.4 SCLC Dependences on Dispersion Parameter σ

Typical calculated CV characteristics for a molecular crystal with dispersion σ as a variable parameter for a $G_g(E)$-type trap distribution (cf. Fig. 3.2) are presented in Fig.5.26. In this case the CV characteristics $j_{SCL} = f(U)_\sigma$ form a common set of curves with a singular point of intersection M. At point M we have $U = U_M$ and $(j_{SCL})\sigma_i = (j_{SCL})\sigma_j$. Since in this voltage range approximation (III) is valid, we obtain the following relationship:

$$\left(\frac{eL^2 N_t - \varepsilon U_M}{\varepsilon U_M}\right)^{\alpha_i} = \left(\frac{eL^2 N_t - \varepsilon U_M}{\varepsilon U_M}\right)^{\alpha_j} \quad ,$$ (5.49)

whence

$$U_M = \frac{1}{2} \cdot \frac{eL^2 N_t}{\varepsilon} = \frac{1}{2} U_{TFL} \quad .$$ (5.50)

Thus, point M corresponds to the passage of E_F through the maximum of the Gaussian curve.

As may be seen from Fig.5.26, the slope n of the CV characteristics increases considerably with increasing σ value. For small σ values ($\sigma = 0.01$ eV) j_{SCL} follows a square law, $j_{SCL} \sim U^2$ for voltages up to $1/2\ U_{TFL}$, and the trap distribution may be regarded as being quasi-discrete. The criterion for distinguishing a narrow Gaussian distribution from a genuine discrete trapping level is provided by the temperature dependence $j_{SCL} = f(1/T)$. The $j_{SCL} = f(U)$ equation for a discrete trapping level at E_t may be formally derived as a specific case from the quadratic range formula (I) (Table 5.6) assuming zero value for σ. It follows that in the case of a discrete trapping level the dependence $j_{SCL} = f(1/T)$ should be linear, whereas for a

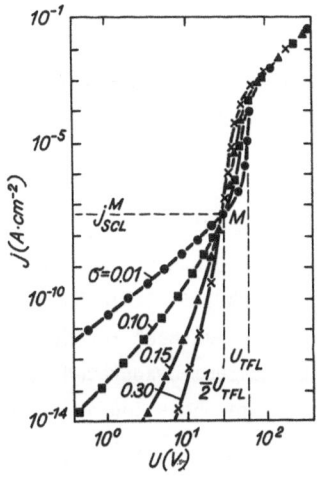

Fig. 5.26. Calculated CV characteristics j_{SCL} = f(U) for molecular crystal with dispersion σ as a variable parameter for $G_g(E)$ type of Gaussian trap distribution. Values of other parameters: E_t = 0.75 eV; T = 300 K; L = 10^{-3} cm; N_t = 10^{16} cm-3; N_e = 4×10^{21} cm-3; μ = 1 cm2/V·s [1.68]

Gaussian distribution the dependence j_{SCL} = f(1/T) should not be linear even for small σ values (σ ≈ kT).

5.9.5 SCLC Temperature Dependences for $G_e(E)$ and $G_g(E)$ Distributions

$G_e(E)$ and $G_g(E)$ trap distributions (cf. Fig.3.2) show pronounced difference in their SCLC temperature (T) characteristics. This is illustrated by calculated CV characteristics for different T values, j = f(1/T) dependences for different U values, and conductivity activation energies E_a = f(U) characteristics for both types of trap distribution (see Figs.5.27,28).

In the case of $G_e(E)$ trap distribution (E_t = 0) the slope n of CV characteristics increases with decreasing T, and they converge at the "half-filled" Gaussian "bell" limit (Fig.5.27a). At the same time, the slopes of j_{SCL} = f(1/T) curves rapidly decrease with increasing U (Fig.5.27b) as E_F approaches the edge of E_e and the corresponding conductivity activation energy E_a approaches zero: $E_a \to 0$ (Fig.5.27c).

In the case of $G_g(E)$ trap distribution (E_t = 0.5 eV) we observe a considerable difference in the behavior of SCLC temperature dependences (Fig.5.28) The increase of n with decreasing T is less pronounced, and CV characteristics converge as U approaches the U_{TFL} values (Fig.5.28a). The j_{SCL} = f(1/T) curves remain practically parallel with increasing U (Fig.5.28b), while E_a reaches the value of E_t ($E_a \to E_t$ = 0.5 eV) as the voltage U approaches 1/2 U_{TFL} (≈ 25V) (Fig.5.28c).

Thus the SCLC temperature dependence, especially E_a = f(U), can be considered as a suitable criterion for distinguishing $G_e(E)$ trap distribution

314

from deep and shallow $G_g(e)$ ones. It also provides a convenient method for E_t determination.

An application of the $E_a = f(U)$ criterion is illustrated by some experimental curves in Fig.5.29. The case in which $E_a \rightarrow 0$, pointing towards a $G_g(E)$ trap distribution, is represented by $E_a = f(U)$ curves for vacuum-evaporated polycrystalline layers of two heterocyclic organic compounds: trans-bisbindonilene (TBB) and betaine-2N-pyridinium-4-aza-indandione-1,3 (N-IPB) (cf. Fig.5.29a). Similar $E_a = f(U)$ dependences with $E_a \rightarrow 0$ exhibit also some other betaine-type derivates of indandione-1,3, e.g., IPB (cf. Table 5.8). The $G_e(E)$ type of Gaussian trap distribution with $E_a \rightarrow 0$ has been recently observed by EIERMANN et al. [3.63] also for amorphous vacuum-evaporated tetracene layers. On the other hand, in case of quasi-amorphous and oriented polycrystalline layers of tetracene and pentacene several sets of shallow $G_g(E)$-type trap distributions have been observed, the shallower one positioned at the depth $E_t \approx 0.1$ eV (cf. Tables 5.8,9, and Fig.5.39).

The case of deep $G_g(E)$-type trap distributions, when E_a approaches a constant value, is illustrated by the $E_a = f(U)$ curves for a tetracene single crystal, for which $E_a \rightarrow E_t \approx 0.6$ eV [3.4], and by the $E_a = f(U)$ curve for an anthracene single crystal for which $E_a \rightarrow E_t \approx 0.5$ eV [5.47] (see Fig.5.29b).

It should be mentioned that such relatively deep trapping levels with E_t value about 0.6 to 0.9 eV and half-width of ca. 0.1 to 0.2 eV have been observed by a number of authors in ultrahigh-purity melt-grown anthracene single crystals (see, e.g. [5.3-5] and Sect.5.12).

Fig. 5.27a,b. Calculated temperature (T) dependences for a $G_e(E)$ type of Gaussian trap distribution. Parameters: $E_t = 0$; $\sigma = 0.2$ eV; $L = 10^{-4}$ cm; $N_t = 10^{16}$ cm-3; $N_e = 4 \times 10^{21}$ cm-3; $\mu = 1$ cm2/V·s. (a) current-voltage characteristics $j_{SCL} = f(U)$ with T as a variable parameter; (b) dependences $j_{SCL} = f(1/T)$ with voltage U as a variable parameter; (c) conductivity activation energy E_a dependence on voltage U [1.68]

Fig. 5.28a,b. Calculated temperature (T) dependences for a $G_g(E)$ type of Gaussian trap distribution. Parameters: $E_t = 0.5$ eV; $\sigma = 0.05$ eV; other parameters as in Fig.5.27. (a) current-voltage characteristics $j_{SCL} = f(U)$ with T as a variable parameter; (b) dependences $j_{SCL} = f(1/T)$ with voltage U as a variable parameter; (c) conductivity activation energy E_a dependence on voltage U [1.68]

Fig. 5.29a,b. Experimental conductivity activation energy (E_a) dependences on voltage U: (a) for $G_e(E)$ type of Gaussian trap distribution in vacuum-evaporated sandwiched thin-layer cells of transbisbindonilene (TBB) (I) (curve 1) and betaine-2-N-pyridinium-4-aza-indandione-1,3 (N-IPB) (II) (curve 2), according to [1.68]; (b) for $G_g(E)$ type of Gaussian trap distribution in a tetracene single crystal (I) (curve 1), according to [3.4], and in an anthracene single crystal (II) (curve 2), according to [5.47]

315

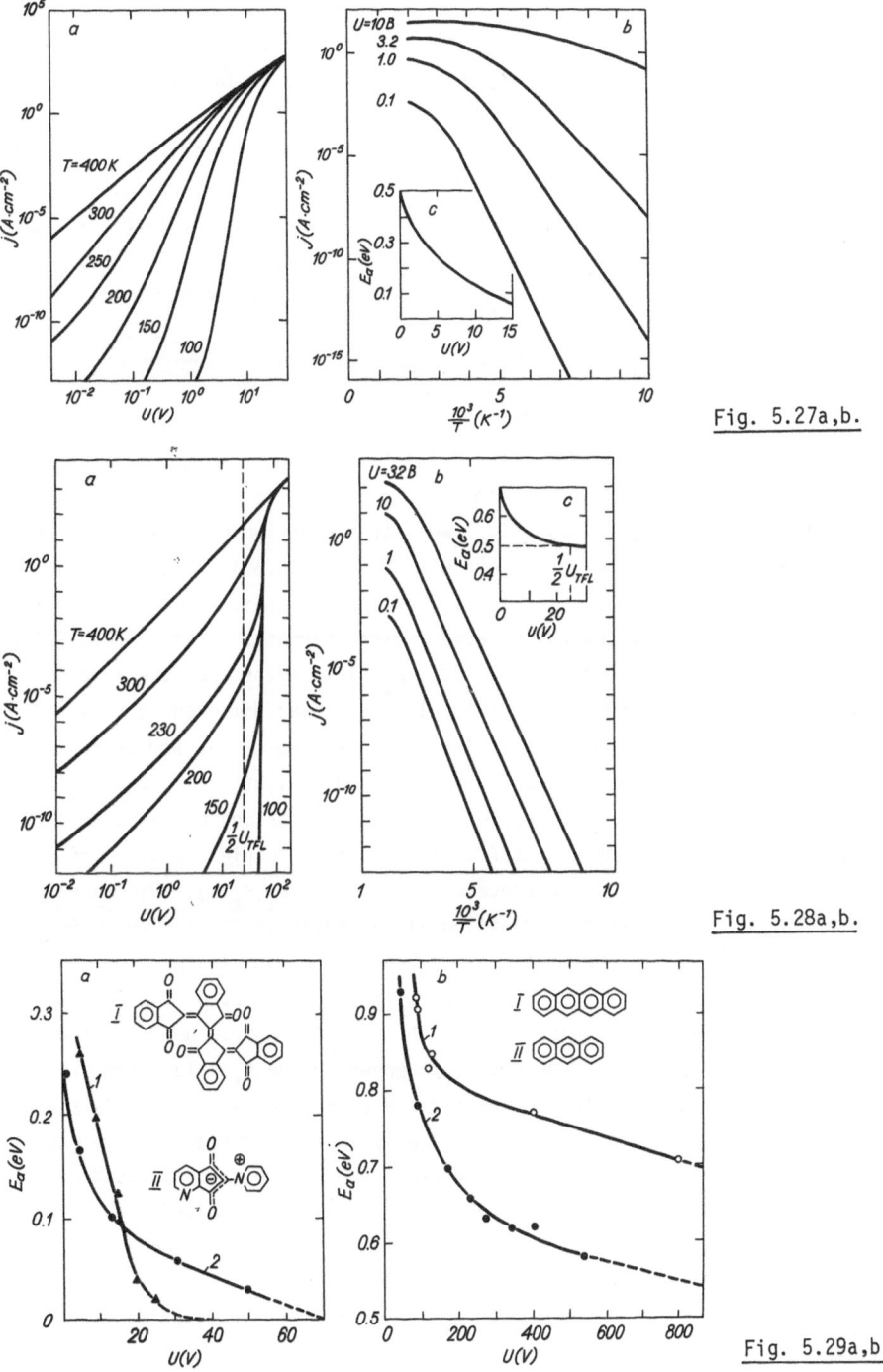

Fig. 5.27a,b.

Fig. 5.28a,b.

Fig. 5.29a,b.

The data show that vacuum-evaporated amorphous, quasi-amorphous and poly-crystalline layers exhibit only either $G_e(E)$ or shallow $G_g(E)$ types of trap distribution, whilst in single crystals, especially those obtained from melt, deep traps of $G_g(E)$ distribution dominate.

This is in good agreement with our present knowledge about the nature of deep structural trapping centers in molecular crystals (cf. Sects.3.2 and 5.7). Only in single crystals such collective defective structures as dislocation aggregations and ensembles may be formed (cf. Sect.3.7), being responsible for the creation of deep structural trapping centers for charge carriers.

5.9.6 SCLC Dependences on E_t Value

Important conclusions for the correct interpretation of experimental data may be drawn from an analysis of the set of the CV characteristics, calculated for different E_t values (Fig.5.30). If a Gaussian distribution curve, having a definite σ value (e.g., E = 0.1 eV in Fig.5.30) is shifted downwards from the value $E_t = 0$ to $E_t = 1.0$ eV in the forbidden energy gap (Fig. 5.30b), the slope n of the CV characteristics increases steadily (Fig.5.30a), especially for higher voltages $U \gtrsim 1/2\ U_{TFL}$. This means that the same Gaussian distribution, having a definit σ value, may have different n values, depending on the position of E_t in the forbidden energy gap and on voltage U. Consequently, the n value for a Gaussian trap distribution cannot be, in principle, used as a direct measure for the determination of the trap distribution parameter σ. As a matter of fact, n is a complex function of σ, E_t, and N_t in a given voltage range, i.e., $n = f(\sigma, E_t, N_t, U)$. This conclusion is in contradiction with the traditionally accepted assumption which is an unjustified generalization of the widely used exponential approximation. (We wish to recall that for an exponential trap distribution the n value is really unambiguously determined by the characteristic trap distribution energy kT_c : $n = (kT_c/kT) + 1$ and does not depend on the E_t value) [cf. (5.33)].

Thus, a high-slope CV characteristic may be caused by a narrow Gaussian distribution, positioned sufficiently deeply in the forbidden energy gap, and it would be a fatal error to interpret such high n value as evidence of a wide trap distribution. If, for example, the CV characteristic for a $G_g(E)$ trap distribution with $\sigma = 0.1$ eV and $E_t = 1.0$ eV would have been approximated exponentially at $U > 1/2\ U_{TFL}$, the corresponding value n = 18 (Fig. 5.30a) would give $kT_c = 0.5$ eV, i.e., a very broad exponential distribution of traps.

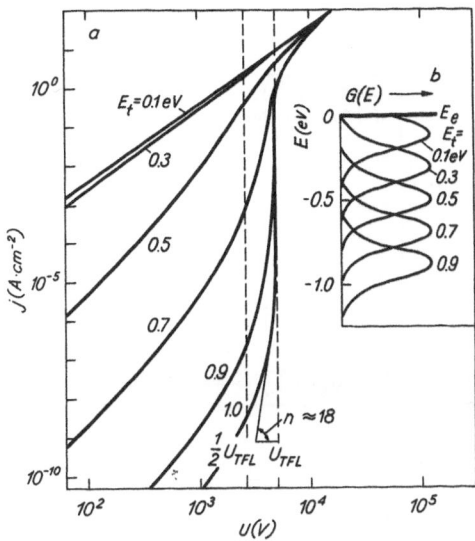

Fig. 5.30. (a) Calculated current-voltage characteristics $j_{SCL} = f(U)$ with Gaussian distribution maximum energy E_t as a variable parameter. Other parameters: $\sigma = 0.1$ eV; $T = 300$ K; $L = 10^{-2}$ cm; $N_t = 10^{14}$ cm^{-3}; $N_e = 4 \times 10^{21}$ cm^{-3}; $\mu = 1$ cm^2/V·s; (b) corresponding Gaussian distribution curves $G(E)$ [1.68]

Possibilities of such interpretation errors are not inherent to the Gaussian approach, since the latter does not regard, unlike the exponential one, the slope value n as an indicator of trap distribution.

Instead, according to the Gaussian approach the curvature of the CV characteristic is correlated, in the given voltage range, with two variable distribution parameters, namely, σ and N_t, which are found by approximation of experimental CV curves by means of analytical formulae in Table 5.6 (cf. Sect. 5.10).

In such an approximation procedure the E_t value should be given, i.e., should be determined independently by the activation energy method from $E_a = f(U)$ dependence (cf. Sect. 5.9.5) and preferably checked by some other independent method, such as TSC, PSC, or IDC methods (cf. Sect. 5.12).

5.9.7 Validity Criteria for Exponential and Gaussian Approximations

The main validity criteria for exponential and Gaussian distributions are presented in Table 5.7 [cf. (5.39-41) and formulae in Table 5.6].

An analysis of many reported experimental $j_{SCL} = f(U)$ dependences for different molecular single crystals, as well as for vacuum-evaporated layers,

318

Table 5.7.

Validity criteria

SCLC dependence	Exponential trap distribution $N(E) = \dfrac{N_t}{kT_c} \exp\left(-\dfrac{E}{kT_c}\right)$	Gaussian trap distribution $N(E) = \dfrac{N_t}{\sigma\sqrt{2\pi}} \exp\left(-\dfrac{(E-E_t)^2}{2\sigma^2}\right)$
lg j versus lg U	linear $n = \dfrac{d(\lg j)}{d(\lg U)} = \text{const}$	nonlinear $n = \dfrac{d(\lg j)}{d(\lg U)} \neq \text{const}$
lg j versus lg L	linear $\ell = \dfrac{d(\lg j)}{d(\lg L)} = \text{const}$	nonlinear $\ell = \dfrac{d(\lg j)}{d(\lg L)} \neq \text{const}$

shows that, in the majority of cases, similarly as for the generalized prototype CV characteristic in Fig.5.22, the slope n of log j_{SCL} versus log U
increases monotonously with field strength $\mathscr{E}$ and n actually may be regarded
as constant only for a narrow voltage range. Thus, according to the first
criterion of Table 5.7 the exponential approximation of trapping states has
been, indeed, merely an *ad hoc* assumption which happens to fit the experimental results for only a narrow voltage range.

But nonlinearity of the lg j_{SCL} versus lg U dependence is merely a condition sufficient for excluding an exponential approximation, but not sufficient for proving the Gaussian one. For that it is necessary to use an appropriate analytical approximation from Table 5.6 and to prove agreement
with experimental data (as has been done for a number of CV characteristics
in Sect.5.10).

The second criterion—nonlinearity of the lg j_{SCL} versus lg L dependence
—is obvious from the approximation formula (II). Approximation formula
(III) also gives nonlinear j_{SCL} versus lg L dependences which become more
obvious after some transformations of (III) at U = const

$$j \sim \frac{1}{L^{2\alpha+1}} \frac{1}{\left(1 - \dfrac{\varepsilon U}{eN_t} \dfrac{1}{L^2}\right)^\alpha} \quad . \tag{5.51}$$

Calculated nonlinear lg j versus lg L dependences for a Gaussian trap
distribution are presented in Fig.5.31a (solid lines) which are compared
with a linear lg j versus lg L dependence for an exponential trap distribution. Experimental nonlinear lg j versus lg L dependences for vacuum-evaporated transbisbindonilene (TBB) thin-layer specimens, speaking in favor of
a Gaussian trap distribution, are demonstrated in Fig.5.31b (cf. Sect.5.8.5
and Fig.5.23).

5.9.8 CV Characteristics for Two Sets of Gaussian Trap Distribution

In Fig.5.32a the calculated CV characeristics for single and double sets of Gaussian trap distribution are presented, the latter being shown in Fig. 5.32b. It is rather striking that the CV characteristics calculated for a single $G_g(E)$ distribution (I) practically coincide with the CV characteristic calculated for double $G_g(E)$ (I) and $G_e(E)$ (II) distributions (Fig.5.32). This means that for a relatively high total density of traps in the first (I) $G_g(E)$ trap distribution ($N_t = 10^{16}$ cm^{-3}) its trap-filled limit U_{TFL} could not be reached in a reasonable, physically possible voltage range, and the second (II) $G_e(E)$ trap distribution does not influence at all the shape of the CV characteristic. Such a situation is rather common for many reported experimental CV characteristics having $G_g(E)$-type trap distribution of high trap density N_t or considerable sample thickness L in case of single crystal specimens.

For overlapping $G_g(E)$ and $G_e(E)$ (III) distributions (Fig.5.32b, Curve IV) the CV characteristic reaches higher n values (Fig.5.32a, Curve 2) and may be easily misinterpreted as caused by a wide distribution "bell". Actually, the CV curve (2) covers only the filling of the narrow $G_g(E)$ distribution "bell", centered at $E_t = 0.5$ eV (Fig.5.32b, Curve 1), and is only slightly influenced by the "tail" of the wide $G_e(E)$ distribution curve (Fig.5.32b, Curve III).

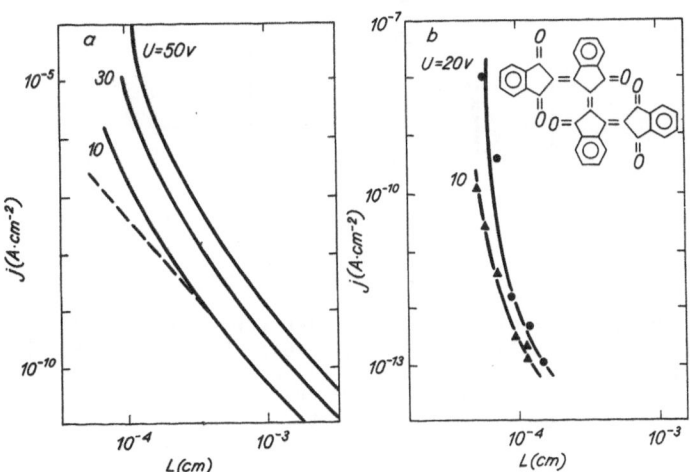

Fig. 5.31. (a) Calculated current dependences on sample thickness $j_{SCL} = f(L)$ for Gaussian (solid line) and exponential (dotted line) trap distributions; (b) experimental $j_{SCL} = f(L)$ dependences for a thin layer cell of transbisbindonilene (TBB) Al/TBB/Au [1.68]

A single wide $G_e(E)$ trap distribution curve (σ = 0.3 eV) (Fig.5.32b, Curve III) displays a very steep CV characteristic (Fig.5.32a, Curve 3) which represents only the superquadratic region.

An interesting situation may arise in the case of two or more narrow sets of $G_g(E)$ low-density trap distributions. Such a situation is shown in Fig. 5.33, which represents a calculated CV characteristic for a double set of $G_g(E)$-trap distribution. In this case the U_{TFL} of the deeper narrow $G_g(E)$ distribution appears as fine structure on the CV characteristic (Fig.5.33, Curve 1). This fine structure becomes more obvious on the differential CV curve (Fig.5.33, Curve 2). If no notice is taken of the fine structure, the CV characteristic in Fig.5.33 can easily be mistaken for a single exponential trap distribution with a constant value of n = 6 and kT_c = 0.15 eV.

As an illustration some typical experimental step-like CV characteristics are shown in Fig.5.34, first obtained by SILINSH et al. [1.67,68] for vacuum-evaporated thin-layer specimens of quasi-amorphous tetracene and pentacene and polycrystalline perylene. These step-like CV characteristics (cf.

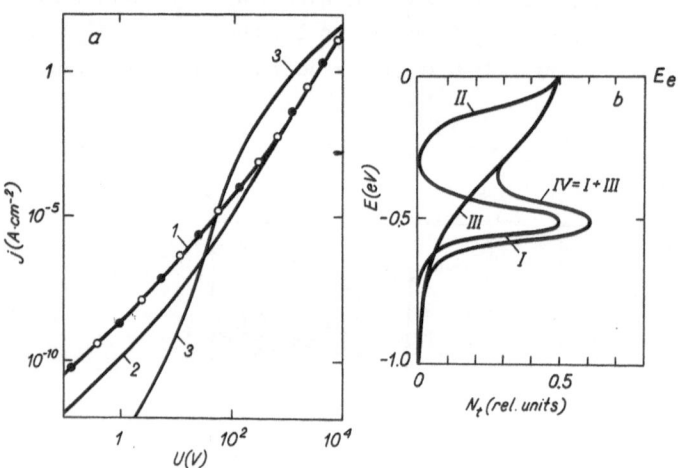

Fig. 5.32. (a) Calculated current-voltage characteristics j_{SCL} = f(U) for single and double sets of Gaussian trap distribution: (1) • - for a single $G_g(E)$ distribution (I) (E_t = 0.5 eV; σ = 0.06 eV); • - for double Gaussian distributions: $G_g(E)$ (I) (E_t = 0.5 eV; σ = 0.06 eV) and $G_e(E)$ (II) (E_t = 0 eV; σ = 0.1 eV); (2) for double overlapping (IV) Gaussian distributions: $G_g(E)$ (E_t = 0.5 eV; σ = 0.06 eV) (I) and $G_e(E)$ (E_t = 0; σ = 0.3 eV) (III); (3) for a single $G_e(E)$ distribution (III) (E_t = 0; σ = 0.3 eV); other parameters: T = 300 K; L = 10^{-4} cm; N_t = 10^{16} cm^{-3}; N_e = 4 × 10^{20} cm^{-3}; μ = 1 cm^2/V · s. (b) Corresponding Gaussian trap distribution curves G(E) in the forbidden energy gap below the electron conductivity level E_e; N_t = relative density of traps [1.68]

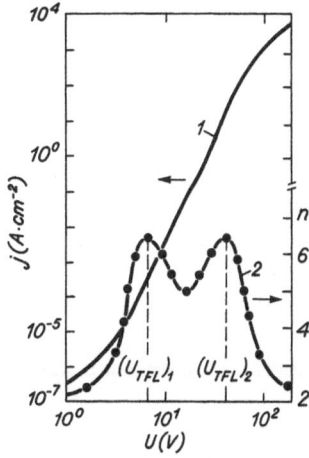

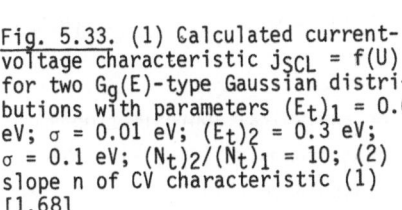

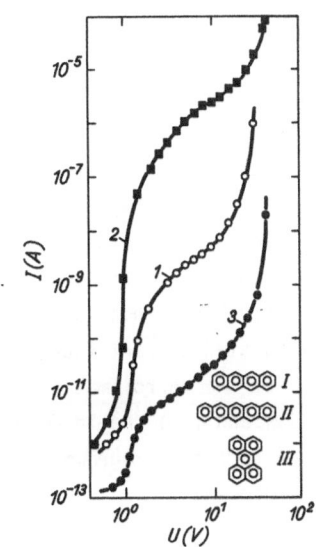

Fig. 5.33. (1) Calculated current-voltage characteristic $j_{SCL} = f(U)$ for two $G_g(E)$-type Gaussian distributions with parameters $(E_t)_1 = 0.6$ eV; $\sigma = 0.01$ eV; $(E_t)_2 = 0.3$ eV; $\sigma = 0.1$ eV; $(N_t)_2/(N_t)_1 = 10$; (2) slope n of CV characteristic (1) [1.68]

Fig. 5.34. Experimental current-voltage characteristics $j_{SCL} = f(U)$, interpreted as a representation of a double set of Gaussian trap distribution: (1) quasi-amorphous layers of tetracene (I); (2) quasi-amorphous layers of pentacene (II); (3) polycrystalline layers of perylene (III) [1.68]

also Fig.5.39) are interpreted as a representation of a double set of Gaussian trap distributions and have been approximated by analytical formulae (III) and IV) (Table 5.6) (cf. Sect.5.10). Similar step-like CV characteristics have been recently observed in vacuum-evaporated thin-layer systems for a number of aromatic and heterocyclic molecular crystals [5.42] (cf. Table 5.8).

5.10 Gaussian SCLC Approximations of Experimental CV Characteristics

5.10.1 Analytical Approximations

In the case of a single set of Gaussian traps the corresponding CV characteristics exhibit, under SCLC conditions, a steadily increasing slope value $n = d(\lg j)/d(\lg U)$ in the voltage range $U \lesssim U_{TFL}$, and an S-shaped curve for

voltages $U > U_{TFL}$ (cf. Fig.5.25). If several sets of Gaussian trap distribution are trap filled, a step-like CV characteristic emerges (see Figs.
5.34,39).

For the approximation of experimental CV characteristics with a monotonously increasing slope value, the analytical formulae (II) and (III) can
be used. In the approximation procedure the unknown parameters σ, N_t, and
N_e are found. The parameter E_t should be determined by an independent method, e.g., from dark conductivity activation energy E_a dependence on voltage
U under SCLC condition: $E_a = f(U)$ (cf. Sect.5.9.5) or by thermo- or photostimulated current techniques (cf. Sect.5.12).

For the approximation of the S-shaped CV characteristics both formulae
(III) and (IV) should be used (cf. Sect.5.9.3).

A computer-programmed least-squares method, based on minimization of the
expression

$$\Phi = \sum_{i=1}^{n} [\lg j_{SCL}^{exp} (U_i) - \lg j_{SCL}^{cal} (U_i, N_t, \sigma, N_e)]^2 \rightarrow \min , \qquad (5.52)$$

has been applied by SILINSH et al. [5.42] for analytical approximation of
the experimental CV characteristics according to formulae (I)-(IV) (Table
5.6). j_{exp} is the experimental and j_{cal} the calculated value of current density at voltage U_i. Consequently, parameters N_t, σ, and N_e should be varia-
ted in order to minimize (5.52). Parameters N_t and σ determine the curvature
of the CV characteristic, whereas parameter N_e determines its linear shift
in $\lg j$, $\lg U$ coordinates.

Typical minimization curves $\Phi = \Phi (N_t)$, according to (5.52), for a single
crystal (SC) and oriented polycrystalline layers (OL) of tetracene (Tc) are
presented, according to [5.42], in Fig.5.35. As can be seen from Fig.5.35,
the $\Phi = \Phi (N_t)$ dependence exhibits a very distinct minimum. Typical σ and
N_e dependences on N_t are also shown in Fig.5.35.

In the case of satisfactory approximation the value of Φ/n, after minimization, should be less than 10^{-4} where n is the number of experimental
points on the CV characteristic used in the approximation procedure (cf.
Fig.5.35).

We have approximated a large number of our own, as well as of reported
experimental CV characteristics under SCLC conditions and determined, applying analytical approximation formulae (II)-(IV) (Table 5.6), Gaussian distribution parameters for a variety of organic molecular crystals. Typical values of Gaussian trap distribution parameters obtained for a number of aroma-

Table 5.8. Typical values of Gaussian trap distribution parameters E_t, σ, N_t and N_e in a number of aromatic and heterocyclic molecular crystals of different crystalline state [5.42]

Compound	Crystalline state	E_t [eV]	σ [eV]	N_t [cm^{-3}]	N_e [cm^{-3}]
(Ac)	SC	0.5-0.8	0.03-0.13	$2 \times 10^{11} - 8 \times 10^{13}$	$10^{16} - 10^{20}$
(Tc)	SC	0.6	0.05-0.14	$10^{13} - 10^{14}$	$10^{18} - 10^{21}$
	OL	0.3-0.4	0.05-0.14	$6 \times 10^{13} - 2 \times 10^{14}$	
		0.1	0.07-0.13	$4 \times 10^{14} - 6 \times 10^{14}$	$10^{12} - 10^{15}$
	QA	0.26	0.11	2×10^{15}	
		0.1	0.05	4×10^{15}	
(Pc)	OL	0.35	0.11	8×10^{13}	$10^{14} - 10^{15}$
		0.08	0.07	8.5×10^{14}	
	QA	0.35	0.06	8×10^{15}	$10^{14} - 10^{15}$
		0.04	0.07	7×10^{13}	
(Pl)	PC	0.4	0.03	4×10^{13}	$10^{11} - 10^{15}$
		0.3	0.05-0.09	$(4-7) \times 10^{15}$	
(TTT)	PC	0.33	0.07-0.11	$6 \times 10^{14} - 3 \times 10^{15}$	$10^{16} - 10^{17}$
		0.2	0.06-0.08	$(4-8) \times 10^{15}$	
(IPB)	PC	0	0.08-0.12	$5 \times 10^{15} - 2 \times 10^{16}$	$10^{10} - 10^{11}$
(N-IPB)	PC	0	0.12-0.15	$(4-8) \times 10^{15}$	$\sim 10^{11}$
(TBB)	PC	0	0.05-0.14	$4 \times 10^{15} - 4 \times 10^{16}$	$10^{10} - 10^{13}$

Notation: SC - single crystal; OL - oriented polycrystalline layers obtained by evaporation in vacuo on substrate at T = 300 K (average grain dimensions: d = 0.1-1 μm); QA - quasi-amorphous layers obtained by evaporation in vacuo on cooled substrate at T ≈ 200 K (d ≲ 500 Å); PC - polycrystalline layers obtained by evaporation in vacuo at T = 300 K.

tic and heterocyclic molecular crystals of different crystalline states are shown in Table 5.8.

The approximation procedure is illustrated in Fig.5.36 for experimental CV characteristics of a tetracene single crystal as reported by BAESSLER et al. [3.4]. The CV characteristic in Fig.5.36 was approximated by formulae (II) and (III) (Table 5.6). σ and N_t values obtained by either of these formulae are usually rather close, although, in the case of formula (II), the obtained parameters σ and N_t are slightly lower (see Fig.5.36). For the approximation data presented in Table 5.8, formula (III) was mainly used.

Figure 5.37 shows approximation of an experimental CV characteristic for an anthracene single crystal, reported by MANY et al. [5.36] using both approximation formulae (II) and (III) (Table 5.6). In Figs.5.36,37 typical CV characteristics of molecular single crystal samples are presented, exhibiting monotonously increasing slope n value. Such CV curves correspond to incomplete filling of deep traps situated at ca. 0.5-0.8 eV (cf. Table 5.8), well below trap-filled-limit voltage U_{TFL}. Unfortunately, for relatively thick single crystal samples with L = 50-100 μm it is impossible to reach within physically accessible voltage range the trap-filled limit and to look "behind" the set of deep trapping states, using SCLC technique.

Figure 5.38 demonstrates the approximation of a typical experimental CV characteristic by formula (III) (Table 5.6) in case of a vacuum-evaporated thin-layer specimen of indandione-1,3-pyridinium betaine (IPB) Au/IPB/Al having $G_e(E)$ type of shallow trap distribution with $E_t \approx 0$ (cf. Table 5.8).

In case of two or more sets of Gaussian trap distribution a step-like CV characteristic in SCLC regime should emerge (cf. Sect.5.9.8).

Such CV characteristics can be safely observed experimentally, however, only in the case of thin-layer systems, at sample thickness L about 0.5-3 μm and, additionally, at not too high total trap density of the more deeply situated distribution set (of an order of 10^{14}-10^{15} cm^{-3}, cf. Table 5.8). Under such conditions it is possible to fill completely the deepest set of traps within accesible voltage range, reach the corresponding trap-filled limit U_{TFL} and, subsequently, shift the quasi-Fermi level to the second shallower set of traps. Thus it is possible to scan both sets of traps during one experimental run which results in a step-like CV characteristic.

The possibility of step-like CV characteristic formation as a result of the presence of several sets of quasi-discrete traps inside the energy gap of an anthracene crystal has been discussed already by POPE and KALLMANN [5.48].

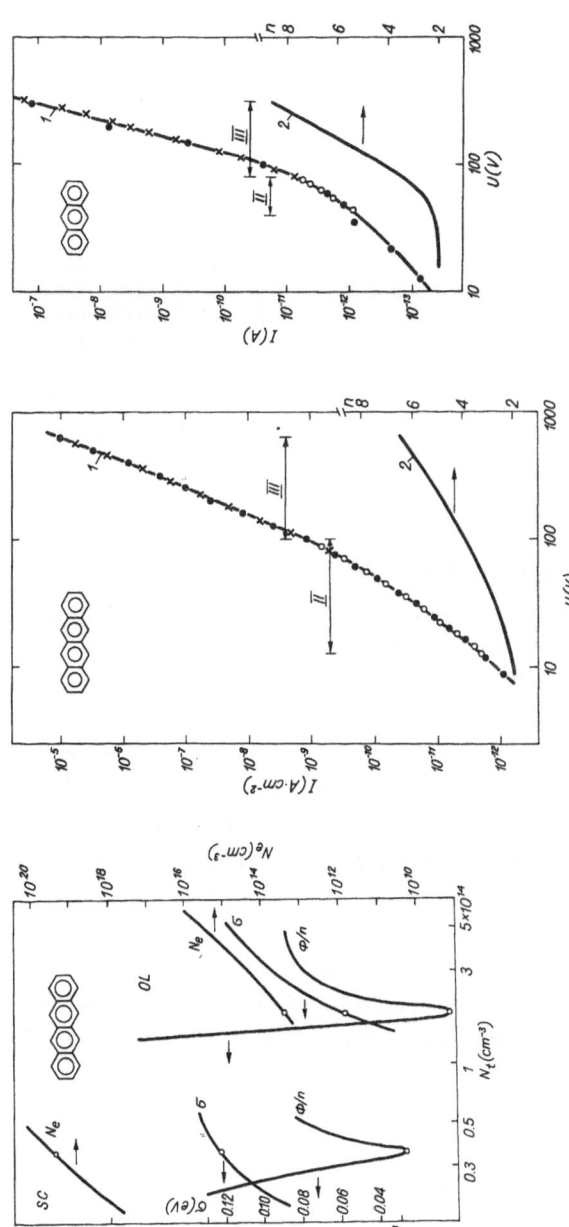

Fig. 5.35. Typical minimization curves $\Phi = \overline{\Phi(N_t)}$ and $\sigma = \sigma(N_t)$ and $N_e = N_e(N_t)$ dependences for single crystal (SC) and oriented polycrystalline layers (OL) of tetracene [5.42]

Fig. 5.36. Approximation of experimental CV characteristic of tetracene single crystal (1), according to [5.42]: (●) averaged experimental points, according to [3.4]; (○) approximation by formula (II) yielding $\sigma = 0{,}11$ eV and $N_t = 5 \times 10^{13}$ cm⁻³; (×) by formula (III) yielding $\sigma = 0{,}14$ eV and $N_t = (1\pm0{,}5) \times 10^{14}$ cm⁻³; 2-value of slope n for CV characteristic (1)

Fig. 5.37. Approximation of experimental CV characteristic for an anthracene single crystal (1): (●) averaged experimental points, according to [5.36]: (○) approximation by formula (II); (×) approximation by formula (III) yielding $\sigma = 0{,}13\pm0{,}03$ eV and $N_t = (7-8) \times 10^{13}$ cm⁻³; 2-value of slope n for CV characteristic (1)

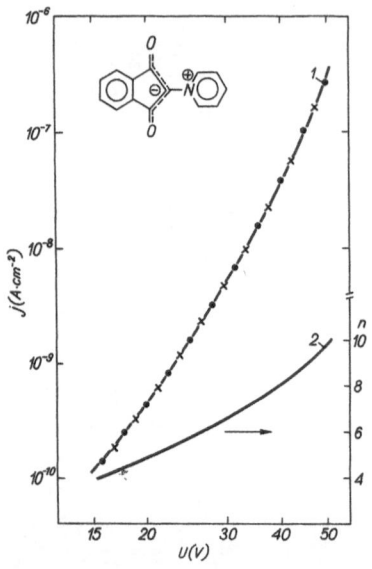

Fig. 5.38. Approximation of the experimental CV characteristic for a vacuum-evaporated thin-layer indandione-1,3-pyridinium betaine (IPB) cell Au/IPB/Au (1): (•) averaged experimental points; (×) approximation by formula (III) yielding σ = 0,12 eV and N_t = 2×10^{16} cm^{-3}; 2-value of slope n for CV characteristic (1)

There have been earlier reports on fine step-like structure of experimental CV curves of molecular crystals. Thus, LAMPERT and MARK [Ref. 1.76, p.110] presented a CV characteristic for vacuum-evaporated copper phthalocyanine layers, showing distinguishable "steps", similar to those of calculated CV characteristic for two sets of Gaussian distribution of traps (cf. Fig.5.33). These CV curves were obtained by SUSSMAN and had not been previously published. In his paper SUSSMAN [3.3] treated such CV curves, however, as linear ones and applied the exponential distribution model. Small "steps" on the CV curves for an anthracene single crystal have also been detected by SWORAKOWSKI [5.2]. These "steps" are, however, just about within the resolution limit of the SCLC method.

Clearly discernible and reliable step-like CV characteristics were first observed in thin-layer vacuum-evaporated specimens of tetracene, pentacene and perylene by BALODE et al. [1.104] and SILINSH et al. [1.67,68] (see Figs.5.34,39).

Figure 5.39 demonstrates the approximation procedure of a step-like CV characteristic for polycrystalline layers of tetracene obtained by evaporation *in vacuo* on a substrate at 300 K (average grain dimensions: d = 0.2-1 μm), according to SILINSH et al. [5.42]. This step-like CV characteristic was approximated by two Gaussian distribution sets using formulae (III) and (IV) (Table 5.6). The $(U_{TFL})_1$ and $(U_{TFL})_2$ are the trap-filled-limit voltages

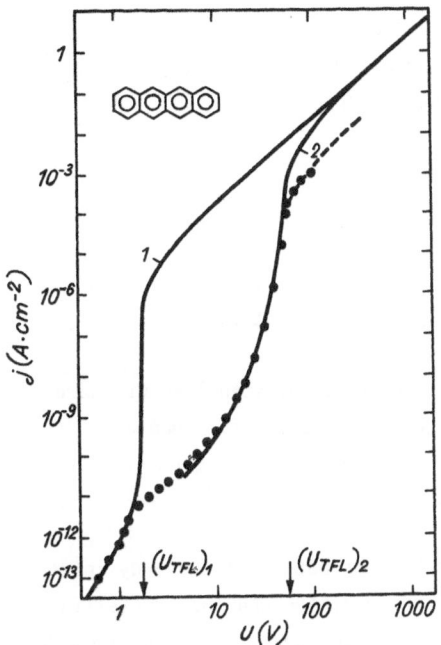

Fig.5.39. Approximation of experimental step-like CV characteristic of oriented polycrystalline tetracene (Tc) layers by formulas (III) and (IV) (Table 5.6). (1) for first, deeper set of trap distribution (E_t = 0.35 eV, σ = 0.05 eV, N_t = 1.3 × 10^{14} cm^{-3}); (2) for second, shallow set of traps (E_t ≈ 0.1 eV; σ = 0.09 eV; N_t = 5.8 × 10^{15} cm^{-3}) [5.42]

for the first (I) deepest and the second (II) shallower distribution sets, respectively. The corresponding trap distribution parameters are given in the caption to Fig.5.39. If there had been only one (the deepest) trap distribution, only one S-shaped CV characteristic would emerge (Fig.5.39, Curve 1). The presence of the second (shallow) trap distribution produces the step-like CV characteristic which may be approximated by the second S-shaped curve (Fig.5.39, Curve 2).

The approximation procedure of such two-step CV characteristics is as follows.

First, a rough estimate of total trap densities $(N_t)_I$ and $(N_t)_{II}$ of both distributions is made by expression (5.47). The rough N_t estimates are further used as initial data for minimization procedure according to (5.52). At first approximation is performed using the analytical formula (III) for slope value range 3 < n < 8 (cf. Table 5.6) and below the U_{TFL} voltage. The corresponding σ, N_t and N_e values are obtained for both distributions. Then these N_t values are inserted in formula (IV), and approximation is performed for the whole S-shaped CV characteristic including U ≥ U_{TFL} region.

The E_t value for the second shallower trap distribution, viz. $(E_t)_{II}$ can be determined from conductivity activation energy E_a dependence on voltage

$E_a = f(U)$ (cf. Sect.5.9.5). Knowing $(E_t)^{II}$ one can evaluate the $(E_t)^I$ value for the first deepest distribution, using experimental $j_M^{(I)}$, $j_M^{(II)}$, $U_M^{(I)}$ and $U_M^{(II)}$ values according to the following expression:

$$E_t^{(I)} = E_t^{(II)} + kT \left(\ln \frac{j_M^{(II)}}{j_M^{(I)}} - \ln \frac{U_M^{(II)}}{U_M^{(I)}} \right) , \qquad (5.53)$$

where $U_M = 1/2\, U_{TFL}$ [cf. (5.50)] and $j_M = j(U_M)$. The indices (I) and (II) correspond, respectively, to the first (I) deepest and the second (II) shallower trap distribution.

The above procedure, applied for approximation of step-like CV characteristics for vacuum-evaporated thin layers of tetracene (Tc), pentacene (Pc), perylene (Pl) and tetrathiotetracene (TTT), yielded corresponding trap distribution parameters for a double set of Gaussian traps which are presented in Table 5.8.

Table 5.8 demonstrates distinct correlations between the energy spectra and distribution parameters of local trapping states and the crystalline structure of samples. These correlations will be subject to special discussion in Sect.5.14.

5.10.2 Differential Method of Analysis of CV Characteristics

NEŠPUREK and SWORAKOWSKI [5.49] have proposed a simple differential method for the analysis of CV characteristics under SCLC conditions. The method permits one to estimate both depths and densities of traps from a single CV characteristic measured at constant temperature.

The following procedure of estimating these parameters is suggested, according to [5.49]:

a) depths of traps, scanned by the quasi-Fermi level E_F during an experimental run, can be estimated from the maximum on the $d[1/(n-1)]/d(E_F/kT)$ versus E_F curves [or, alternatively, from the inflection points on the $1/(n-1)$ versus E_F curves] where n is the slope of the CV characteristic on log-log plots.

b) densities of the traps can be either calculated according to the conventional trap-filled-limit formula [cf. (5.47)], or the relative trap densities can be found from the areas under $1/(n-1)$ versus E_F curves.

However, the method was shown to give correct results only for discrete traps and narrow trap distributions with σ value not exceeding kT, viz. $\sigma \lesssim kT$, whereas it yields overestimated values of trap depth if applied to broader distributions [5.49,50]. It has also been shown by the authors of [5.49] that the trap-filled-limit voltage U_{TFL} for distributed traps depends on σ value, namely

$$U_{TFL} = U_{TFL,discr.} \; f(\sigma) \tag{5.54}$$

The function $f(\sigma)$ amounts to unity for $\sigma \to 0$ and decreases for increasing σ values: $1 \geq f(\sigma) \geq 1/2$.

This important result shows that the conventional trap-filled-limit voltage formula (5.47) can be safely used only for discrete traps. In the case of Gaussian distribution of traps it can be used only for rough approximation (cf. Sect.5.10.1).

Recently NEŠPUREK and SWORAKOWSKI [5.50] have extended the differential method of analysis of CV characteristics under SCLC conditions to any arbitrary distribution of traps. The extended method is based on the assumption that on changing the voltage one changes the injection level, and consequently the quasi-Fermi level E_F is shifted within the forbidden energy gap. If one assumes the low temperature approximation to be valid, E_F constitutes a demarcation level between empty and filled traps. Hence, in principle, one can determine distributions of traps scanned by the quasi-Fermi level during an experimental run.

As shown in [5.50], one may determine the function characterizing an energy distribution of traps, $h(E)$, from the equation

$$h(E_F) = \alpha \, \frac{1}{n-1} \, \exp(- \frac{1}{kT} \int_{E_0}^{E_F} \frac{1}{n-1} \, dE_F) \tag{5.55}$$

n being the slope of the CV characteristic in log-log plots; E_F denoting the position of quasi-steady-static Fermi level at the rear (noninjecting) electrode; E_0 standing for the depth of the deepest traps scanned during an experimental run, and α being a normalization parameter. The position of E_F can be determined from the relationship

$$E_F(U) = kT \, \ln\frac{x_1 N_e e\mu}{L} + kT \, \ln\frac{U}{J} \quad , \tag{5.56}$$

where μ is the carrier mobility, L the electrode spacing, j stands for the current density, N_e for effective density of conductivity states, and x_1 is a parameter ($1 \leq x_1 \leq 2$).

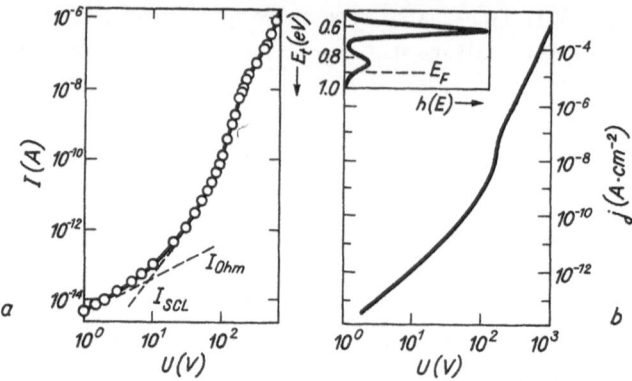

Fig. 5.40. Experimental (a) and computed (b) CV characteristics of a DPA single crystal. Parameters used in calculations: $E_t(I) = 0.82$ eV, $N_t(I) = 2.7 \times 10^{12}$ cm^{-3}; $\sigma(I) = 0.044$ eV; $E_t(II) = 0.64$ eV; $N_t = 9.7 \times 10^{12}$ cm^{-3}; $\sigma(II) = 0.028$ eV; $N_e = 4 \times 10^{21}$ cm^{-3}; $\mu = 10^{-2}$ cm^{2}V^{-1}s^{-1}; T = 300 K. Inset: h(E) function calculated according to (5.55) showing the proposed energy distribution of traps [5.51]

The differential method described was used by NEŠPUREK et al. [5.51] for the investigation of trap distribution in sublimation-grown N,N'-diphenyl-p-phenylendiamine (DPA) single crystals by SCLC technique. Experimental and computed CV characteristics of a DPA crystal are shown for comparison in Fig.5.40. The computations were carried out employing the exact SCLC theory for Gaussian trap distribution using the parameters contained in the caption to Fig.5.40. The parameters used in computations were deduced from the analysis of differential curves. Such analysis led the authors to the conclusion that the experimental step-like CV characteristic is controlled by two sets of narrow trap distributions centered at 0.64 and 0.82 eV, their distribution parameters not exceeding 0.03 and 0.05 eV, respectively. The h(E) function, as calculated from (5.55), is shown in the inset of Fig.5.40.

An obvious disadvantage of the differential method, as presented in [5.50,51], is the fact that the parameter N_e of the effective density of conductivity states should be previously known in order to find the correct E_F value according to (5.56). The authors of [5.51] used the value of $N_e = 4 \times 10^{21}$ cm^{-3}, as recommended by HELFRICH [1.75], which actually is the density of molecules per cm^{-3}. The effective density of conductivity states N_e in molecular single crystals is lower and lies somewhere between 10^{19} and 10^{21} cm^{-3}, as has been estimated by analytical approximation (cf. Sect.5.10.1) which yields the effective N_e value as a variable parameter (cf. Table 5.8). Therefore the E_t values proposed in [5.51] seem to be overestimated. This conclusion has been partially confirmed by a rough estimate of trap depth in DPA crystals by the method of photocurrent decay curves measured in [5.51], and yielding a value of 0.56 eV.

Thus it may be concluded that the differential method of CV-curve analysis provides an adequate relative trap distribution spectrum but should be used in combination with the method of analytical approximation or other independent measurement techniques in order to check the absolute values of trap depth E_t.

5.11 SCLC Theory for Spatially Nonuniform Trap Distribution

In phenomenological SCLC theories for insulators it is commonly assumed that the trap distributions vary only in energy but are spatially uniform throughout the bulk of the sample. This assumption has been questioned by a number of authors [5.20,24,52-58] who have discussed a variety of spatial distributions of trap density in insulators in terms of diffusion-free (high-voltage) SCLC theory [5.20,24,55].

NICOLET et al. [5.52,53] have considered a model with uniform shallow trap distribution in a narrow region near the injecting electrode, and a trap-free interior.

DELANNOY et al. [5.57] have extended the model including finite trap density also in the interior of the sample and have discussed its possible application to tetracene crystals.

SWORAKOWSKI [5.54] has given a general formulation for an arbitrary spatial variation of trap density for shallow and exponential traps. He considered the case when the trap density decreases away from the sample surface as $\exp(-x/r)$ or as $\left|1 + \exp[(x-a)/r]\right|^{-1}$, where r and a are parameters. The latter case is similar to the step-function model of [5.52,53,57].

HWANG and KAO [5.56] have given general formulations identical to those of SWORAKOWSKI [5.54] for shallow, exponential, as well as energetically uniform distributions. According to [5.56] the distribution density of traps $h(E,x)$ is a function of their energy E spectrum and spatial distribution dependence on the distance x from the injecting electrode

$$h(E,x) = N_t(E)S(x) \quad .$$

(5.57)

The authors of [5.56] have also calculated the most probable spatial dispersion function $S(x)$ of traps

$$S(x) = 1 + B \left[\exp\left(-\frac{x}{x_0}\right) + \exp\left(-\frac{L - x}{x_0}\right) \right] \quad , \tag{5.58}$$

where B and x_0 are constants.

As can be seen from (5.58), the trap density decreases exponentially with distance from both sample surfaces, having maximum density at $x = 0$ and $x = L$.

This corresponds to the widely recognized assumption that the spatial trap density is likely to be much higher near the insulator surface than in the bulk [5.20].

Later HWANG and KAO [5.58] have applied a similar treatment to a wide and narrow Gaussian (in energy) trap distribution and have also introduced a Gaussian spatial dependence of trap density.

The main conclusions of all these above-mentioned treatments of diffusion-free (high-voltage) SCLC theories of spatially nonuniform trap distribution are as follows [5.20]:

a) a high trap density near the rear (noninjecting) electrode has a negligible effect on the current;

b) a high trap density near the injecting electrode lowers the current at a given voltage, but does not alter the shape of the CV curve. At low voltages such an electrode acts as a quasi-blocking one and produces a rectifying effect (cf. Sect.5.8.2);

c) the reduction in current can be described in terms of an increased effective thickness of the insulator, viz. $L_{ef} > L$. This is sometimes convenient but, as NICOLET [5.52] has shown, it is more appropriate to consider the current as SCLC through the region of high trap density near the injecting electrode, driven by an effective voltage U_{ef}. The latter is lower than the applied voltage U by the ratio between the thickness of the high trap-density region L' and the total thickness L of the sample ($U_{ef} = (L'/L).U$).

At low voltages diffusion must be taken into account, as has been done by ROZENTAL and PARITSKII [5.55] who considered shallow traps with an exponential spatial distribution. Their results, however, do not lend themselves to generalized analyses being entirely numerical. BONHAM [5.20,24] has recently extended the analytical SCLC theory, including diffusion effects, for shallow and exponential distribution of traps whose concentration varies exponentially [5.24] or step-wise [5.20] with distance.

Two simple step-function spatial distributions are considered — one symmetrical and the other asymmetrical, with higher trap density near the insulator surface. Although these models of [5.20] seem to be artificial and oversimplified they nevertheless have considerable predictive power, at least qualitatively. Thus, for instance, an analysis of the symmetric model shows that such step-like spatial trap distributions may give rise to fine structure in the CV characteristics, which is not predicted by the diffusion-free theory. Therefore one should consider the possibility that the experimental step-like CV characteristics (cf. Fig.5.34,39) might be caused, or at least influenced, by spatial variations of trap density. Such possibility, however, is practically negligible.

First, recently EIERMANN et al. [3.63] used an independent method of thermostimulated conductivity (cf. Sect.5.12) for studying quasi-amorphous thin layers of tetracene, obtained by vacuum evaporation on cooled substrate under conditions very similar to those used by SILINSH et al. [5.42]. They observed several sets of Gaussian trap distributions which are in good agreement with those observed in [5.42] by SCLC technique (cf. Sect.5.12). These data provide additional support to the validity of the interpretation of step-like CV characteristics as caused by energy distribution of Gaussian traps (cf. Sect.5.10.1). Secondly, significant fine structure of CV characteristics may be caused, according to [5.20], only by nonmonotonous, step-function spatial variations of trap density. In real situations, however, one should expect much more monotonous spatial variations of traps, the most probable being exponential decrease of trap density away from the sample surface [cf. (5.58)].

Thirdly, experimental CV characteristics are usually measured at sufficiently high voltages, especially for the second, shallow trap distribution (cf. Figs.5.34,39), in the voltage range at which the diffusion-free SCLC theory is valid (or almost valid). Therefore the main conclusions of the diffusion-free SCLC theory for spatially nonuniform trap distribution, as cited above, can also be regarded as valid. The most important conclusion in this case is that higher trap density near the injecting electrode only lowers the current at a given voltage (which may be taken into account using reduced effective voltage, viz. $U_{ef} = U - \delta U$), but *does not alter the shape* of the CV characteristic.

Empirical experience gained by analytical approximation of step-like CV characteristics (cf. Sect.5.10.1) indicates, however, that there may be a slight effect of spatially nonuniform trap distribution on the shape of CV curves. Thus, analytical approximations of experimental step-like CV characteristics are performed separately for different portions of the CV curve, for which the approximation formulae (III) and (IV) (Table 5.6) hold (cf. Sect.5.10.1 and Fig.5.39). When, however, one tries to approximate the whole step-like CV characteristic numerically, using the basic SCLC equations (5.43-45), a satisfactory agreement between calculated and experimental CV curves cannot be reached. This is apparently due to slight shifts of current-voltage "steps" relatively to each other by a value of δU owing to nonuniform trap distribution near the injecting electrode. Such spatial nonuniformities of trap distribution are probably indicated also by reduced trap-filled-limit current value in some cases, as compared to the calculated one (cf., e.g., Fig.5.39).

Another important result from BONHAM's [5.20] analysis of the asymmetric step-function trap distribution model is the concept of quasi-blocking rectifying contact, caused by spatially nonuniform very high trap density near the injecting electrode, already discussed in Sect.5.8.2.

Concluding the paper [5.20] BONHAM emphasized a significant fact, viz. that space-charge effects caused by spatially nonuniform trap distribution are important only near the injecting electrode. The form of spatial trap distribution in the bulk is not very important, and its influence on the shape of the CV curve may be regarded as negligible.

5.12 Investigation of Local Trapping States by Thermally Activated Spectroscopy Techniques

Apart from SCL current methods various thermally activated spectroscopy techniques are widely used for investigating energy spectra of local charge carrier trapping states. Thermally activated spectroscopy is especially convenient for studying deep quasi-discrete trapping states with E_t values exceeding 0.6-0.7 eV, not accessible to SCLC techniques. Among other thermally activated spectroscopy techniques the thermally stimulated current (TSC) method has become the most popular one in molecular crystal studies.

The essence of the TSC method consists in filling the given trapping states with charge carriers by means of photoexcitation or field injection at low temperatures. After that the temperature of the sample is raised according to a definite law, and the charge carriers which have been thermally released are observed by means of peaks of thermally stimulated conductivity $j = f(T)$ in an external electric field $\mathscr{E}$ (cf. Figs.5.41,42).

According to conventional TSC techniques the trapping states are preliminarily filled by charge carriers through intrinsic photoionization of the
crystals upon illumination with UV light [3.7,5.59-61]. This leads, however, to bipolar filling of traps both by electrons and holes, and it is
difficult to digstinguish between TSC peaks due to electrons and those due
to holes upon thermal release of charge carriers. A more selective method of
trap filling consists in monopolar field injection of charge carriers [5.62-
64]. It is then possible to fill the traps either only with electrons or
only with holes and thus study the corresponding trapping states [5.3]. A
comprehensive TSC theory in case of optical or electric trap filling has
been developed by SIMMONS et al. [5.65] for arbitrary trap distributions
including discrete, exponential and Gaussian ones. TSC theory applications
in case of monopolar trap filling has been specially analyzed by SAMOC et
al. [5.66]. Under the conditions for which the SIMMONS et al. [5.65] model
is valid the TSC spectrum j = f(T) directly reflects the density profile of
the occupied traps

$$j(T) = 0.6eL \ \beta(E/T) \ f(E) \ N(E) \quad , \qquad (5.59)$$

where L is the thickness of the sample, N(E) is the trap density per unit
volume and energy (in eV), and f(E) is the probability that a trap at an
energy E is occupied.

The energy depth E_t of a charge carrier trapping state is related to the
temperature T at which the trap becomes emptied by

$$E_t = T \left(A + \ln \frac{\nu}{\beta} + B \right) + C \quad , \qquad (5.60)$$

where β is the linear heating rate in Ks^{-1} and ν is the attempt-to-escape
frequency (frequency factor). A,B,C are constants. (The ν value is usually
of the order of 10^{10} to $10^{12} \ s^{-1}$ according to [5.65]).

The ν value can be determined from the shift of the TSC signal temperature with varying heating rate [3.63]

$$\log \nu = \frac{(T_2 \log \beta_2 - T_1 \log \beta_1)}{T_2 - T_1} - 1.66 \quad . \qquad (5.61)$$

Figures 5.41,42 present typical TSC curves for ultrapure anthracene single crystals according to PARKINSON et al. [5.3]. Monopolar injection of
holes into the anthracene crystal was effectuated from a KJ/J_2 electrode,
but that of electrons from a metallic electrode of Na und K amalgam. The β

value was altered within the range of 10^{-2} to $1 \text{ K} \cdot \text{s}^{-1}$, the temperature from 220 to 340 K. The following characteristic E_t values of deep trapping states in single crystal samples of anthracene were obtained from an analysis of typical TSC curves (cf. Figs.5.41,42), according to (5.60) [5.3]

$$\text{for holes } E_t = 0.75; \ 0.85 \text{ and } 0.95 \text{ eV} \qquad\qquad (5.62)$$
$$\text{for electrons } E_t = 0.75 \text{ and } 0.85 \text{ eV} \quad . \qquad\qquad (5.63)$$

MAETA and SAKAGUCHI [5.66a] have used computer-simulated analysis of TSC spectra obtained for anthracene single crystals in the 56 to 300 K temperature range. Authors have detected ten closely spaced shallow trapping levels in the 0.03 to 0.34 eV energy interval which form a wide overlapping TSC spectrum band, as well as two distinct deeper levels at 0.51 and 0.60 eV.

Employing TSC technique EIERMANN et al. [3.63,5.67] have recently established the existence of three distinct sets of Gaussian trap distributions in vacuum-evaporated tetracene layers formed on a cooled substrate. It has been shown that at substrate temperatures T_f below 180 K the tetracene layers possess amorphous structure, whilst in the temperature range $T_f = 180$-200 K quasi-amorphous layers consisting of very small microcrystallites were obtained. The authors of [3.63] have established a very distinct correlation between distribution parameters E_t and N_t of the first set of shallow traps, and the formation temperature of the layer, which apparently determines its crystalline structure (see Table 5.9). In layers formed at $T_f < 130$ K the

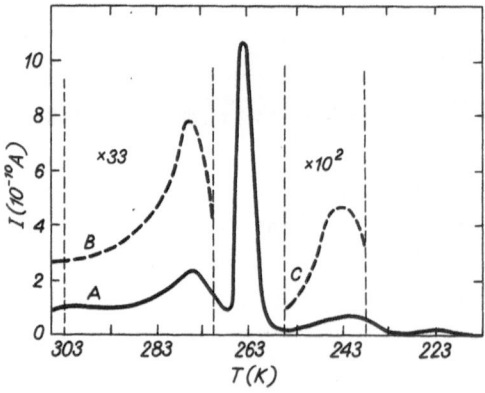
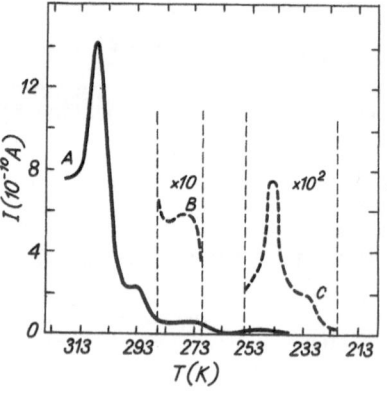
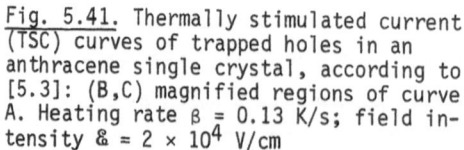

Fig. 5.41. Thermally stimulated current (TSC) curves of trapped holes in an anthracene single crystal, according to [5.3]: (B,C) magnified regions of curve A. Heating rate $\beta = 0.13$ K/s; field intensity $\mathcal{E} = 2 \times 10^4$ V/cm

Fig. 5.42. Thermally stimulated current (TSC) curves of trapped electrons in an anthracene single crystal, according to [5.3]: (B, C) magnified sections of curve A. β and $\mathcal{E}$ values as in Fig.5.41

hole conductivity level E_h splits into $G_e(E)$ type of distribution of local states with $E_t = 0$, distribution parameter $\sigma \approx 0.06$ eV, and local state density $N_t = 2 \times 10^{21}$ cm^{-3}, which is practically equal to molecular density. With rise in T_f from 130 to 180 K, both N_t and σ decrease, indicating a decrease of structure disorder (cf. Table 5.9). At $T_f > 180$ K, when quasi-amorphous structure emerges, the distribution maximum is shifted to $E_t = 0.07$ eV above the hole conductivity level E_h and the σ and, especially, N_t values decrease still further.

Trap distribution sets centered near 0.4 and 0.7 eV with values of σ 0.07 and 0.1 eV, respectively, and containing ca. 10^{15} states/cm^3 are almost independent of film formation conditions and are supposed to be most probably also of structural origin. (It is suggested by the authors that the distribution set a $E_t = 0.7$ eV may be caused by trapping centers formed by pre-dimer molecular pairs responsible also for excimer emission observed in both amorphous and quasi-amorphous tetracene layers [3.61]).

The Gaussian nature of the first shallow distribution set has been additionally confirmed by TSC peak versus energy-square linear dependence as shown in Fig.5.43. Comparison of the experimental values of Gaussian trap distribution parameters E_t, σ and N_t for shallow and deep sets of trapping states obtained by independent TSC and SCLC methods in quasi-amorphous tetracene layers, formed at similar conditons, demonstrates a satisfactory agreement (see Table 5.10). These results can serve as additional evidence of the existence of several sets of Gaussian trap distributions in molecular

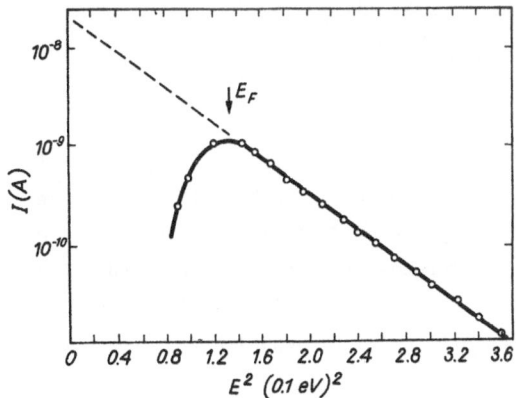

Fig.5.43. Gaussian analysis of the high-energy wing of a TSC peak of a vacuum evaporated tetracene layer formed at substrate temperature T = 160 K according to [3.63]. E_F denotes the position of the quasi-Fermi level for trapped holes above the hole conductivity state

Table 5.9. Experimental values of Gaussian trap distribution parameters σ, E_t and N_t [cf. (3.5)] as well as density of occupied traps n_t as a function of formation temperature T_f of vacuum evaporated tetracene layers determined by TSC technique [3.63]

Substrate temperature T_f [K]	n_t [cm^{-3}]	N_t [cm^{-3}]	σ [eV]	E_t [eV]	Crystalline state
80			> 0.06	0	
130	2.5×10^{19}	2×10^{21}	0.060 ± 0.005	0	
140	2.3×10^{18}	1×10^{20}	0.051 ± 0.003	0	amorphous
160	2.5×10^{17}	5×10^{18}	0.049 ± 0.003	0	
180	1.0×10^{17}	4×10^{17}	0.033 ± 0.003	0.07 ± 0.01	quasi-amorphous
200	2.4×10^{15}	5×10^{16}	0.033 ± 0.003	0.07 ± 0.02	

Table 5.10. Comparison of experimental values of Gaussion trap distribution parameters E_t, σ and N_t for shallow and deep sets of trapping states for quasi-amorphous tetracene layers evaporated on cooled substrate at 200 K, determined independently by TSC and SCLC techniques

Gaussian trap distribution sets	Determined by TSC technique according to EIERMANN et al. [3.63]			Determined by SCLC technique according SILINSH et al. [5.42]		
	E_t [eV]	σ [eV]	N_t [cm^{-3}]	E_t [eV]	σ [eV]	N_t [cm^{-3}]
I set of shallow traps	0.07 ± 0.02	0.033 ± 0.005	5×10^{16}	0.1	0.05	4×10^{15}
II set of deep traps	0.4	0.07	$\sim 10^{15}$	0.26	0.11	2×10^{14}
III set of the deepest observed traps	0.7	0.1	$\sim 10^{15}$	not detected		

crystal layers and confirm the validity of SCLC theory in the interpretation of step-like CV characteristics (cf. Sect.5.10.1). SAKURAI [5.67a] has recently reported observation of two deep trapping levels at 0.55 and ca. 0.9 eV, detected by TSC method in quasi-amorphous tetracene layers evaporated on a cooled substrate at $T_f \approx 180$ K. These trapping levels disappear after thermal treatment of the layers and are therefore supposed to be of structural origin.

As an example of combined application of SCLC and TSC methods for studying local states in organic dye layers the paper by BLAGODAROV et al. [5.68] may be mentioned.

Another method of thermally activated spectroscopy of local trapping states is a variant of TSC technique known as termally stimulated depolarization (TSD) of preliminarily filled traps. The TSD method uses either linear heating [5.61], or fractional thermally stimulated depolarization (FTSD) [5.69]. The general theory of thermally activated spectroscopy shows that fractional heating provides, in principle, higher resolution than the linear one. Methods of interpretating FTSD curves are described in [5.70].

An experimentally convenient and widespread method is also that of isothermal decay current (IDC) technique. It is frequently used for independent checking of data obtained from TSC measurements (cf., e.g. [5.2,4,5]). The theory of isothermal currents for direct determination of trap parameters of arbitrary, including Gaussian, distribution is discussed in great detail by SIMMONS and TAM [5.71] (see also [5.66]).

The trapping state energy E_t is described, according to IDC theory, by a simple expression [5.66]).

$$E_t = kT_m \ln(\nu t) \quad , \tag{5.64}$$

where T_m is the temperature at which the IDC curve is obtained, the curve itself being construed in $It = f(\log t)$ coordinates (cf. Fig.5.45).

It is thus possible, using (5.64), to determine E_t, as well as the frequency factor ν values from a family of IDC curves obtained at various temperatures T_m [5.4]. A comparison of both methods shows that the best experimental conditions are obtained when IDC curves are plotted for temperatures close to those at which TSC reaches peak value [5.4].

For comparison of data, obtained independently by TSC and IDC methods, TSC spectra (a) and temperature dependence of IDC curve peaks (b) for two different anthracene crystals A and B, according to SAMOC et al. [5.4], are presented in Fig.5.44. It can be seen from Fig.5.44a,b that both methods yield close E_t values for these crystals.

Comparison of independent TDS and IDC measurements under conditions of monopolar charge carrier injection was used by SAMOC et al. [5.5] for determining characteristic depth E_t of $G_g(E)$-type distributions in an anthracene single crystal

for holes E_t = 0.4; 0.5; 0.6 and 0.85 eV $\tag{5.65}$

for electrons E_+ = 0.5 and 0.8 eV . $\tag{5.66}$

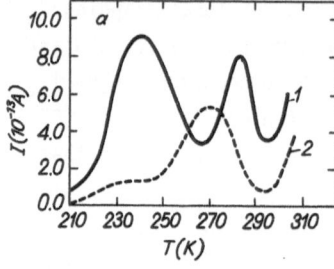

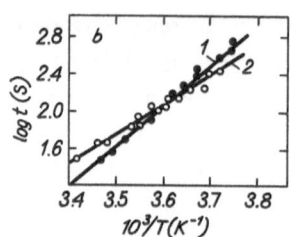

Fig. 5.44a,b. Thermally stimulated current (TSC) spectra for two anthracene crystals (A and B), according to [5.4] (a) 1 - crystal A, heating rate $\beta = 0.038$ K/s; E_t determined from peak at 283 K equals 0.83 eV; 2 - crystal B, $\beta = 0.032$ K/s, E_t determined from peak at 272 K equals 0.62 eV, and temperature dependences of isothermal decay current (IDC) peak positions (b) 1 - crystal A, $E_t = 0.87$ eV, $\nu = 5 \times 10^{13}$ s^{-1}; 2 - crystal B, $E_t = 0.61$ eV; $\nu = 1 \times 10^9$ s^{-1}

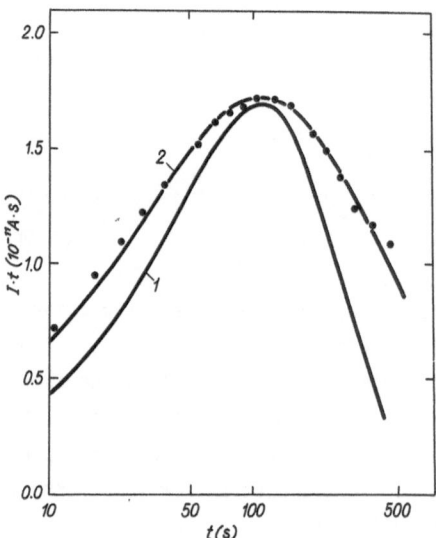

Fig. 5.45. Comparison between experimental curve of isothermal decay current (IDC) (dots) and calculated IDC curves, according to [5.5]: 1 - calculated IDC curve for a discrete trapping level; 2 - calculated IDC curve for Gaussian trap distribution with parameter $\sigma = 0.05$ eV. Experimental IDC curve was obtained at T = 288 K

Possible causes of asymmetry in trap depth E_t values for holes and electrons in trapping on centers of presumably structural origin will be discussed in Sect.5.15.

Using the theory evolved in [5.66], a numerical analysis was performed by SAMOC et al. [5.5] in order to test the validity of the Gaussian distribution model for describing experimental IDC curves. A comparison between calculated and experimental data (see Fig.5.45) shows that the IDC method unambiguously confirms Gaussian $G_g(E)$-type distribution of traps in the forbidden gap of an anthracene crystal. These data also lend additional support to the validity of the Gaussian model of local state distribution in molecular crystals.

It was found from analysis of experimental IDC curves that Gaussian $G_g(E)$-distributed states in anthracene, with E_+ values according to (5.65,66), have a distribution parameter σ = 0.04-0.07 eV. The TSC curves yield a smaller value: σ = 0.02-0.03 eV. The discrepancy between the data obtained by both methods is not clear [5.5]. According to the authors of [5.5], the results obtained from IDC measurements, however, are more reliable than those obtained by the TSC method. As can be seen from Table 5.8 the typical σ values for deep trapping states in anthracene crystals, obtained by SCLC technique, are of the same order of magnitude as those obtained by IDC method in [5.5].

Another promising but less popular method of thermally stimulated spectroscopy of local states ought to be mentioned, namely the method of thermoluminescence which has also been applied for investigation of molecular crystals (cf. [1.71]). In this case the crystal is excited by ionizing radiation (UV or X-rays) at low temperature (T = 4 or 77 K). After that the trapped carriers are liberated by linear or fractional heating [5.72], and their recombination luminescence I = f(T) is recorded. This method does not, however, permit separation of electron and hole trapping states, as in the case of TSC and IDC methods. Thermoluminescence can, however, be used as an additional method for studying shallow, quasi-continuously distributed traps in molecular crystals (cf., e.g. [5.73-75].

5.13 Other Experimental Methods for Local Trapping State Study

1) Among other experimental methods one should mention first the photodielectric effect (PDE) which opens certain new promising avenues for studying local states in molecular crystals. Thus, ROSENSTEIN et al. [5.76,77] have shown that a comparison of temperature dependence of PDE and alternating current TSC, and the temperature dependence of rate of growth and decrease of PDE leads to the conclusion that PDE of the first kind is caused by those charge carriers in organic dyes, which effectuate active alternating current conductivity. This is additional evidence in favor of partial localization of charge carriers on the conductivity levels (cf. Sect.1.5). According to such a localization model the active conductivity proceeds by means of activationless hopping of charge carriers between localization centers [5.77].

2) An important method of studying local states is undoubtedly that of photostimulated conductivity (PSC).

The general theory of PSC under SCLC conditions in molecular crystals was developed by HELFRICH [1.75]. He obtained the following generalized formula for photostimulated current j_{ph} created by release of trapped charge carriers either by excitons or as a result of direct photoionization

$$j_{ph} = \frac{9}{8} \frac{\epsilon\mu}{N_t} N_e^{1/m} \left(\frac{A(\lambda)}{vs}\right)^{(1-1/m)} I^{(1-1/m)} \frac{v^2}{L^3} \quad , \tag{5.67}$$

where $m = T/T_c$; T_c is the parameter of exponential trap distribution; $A(\lambda)$ is a constant depending on the wavelength of the exciting light; v is the mean thermal velocity of the charge carrier; s is the cross section of the trapping center; I is the light intensity.

Photostimulated conductivity, especially in the case of trapped charge carrier release by excitons has been studied by many authors (see, e.g. [1.75,2.134,3.1,5.36,78-82]). Since, however, (5.67) has been obtained for an exponential distribution model it is not valid in the case of the more feasible Gaussian distribution.

Direct photoionization of traps with quasi-continuous energy spectrum under SCLC conditions can yield valuable information on photoionization processes of traps, but very little on the distribution parameters of the traps themselves (see, e.g. [2.133]).

Only in the case of deep narrow trapping states of $G_g(E)$ type valuable information can sometimes be gained with respect to E_t values of these distributions. Figure 5.46 shows, as an example, the dependence $j_f = f(h\nu)$ of photocurrent under SCLC conditions upon direct photoionization of hole trapping states, according to PARKINSON et al. [5.3]. In this case peaks of photocurrent can clearly be discerned, coinciding, within limits of error, with E_t values obtained by the TSC method [cf. (5.62)].

The applicability of PSC technique for determining deep quasi-discrete trapping states in anthracene single crystals has been also demonstrated by BRODRIBB et al. [5.83]. The authors have observed deep electron and hole traps in the energy range 0.8 to 1.35 eV.

3) Rather promising although less frequently used are ac conductivity and dielectric dispersion techniques. This is mainly due to the interpretation difficulties of the obtained experimental data. The potential possibilities of these methods have been demonstrated by PETHIG et al. [5.84,85], ROSENSTEIN et al. [5.76,77] and others. Thus PETHIG et al. [5.84,85] have shown that ac conductivity and dielectric dispersion measurement data in such organic solids as anthracene and β carotene, perylene-chloranil com-

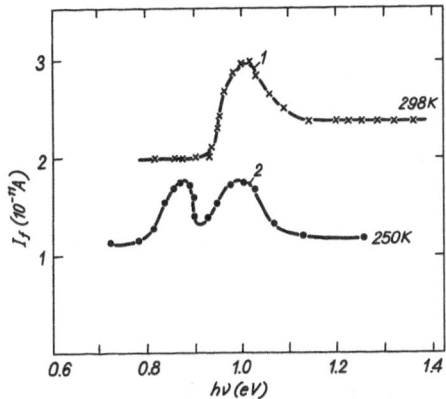

Fig. 5.46. Photocurrent $j_f = f(h\nu)$ dependence for direct photoionization of local hole trapping states filled under conditions of SCLC, according to [5.3]: (1) at T = 289 K; (2) at T = 250 K

plexes, as well as in a number of biopolymers, can be interpreted in terms of hopping electron transport. An analysis of the temperature dependences of ac measurements is considered to lead to knowledge of the energy distribution of the electron trapping states [5.84,85]. TWAROWSKI and ALBRECHT [5.86] have recently carried out low-frequency (f = 10^{-2}-10^{-4} Hz) capacitance measurements in polycrystalline tetracene layers. The authors show that this method allows one to estimate the density n_t of charge carriers trapped in shallow trapping states. The n_t values obtained from these measurements are of the order of $n_t = (0.6-1.4) \times 10^{16}$ cm^{-3} in Al/Tc/Au and Al/Tc/Nesatron sandwich-type cells. On the other hand, the density n of free charge carriers, as estimated from conductivity and mobility data, does not exceed 10^7 cm^{-3}. BAK et al. [3.80] have also recently measured the ac conductivity in polycrystalline tetracene layers. The authors suggest that at low frequencies (f < 1 kHz) the ac conductivity in the given Tc layers is controlled by intergrain barriers, having a height of ca. 0.07 eV. At high frequencies (f > 1 kHz) the ac conductivity is interpreted in terms of hopping transport via localized states inside separate crystallites.

4) For the study of electron trapping states of quasi-discrete and narrow $G_g(E)$ distribution, photoelectron emission technique can be applied [1.144, 2.136]. This technique is especially effective for the investigation of deep electron trapping states of impurity origin when their occupancy exceeds 10^{16}-10^{17} cm^{-3} (see [1.144] and references therein).

5) Pulsed photoconductivity methods (see, e.g. [1.45,75,5.30]) are rather sensitive to the presence of local trapping states for charge carriers. The shape of the conductivity pulse is, however, on principle incapable of yielding quantitative information on the energy spectrum of local trapping states.

One can make use, however, of the elegant method of drift-current technique for the measurements of trapping parameters of guest molecules in a doped molecular crystal, as demonstrated by KARL et al. (see [1.45] and references therein).

Thus, the E_t value of an impurity center can be estimated from the temperature dependence of effective charge carrier mobility μ_{eff} in a doped crystal according to [1.45]

$$\mu_{eff} = \mu_0 \left[1 + \frac{N_t}{N_e} \exp \left(\frac{E_t}{kT} \right) \right]^{-1} , \qquad (5.68)$$

from the slope value of $\mu_{eff}(T)$ dependence in $(\mu_{eff}, 1/T)$ coordinates (see [Ref. 1.45, p.278]). Most trapping energies E_t of guest molecules in a doped anthracene crystal shown in Fig.5.50 have been obtained by this experimental technique (see [1.45] and references therein).

6) Additional information on the occupancy of local trapping states can be obtained by means of the photomagnetic effect caused by interaction between triplet excitons with occupied paramagnetic trapping centers (see, e.g. [4.1,2,5.79,80]). In case of the presence of two types of paramagnetic centers their relative occupancy ratio can determine the sign of the photomagnetic effect, as shown by KAULACH and SILINSH [5.87].

7) Recently CAMPOS et al. [5.88] have estimated deep trapping states in naphthalene from long-time decay ($\sim$ 250 h) of the surface potential of corona-charged crystals. This rather exotic technique yields two sets of deep trapping states in naphthalene crystals: the first quasi-discrete one at $E_t = 0.6$ eV with density $N_t \approx 10^{17}$ cm^{-3}, and the second statistically distributed one at $E_t = 1.1$ eV with $N_t \approx 10^{13}$ cm^{-3}.

5.14 Correlations Between Distribution Parameters of Local Trapping States and Crystalline Structure

As already repeatedly emphasized (cf., e.g., Sects.5,2,7,10,12), deep trapping states of $G_g(E)$-type distribution play a dominant part in local trapping of charge carriers in molecular single crystals. The structural origin of these centers has been convincingly proved at present, both by deep purification of crystals, excluding the presence of possible impurity centers, as well as through studies of the effects of crystal growing technique, thermal treatment, and mechanical deformation of the crystal on density and spectral distribution of the given local states. Very informative, in this case is the comparison between corresponding trapping states of structural origin for charge carriers and excitons (cf. Sects.4.3 and 5.2). On the other hand, shallow traps of $G_e(E)$ and $G_g(E)$ types of distribution are dominant in thin layers of vacuum-evaporated samples of a number of aromatic and heterocyclic compounds (cf. Sects.5.10,12).

The influence of crystalline structure on energy spectra and distribution parameters is most clearly seen in the case of vacuum-evaporated layers. It has been found that the values of parameters E_t, σ and N_t of evaporated layers of aromatic and heterocyclic compounds are rather sensitive to the conditions of their preparation, such as substrate temperature, evaporation rate, thickness of layers and other factors, which either directly or indirectly influence the crystalline structure of the sample (cf. Tables 5.8 and 5.9).

The most direct evidence of such influence is demonstrated by data in Table 5.9. The distribution parameters of these states can also be easily changed by subsequent thermal treatment of the layers obtained. One can safely say that there is, at present, abundant evidence speaking in favor of the structural nature of these local charge carrier trapping states.

In recent years a lot of additional information has been obtained on real crystalline microstructure of vacuum-evaporated layers of a number of aromatic hydrocarbons, especially tetracene and pentacene.

KAMURA et al. [1.97], and MARUYAMA and IWASAKI [1.98] showed already in 1974 that the crystalline structure of vacuum-deposited layers of tetracene (Tc) and pentacene (Pc) depends on the temperature of the substrate. Thus, e.g., at substrate temperature of 300 K layers of oriented Pc crystallites are formed, as ascertained by X-ray diffraction and appearance of Davydov doublets in optical absorption spectra (cf. Fig.4.1). At substrate temperature from 210 K to 90 K, presumably, amorphous layers of Pc were formed, according to [1.97,98], judging by X-ray data and absence of Davydov splitting in absorption spectra (cf. Fig.4.2).

In 1974-1978 a series of papers were published, reporting the results of studies on the effect of crystalline structure of evaporated Tc and Pc layers on their optical [1.100-103,2.130,138,3.63,5.67,89-91] and electrophysical [1.67,68,104,2.121,153,3.63,5.42,67] properties. Detailed studies of local trapping state energy spectra and distribution parameters in vacuum-evaporated Tc and Pc layers have been carried out by SILINSH et al. [1.67, 68,104,5.42] and EIERMANN et al. [3.63,5.67].

It can be considered as an established fact, at present, that deposition *in vacuo* of Tc and Pc layers on an uncooled substrate at 300 K leads to the formation of a structure consisting of oriented planar crystallites. Electron microscope studies show that Tc and Pc layers obtained by this method consist of planar oriented crystalline grains of ca. 0.2-1.0 μm in diameter [1.104,2.153]. The ab plane of the crystallites is oriented parallel to the substrate, thus forming an anisotropic texture with a dominant orientation of planar crystallites. Electron diffraction patterns of these layers show sharp reflection indicating a dominant orientation of crystallites and confirming texture-like structure of these layers [1.104,2.153].

The appearance of well-defined Davydov splitting in the absorption spectra of the given Tc and Pc layers (cf. Figs.2.12,16 and 4.1) also speaks in favor of dominant orientation of the ab plane of microcrystallites parallel to the plane of the base.

On the other hand, as reported by BALODE et al. [1.104], GRECHOV et al. [2.153] and EIERMANN et al. [3.63], Tc and Pc layers formed on a cooled substrate at T = 180-200 K cannot be considered as amorphous, as earlier suggested in [1.97,98]. Electron microscope and electron diffraction studies show [1.104,2.153,3.63] that these layers actually consist of small and random-distributed microcrystallites of 30-40 Å [1.104] and larger size but not exceeding 500 Å [2.153]. Therefore it is more convenient to call such layers quasi-amorphous [1.104,3.63].

More detailed studies of electronic absorption spectra of such quasi-amorphous Pc layers by VERTSIMACHA et al. [2.130] showed that they also reveal Davydov splitting, partly concealed by effects of light dispersion in the fine-grained structure (cf. also [1.100]). This means that the microcrystallites of such a quasi-amorphous layer also possess distinct orientation of the ab plane parallel to the substrate. As shown by EIERMANN et al. [3.63], only at substrate temperatures T_f below 170 K genuinely amorphous layers of Tc can be obtained, exhibiting an essentially diffuse electron diffraction pattern. The transition from quasi-amorphous to amorphous structure in Tc layers can also be seen from E_t value shift to $E_t = 0$ below 180 K (cf. Table 5.9).

It is interesting to mention that anthracene forms, as observed from electronic spectra by MARUYAMA et al. [5.90], pronounced oriented structures, even if deposited on a substrate cooled to 90 K. In the case of such aromatic compounds as perylene and coronene deposition on a cooled substrate at first creates quasi-amorphous structure which is rapidly transformed into oriented crystallite structure with rise in temperature [5.91]. It appears that vacuum-evaporated aromatic hydrocarbon molecules have a strong tendency towards formation of oriented crystallite structures if deposited on nonmetallic [1.104,3.63,5.42], as well as on metallic [2.153] substrates.

The above-mentioned data have been confirmed independently by low-energy electron diffraction (LEED) studies of thin layers of naphthalene crystals grown by vapor-phase epitaxy on a clean Pt (111) substrate, performed by FIRMENT and SAMORJAI [3.82]. The authors show that at deposition temperatures below 105 K the naphthalene layers are disordered, whilst in the temperature range of 105-200 K diffraction patterns charcteristic of well-ordered surface structures persisted. Diffraction spot sizes indicate domain dimensions of ca. 70 Å over the range of layer thickness ($\sim$ 1000 Å) observed. The symmetry of the diffraction pattern suggests that the naphthalene film consists of a large number of crystallites with ab plane parallel to the substrate surface, oriented in six equivalent directions. Grain boundaries separate the crystallites [3.82].

Tables 5.8,9 and Figs.5.47,48 (see also Sects.5.10,12) illustrate the basic correlations between the distribution parameters of local trapping states (E_t, σ, N_t) as well as of N_e (see Fig.5.48) and crystalline structure of the samples [single crystals (SC), oriented (OL), quasi-amorphous (QA) and amorphous (A) layers].

As can be seen, the density N_t of local trapping states is most distinctly dependent on the crystalline state. This is particularly marked in the case of tetracene for which abundant reported data are available. In Tc single crystals the density of deep trapping states is of the order of $N_t = 10^{13}$-10^{14} cm^{-3}; in oriented Tc layers $N_t = 10^{14}$-10^{15} cm^{-3} for deep traps (DT) and $N_t = 10^{15}$-10^{16} cm^{-3} for shallow traps (ST); in quasi-amorphous Tc layers N_t increases to $N_t = 5 \times 10^{15}$-5×10^{17} cm^{-3} for shallow traps (ST) and reaches the value of 10^{19}-10^{21} cm^{-3} for amorphous Tc layers (see Tables 5.8, 9 and Fig.5.47).

As can be seen from Table 5.9, the N_t value for shallow traps in quasi-amorphous and amorphous Tc layers also exhibits distinct dependence on the formation temperature T_f of the layer: N_t increases with decreasing formation temperature T_f indicating increase in structural disorder. At T_f below 130 K the density N_t of local states in amorphous Tc layers practically reaches the value of the density of molecules, i.e., each molecule acts as a trapping center for a charge carrier.

Less distinct is the σ dependence on crystalline state. σ may have values within the range of 0.03 to 0.14 eV in single crystals, as well as in oriented and quasi-amorphous layers (cf. Tables 5.8,9 and Sects.5.10,12). Only

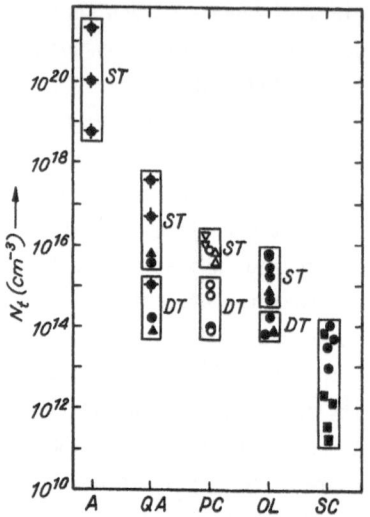

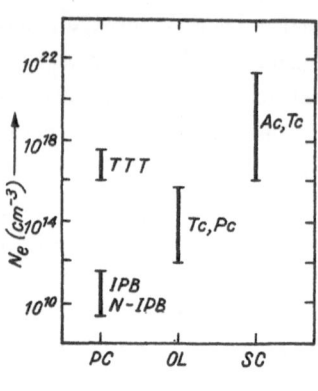

Fig. 5.47. Correlations between the density of trapping states N_t and crystalline structure in molecular crystals. (A) amorphous, (QA) quasi-amorphous, (PC) polycrystalline, (OL) oriented layers; (SC) single crystals, (■) anthracene (Ac), (•) tetracene (Tc), (▲) pentacene (Pc), (○) tetrathiotetracene (TTT), (▽) IPB, (△) N-IPB (see Table 5.8), (◆) tetracene (Tc), according to [3.63] (cf. Table 5.9); (DT) deep traps; (ST) shallow traps [5.42]

Fig. 5.48. Typical dependence of effective density of states at the conductivity level N_e on crystalline structure in molecular crystals (Notation as in Fig.5.47) [5.42]

in polycrystalline layers of some heterocyclic compounds, such as IPB and N-IPB (cf. Table 5.8), σ reaches the typical value of 0.10 to 0.15 eV in the case of shallow traps of $G_e(E)$ distribution, thus indicating that this set of shallow traps may be regarded as quasi-continuous. More distinct dependences of σ on the crystalline state are observed, according to [3.63], in amorphous and quasi-amorphous layers of Tc (cf. Table 5.9). With decreasing formation temperature T_f of the layers, i.e., with increasing disorder, σ steadily increases from 0.03 eV to ~0.06 eV.

The energy spectra of trap distribution, as characterized by the set of E_t values, also strongly depend on the crystalline state of the samples. Thus, as repeatedly stressed, only deep trapping states of the order of $E_t = 0.5$-0.9 eV have been detected in single crystal samples (cf. Sects.5.10 12). In oriented and quasi-amorphous layers of aromatic hydrocarbons (tetracene, pentacene), at least two Gaussian trap distributions are observed, a shallow one (E_t = 0.1 eV) and a more deeply situated one (E_t = 0.3-0.4 eV) (cf. Table 5.8 and Fig.5.47). The position of these distribution maxima, especially of the shallow one, also depends on the crystalline state (cf. Tables 5.8,9).

In evaporated polycrystalline layers of such heterocyclic compounds as IPB, N-IPB and TBB only shallow wide $G_e(E)$ trap distribution, centered at the conductivity level ($E_t \approx 0$), is observed, but in TTT layers two $G_g(E)$ trap distributions emerge, one at $E_t \approx 0.2$, the second at $E_t \approx 0.3$ eV (see Table 5.8).

It has been also established, according to [5.42], that the effective density of states at the conductivity level is strongly influenced by the crystalline structure of the samples (see Fig.5.48 and Table 5.8). Thus, in Ac and Tc single crystals N_e is of the order of 10^{16}-10^{21} cm^{-3}; in oriented Tc and Pc layers $N_e = 10^{12}$-10^{16} cm^{-3}, but in less regular polycrystalline layers of IPB and N-IPB the N_e value is rather small: $N_e = 10^{10}$-10^{12} cm^{-3}. As an interesting exception, evaporated polycrystalline layers of TTT should be mentioned, which is an exotic compound having relatively high conductivity ($\sigma = 10^{-10}$-10^{-9} Ω^{-1} cm^{-1}, see [2.141,142,5.18]). As may be seen from Table 5.8 and Fig.5.48, in TTT layers N_e reaches the value of $N_e = 10^{16}$-10^{17} cm^{-3} which equals the typical N_e values for single crystals of Ac and Tc.

It is interesting to notice that the N_t and N_e dependences on the crystalline state of the samples are mutually correlated: N_t steadily increases, whilst N_e decreases with increasing structural disorder of the sample (cf. Figs.5.47,48).

Unlike N_t, it is rather difficult to assign direct physical meaning to the parameter N_e. It can be regarded as an effective quantity characterizing the specific conductivity σ of the sample through the well-known relation $\sigma = e\mu N_e \exp(-E_F/kT)$. It may, however, be questioned whether this relation is valid for a charge carrier hopping model in highly disordered systems. The low effective value of N_e in polycrystalline layers of some heterocyclic compounds, such as IPB, N-IPB and TBB ($N_e \approx 10^{10}$-10^{12} cm^{-3}, cf. Table 5.8) is, however, reasonable. The quasi-Fermi level in such systems may actually be shifted at high injection levels to the very edge of conductivity states, resulting in $E_a \to 0$ (cf. Fig.5.29). In such a case the effective value of N_e determines the upper limit of possible σ value, in accordance with the experimentally observed one. (If in such a situation the N_e value would be of the order of 10^{20}-10^{21} cm^{-3} one should, in such a case, observe metallic conductivity, which obviously does not appear. The resistance of such systems at $E_e = 0$ remains pretty high and can be described by the correspondingly low effective N_e value for the given charge carrier drift mobility μ). It should be mentioned that the value of N_e can be estimated only within the limits of an order of magnitude. The linear shift of the CV characteristic in log-log plots is determined, apart from N_e, also by μ and E_t (cf. Table 5.6) on which reliable data are not always available.

5.15 Local Lattice Polarization by Trapped Charge Carrier in Molecular Crystals

Localization of a charge carrier on a neutral molecule and formation of a molecular ion (cf. Sect.2.7.2) produces a considerable change in the forces of attraction between molecules. Van der Waals forces of the r^{-6} type between neutral molecules (cf. Sect.1.1) are replaced by attraction between an ion and induced dipoles, which follows r^{-4} law [cf. (2.12)]. Typical relaxation time τ_e of lattice polarization in anthracene-type crystals is of the order of 10^{-12}-10^{-11} s (cf. Fig.1.22). Consequently, in this case, lattice polarization cannot take place during typical free charge carrier localization time $\tau_h = (2-5) \times 10^{-14}$ s (cf. Fig.1.22). The situation changes drastically in the case of charge carrier trapping in shallow traps. Localization (trapping) time τ_t of a charge carrier in a trap of depth E_t can be evaluated by the expression

$$\tau_t = \frac{1}{\nu} \exp\left(\frac{E_t}{kT}\right) \quad , \tag{5.69}$$

where ν is the frequency factor.

Typical values of τ_t for various trap depths E_t in an anthracene crystal are presented in Table 5.11.

Table 5.11 shows that local lattice polarization of anthracene-type crystal must take place already at trap depths of $E_t \gtrsim 0.03$-0.05 eV. Such local lattice polarization is, obviously, of "self-trapping" nature, causing the deepening of the potential well of the trapped carrier. The process of local lattice deformation due to interaction between an ion and induced dipoles on surrounding neutral molecules is illustrated in Fig.5.49.

A rough estimate of the self-trapping energy gain ΔE_ℓ through local lattice deformation by a trapped charge carrier (molecular ion) in an anthracene-type crystal has been reported by SEKI et al. [1.159] and SWORAKOWSKI [1.21]. The ΔE_ℓ values obtained are 0.08 eV after [1.159], and 0.09 eV after [1.21]. There is, at present, no direct experimental evidence of the existence of such self-trapping states in molecular crystals. One may, however, suggest that the typical set of shallow trapping states, centered at $E_t = 0.04$-0.1 eV above the conductivity state E_h in oriented and quasi-amorphous tetracene and pentacene layers might be caused by such a "self-trapping" effect shifting the initial $G_e(E)$-type distribution with $E_t = 0$ by the given E_t value.

Table 5.11. Typical charge carrier trapping times τ_t in traps of depth E_t in an anthracene crystal

E_t [eV]	E_t/kT	τ_t [s]
0.026	1	2.72×10^{-11}
0.05	1.92	6.84×10^{-11}
0.10	3.85	4.68×10^{-10}
0.20	7.69	2.19×10^{-9}
0.30	11.54	1.03×10^{-7}
0.40	15.38	4.80×10^{-6}
0.50	19.23	2.25×10^{-4}
0.60	23.08	1.05×10^{-2}
0.70	26.92	0.49
0.80	30.77	23
1.00	38.46	5.05×10^4 s = 14.2 h
1.20	46.15	$\sim 3 \times 10^{15}$ h

Note: For calculating τ_t according to (5.69) the value $\nu = \nu_{max} = kT_D/h = 2.9 \times 10^{12}$ s^{-1} was used, where T_D is the Debye temperature for an anthracene crystal, $T_D = 140.7$ K.

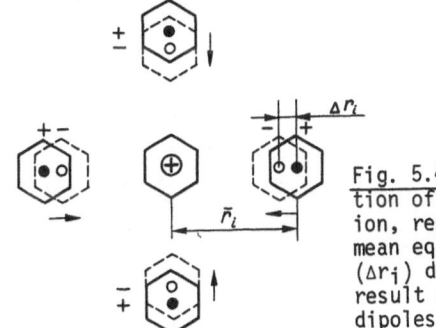

Fig. 5.49. Schematic picture of local deformation of a molecular crystal around a molecular ion, resulting in "self-trapping" effect. ($\bar{r}_i$) mean equilibrium distance of neutral molecules; (Δr_i) displacement of neutral molecules as a result of interaction of the ion with induced dipoles on surrounding neutral molecules

The dynamic aspects of charge carrier interaction with lattice vibrations in molecular crystals has been approached in terms of the small polaron model [1.157,5.92]. However, as shown by VILFAN [1.157], owing to inequality of relaxation times, viz, $\tau_h < \tau_\ell$ (cf. Fig.1.22). the probability of small lattice polaron formation in anthracene-type crystals may be regarded as negligible. Typical temperature dependences of mobility in this case also do not speak in favor of the small lattice polaron model (cf. Sect.1.5). It seems more appropriate to treat electron-phonon interaction in anthracene-type crystals in terms of electronic polarization fluctuations ΔP having a value ca. 0.03 eV [1.157,158].

Table 5.11, also demonstrates the fact that the SCLC method under steady-state conditions can be applied only for studying traps of depth not exceeding $E_t \approx 0.8$ eV.

As already repeatedly stressed, structural traps of polarizational origin must have, in point approximation, equal depth E_t for both electrons and holes. As, however, found in the qualitative analysis by MUNN [5.93], the actual distribution of charges at a positive and a negative ion can lead to asymmetry in lattice polarization by a trapped charge carrier. According to quantum mechanical calculations, the mean electron density on peripheral hydrogen atoms is higher in the case of anions than in the case of cations. Consequently repulsive forces grow faster when a neutral molecule approaches an anion than when it approaches a cation. Therefore the displacement Δr_i of neutral molecules from equilibrium position (cf. Fig.5.49) is larger in the case of local lattice polarization by a cation than in case of polarization by an anion.

There is indirect experimental confirmation of this concept (cf. Sect. 5.12). As can be seen from (5.62,63,65,66), the characteristic depth of traps for holes E_t^h in an anthracene crystal exceeds the possible corresponding value of traps for electrons E_t^e by ca. 0.05-0.1 eV, e.g., actually $E_t^h > E_t^e$.

Since the suggestion of such charge carrier self-trapping asymmetry in aromatic hydrocarbon crystals [5.93] is of essential significance, one should expect more detailed theoretical as well as experimental studies of this phenomenon in the near future.

5.16 Guest Molecules as Trapping Centers in a Host Lattice

Although this chapter has been devoted mainly to charge carrier trapping states of structural origin, we think it appropriate to have a brief discussion here also about local trapping states formed by guest molecules in a doped crystal, as well as by impurity molecules in a real crystal.

The formation of local charge carrier trapping states by guest molecules in a host lattice was considered theoretically already by GUTMAN and LYONS (see [Ref. 1.38, p.368]). The authors showed that the position of a local trapping state in the energy diagram of the host crystal, produced by a guest molecule, is determined by molecular parameters $(I_G)_{guest}$ and $(A_G)_{guest}$

of the guest molecule and electronic polarization energy of the host crystal by a charge carrier localized on the guest molecule P_{guest}. Consequently these local states can be described in the framework of the Lyons polarization model. Later SWORAKOWSKI [3.5] and KARL (see [Ref. 1.45, p.271]) have given similar, more detailed theoretical considerations on energy values of the trapping states formed by guest molecules. According to these approaches the local state for electron trapping $(E_t^e)_{guest}$ formed by a guest molecule in a host lattice can be described by the expression

$$(E_t^e)_{guest} = (A_C)_{guest} - (A_C)_{host} = [(A_G)_{guest} + (P_e)_{guest}]$$
$$- [(A_G)_{host} + (P_e)_{host}] \quad . \tag{5.70}$$

The difference in electronic polarization energy

$$(\Delta P_e)_{guest} = (P_e)_{guest} - (P_e)_{host} \tag{5.71}$$

is, presumably, caused by distortion of the host lattice in the vicinity of the guest molecule. Such distortions have been detected by methods of defect fluorescence (cf. Sect.4.3) and indirectly by drift current technique in doped anthracene crystals (cf. [1.45,135]).

If the lattice around the guest molecule is compressed, $(\Delta P_e)_{guest}$ is positive. If, on the contrary, the surrounding lattice becomes dilated, $(\Delta P_e)_{guest}$ is negative.

The lattice distortion around a single guest molecule, e.g., such as a tetracene molecule in an anthracene host, should not produce, according to analysis presented in previous sections (cf., e.g., Sect.5.2), a change in polarization energy greater than ca. ±0.1 eV.

Consequently, one can assume, as a zero order approximation, that $(P)_{guest}$ in approximately equal to $(P)_{host}$, introducing thus a possible error not exceeding ca. ±0.1 eV. In case of such an approach (5.70) can be simplified as follows:

$$(E_t^e)_{guest} = (A_G)_{guest} - (A_G)_{host} \quad . \tag{5.72}$$

In terms of similar approximation the energy of a local state for hole trapping by a guest molecule $(E_t^h)_{guest}$ is given by

$$(E_t^h)_{guest} = (I_G)_{guest} - (I_G)_{host} \quad . \tag{5.73}$$

Expressions (5.72,73) show that the local state energy for guest molecules

Table 5.12. General conditions for local state formation by guest molecules in a host lattice

	Relations between molecular parameters	Characteristics	Examples
I	$(A_G)_{guest} > (A_G)_{host}$ $(I_G)_{guest} < (I_G)_{host}$	Traps for electrons Traps for holes	Tetracene in anthracene host
II	$(A_G)_{guest} < (A_G)_{host}$ $(I_G)_{guest} > (I_G)_{host}$	Antitraps for electrons Antitraps for holes	Anthracene in tetracene host
III	$(A_G)_{guest} > (A_G)_{host}$ $(I_G)_{guest} > (I_G)_{host}$	Traps for electrons Antitraps for holes	Anthraquinone in anthracene host Cation radical in neutral lattice
IV	$(A_G)_{guest} < (A_G)_{host}$ $(I_G)_{guest} < (I_G)_{host}$	Antitraps for electrons Traps for holes	Phenothiazine in anthracene host Anion radical in neutral lattice

in the given approximation is unequivocally determined by the difference between the corresponding molecular parameters of the guest and host molecules, namely, electron affinity and ionization energy.

Depending on possible relations between the guest and host molecular parameters four types of general conditions for local state formation by guest molecules emerge (cf. Table 5.12).

After the early work by HOESTEREY and LETSON [5.94] a number of authors have studied trapping levels in doped molecular crystals, the most detailed investigations being carried out in last years by KARL and collaborators (see [1.45,135] and references therein).

Some characteristic types of trapping levels in a doped anthracene cysttal, which illustrate well the general conditions for local state formation by guest molecules (cf. Table 5.12), are shown, according to KARL [1.45], in Fig.5.50.

As can be seen from Table 5.12 and Fig.5.50, in the case of the polyacene series higher analogues of the host form, as a rule, traps for both electrons and holes, whilst the lower analogues form only antitrapping local states. From this point of view tetracene, for instance, is a rather undesirable impurity in an anthracene crystal. Anthracene guest molecules in a tetracene crystal, on the contrary, are quite innocent impurities, practically not affecting the transport of charge carriers.

Anthraquinone in an anthracene host, as well as other quinone-type oxidation products in aromatic crystals, form, as a rule, relatively deep trapping centers *only* for electrons.

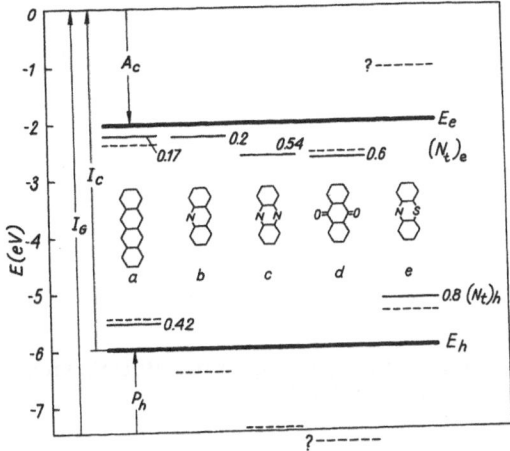

Fig. 5.50. Measured trapping levels in anthracene formed by tetracene (a), acridine (b), phenazine (c), anthraquinone (d) and phenothiazine (e) (solid levels), as compared to the values obtained from (5.72,73) (dashed levels), according to [1.45]

Such deep electron trapping states have been detected by BELKIND et al. (see, e.g. [1.144,2.136] and references therein) in a number of aromatic crystals using photoemission technique (cf. Sect.5.13). Thus in tetracene crystals trapping states were observed of typical depth $E_t = 1.50 \pm 0.20$ eV, which have been identified as quinone-type impurity centers. First, the concentration of these trapping states increases upon photooxidation of the sample. Secondly, similar trapping states emerge with E_t value 1.45 ± 0.1 eV in a dioxitetracenequinone-doped tetracene crystal (cf. [Ref. 1.144, p. 124]).

It is interesting to mention that the trapping states formed by quinone-type impurities are usually deeper than one should expect from (5.72), i.e., $(\Delta P_e)_{guest}$ is not negligible (cf. [Ref. 1.144, p.125]; see also Fig.5.50d). This may be caused by nonuniform distribution on the trapped electron over the quinone-type anion, resulting in an increase in electronic polarization energy.

The quinone-type electron trapping centers usually dominate in real aromatic crystals and cause an asymmetrical energy structure of trapping states (cf. Fig.5.51). If free charge carriers have been photogenerated in such a crystal, the electrons are effectively trapped by deep electron trapping states, while holes are trapped mainly by shallow traps of structural origin (cf. Fig.5.51). As a result the quasi-Fermi level for electrons $(E_F)_e$ remains fixed on the quasi-discrete trapping levels, whereas the quasi-Fermi

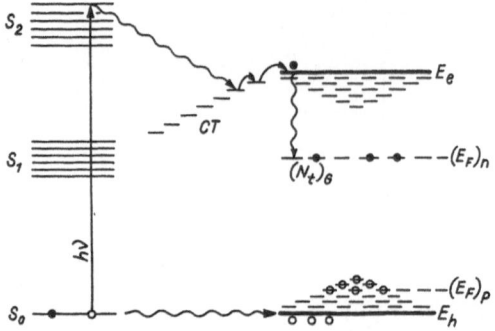

Fig. 5.51. Schematic diagram of an asymmetric trapping state distribution, characteristic for aromatic organic molecular crystals

level for holes $(E_F)_h$ steadily approaches the hole conductivity level E_h with increasing trapped hole concentration.

Such asymmetric trap distribution, apparently, determines p-type conductivity typical for practically all aromatic and a number of heterocyclic organic crystals. Because of this factor it is also extremely difficult to obtain conditions of genuine intrinsic conductivity for the majority of organic molecular crystals. The experimentally estimated activation energy of dark conductivity E_a is usually lower than the value of $\Delta E_G/2$, and only in exceptional cases, under special precautions the value E_a^0 of genuine intrinsic dark conductivity can be obtained (cf. Sect.2.9).

On the other hand, the absence or negligible density of deep impurity centers for hole trapping in the majority of aromatic and a number of heterocyclic molecular crystals allows one to scan by monopolar hole injection the shallow trapping states of structural origin (cf. Sect.5.10).

In some heterocyclic organic crystals, e.g., in tetrathiotetracene (TTT) crystals, deep electron trapping states may be, presumably, formed by cation-radical TTT$^+$ impurities. In this case asymmetric p-type conductivity also dominates, with typical E_a value ca. 0.3 eV. Only under special precautions genuine intrinsic conductivity activation energy $E_a^0 = \Delta E_G/2 = 1.0$ eV can be observed [2.142].

It has been shown by SILINSH et al. [1.66,62] that a guest molecule, having polarizability different from that of the host molecule, may form local " satellite" states of polarization origin for an approaching charge carrier. Indeed, if the molecular polarizability for host molecules is α_h, and for guest molecule α_g, and $\alpha_h \neq \alpha_g$, then for a charge carrier approaching a guest molecule the polarization energy changes with distance r_g, i.e., $P = P(r_g)$.

The difference in polarization energy $\Delta P(r_g)$ caused by the guest molecule equals

$$\Delta P(r_g) = \frac{e^2(\alpha_g - \alpha_h)}{2r_g^4} \quad , \tag{5.74}$$

in terms of simplified approximation of charge carrier-induced dipole inter-action (cf. Sect.2.4.1).

If $\alpha_g > \alpha_h$ and $\Delta P(r_g) > 0$, local polarization "satellite" states are for-med in the vicinity of the guest molecule, acting as charge carrier trapping centers. On the other hand, if $\alpha_g < \alpha_h$, and $\Delta P(r_g) < 0$ local polarization "satellite" states are formed in the vicinity of the guest molecule, acting as antitraps for an approaching charge carrier.

If the guest molecule itself is a charge carrier trapping center, polar-ization "satellite" states may either increase (if $\Delta P(r_g) > 0$) or diminish (if $\Delta P(r_g) < 0$) the effective capture cross section of the guest molecule.

SILINSH et al. [1.62] have evaluated the $\Delta P(r_g)$ values for polarization "satellite" states of some guest molecules in anthracene crystals by SCPF method (cf. Sect.2.4.2). Thus, e.g., for naphthalene as a guest molecule ($\alpha_g = 16.4 \times 10^{-24}$ cm^3) in anthracene host ($\alpha_h = 25.1 \times 10^{-24}$ cm^3) following $\Delta P(r_g)$ values were obtained: if a naphthalene molecule is in the crystal site (0,0,0) and the charge carrier localized on the (1/2, 1/2, 0) site, $\Delta P = -0.019$ eV; on (0,1,0) site $\Delta P = 0.020$ eV; on (1,1,0) site $\Delta P = 0.009$ eV. This means that for a guest molecule, having $\Delta\alpha = \alpha_g - \alpha_h$ ca. 9×10^{-24} cm^3, the nearest-neighbor "satellite" states have $\Delta P(r_g)$ values of kT order at room temperature.

6. Summing Up and Looking Ahead

... sicut lux seipsam et tenebras manifestat,
sic veritas norma sui et falsi est.

Spinoza

To seek the light of truth, while truth the while
Doth falsely blind the eyesight of his look:
Light seeking light, doth light of light beguile;
So ere you find where light in darkness lies,
Your light grows dark by losing of your eyes.

Shakespeare

We have attempted, in the present work, to give not only an account of well-established experimental facts, physical models and theories, but have raised, as the reader might have noticed, also a number of disputable, even controversial problems. This is natural, if we are dealing with a scientific discipline that is still in the early stages of development, as is the case with the physics of organic molecular crystals.

As one of the well-established models we consider that of polarization-caused local states in a molecular crystal. This model has been borne out by a great number of experimental data and theoretical calculations. It may be regarded as generally accepted at present.

The model of Gaussian distribution of the energy spectra of local states of structural origin is also well established, in our view, on the basis of independent experimental studies by SCLC measurements in Gauss approximation, and by thermally activated spectroscopy technique. The statistical interpretation of the energy spectra of local states, on the other hand, cannot, at the present state, be considered as sufficiently supported by adequate models of random distribution of displaced molecules in structural defects and lattice faults in real molecular crystals.

The problem of structural origin of deep trapping centers with narrow Gauss distribution of the energy spectrum also requires further study. Rather interesting empirical correlations have been found between such states and specific structural defects of the crystal. This indicates certain possibilities for evolving molecular models of such defects. The atom-atom potential method seems most promising in this respect. Applied in combination with calculations of electronic polarization, it may provide direct relationships between the coordinates of displaced molecules in defect structures with the corresponding energy spectrum of local states.

It will be also necessary to combine more effectively traditional optical and electrophysical methods of studying the energy spectra of local states with up-to-date techniques of structural microanalysis of defects in molecular crystals.

The concept of electronic polarization of the crystal by a localized charge carrier being the cornerstone of the local state model, it is essential to develop more generalized methods of calculating electronic polarization in the nearest future, thus providing a possibility of approaching more complex molecular systems than an anthracene crystal. It might be viable to work out a "hybrid" approach of calculation, which will combine methods of microelectrostatics and quantum mechanics.

Finally, a more detailed and profound treatment, both theoretical and experimental, ought to be devoted to problems of vibronic relaxation of the molecule in the process of formation of ionic states. The same applies to the problem of local lattice polarization of the crystal upon trapping of a charge carrier. We are convinced that these targets in organic solid-state physics will be reached in the nearest future.

In further perspective we may safely predict a development of a comprehensive dynamic electronic polaron theory, as well as further advance in more generalized vibronic and lattice polaron concepts in application to various classes of organic molecular crystals.

The borderline of science with the unknown is a virgin area where light gradually vanishes into the darkness.

This borderline of light and dark resembles in many ways the *chiroscuro* scenery so reminiscent of Rembrant's canvas or philosophical drawings by Escher, such as "Day and Night".

One of the main dilemmas a scientist confronts at these advanced frontiers has been brilliantly formulated by Spinoza. As light includes in itself the darkness, so truth consists of itself and fallacy as an integral part.

Another borderline dilemma of cognition has been set forth by Shakespeare in the ingenious verse we have chosen as an epigraph. The light of truth might falsely blind the eyesight of a scholar — the light grows dark by losing of his eyes.

The problem of the *chiroscuro* borderline of cognition, raised already by the most brilliant minds of the Renaissance has become even more acute in our days. In this context we should like to quote the words of one of the founders of quantum mechanics, the French physicist Louis de Broglie: "We must never forget (and the history of science proves it) that every success in our cognition sets more problems than it solves, and every newly discovered land lets sense the existence of new vast continents, as yet unknown to us".

References

1.1   J.D. Cox, G. Pilcher: *Thermochemistry of Organic and Organometallic Compounds* (Academic Press. New York 1970)
1.2   A.I. Kitaigorodsky: Molecular Crystals and Molecules (Nauka, Moscow 1971) (in Russian) [English transl.: (Academic Press, New York 1973)]
1.3   *Handbook of Chemistry and Physics,* 54th ed. (CRC Press, Boca Raton, Florida 1973-1974)
1.4   F. London: Z. Phys. Chem. *11*, 222-251 (1930)
1.5   R.A. Buckingham: Proc.Roy.Soc. A *168*, 264-283 (1938)
1.6. R.A. Buckingham, J. Corner: Proc. Roy. Soc. A *189*, 118-129 (1947)
1.7   J.E. Lennard-Jones: Proc. Cambr. Philos. Soc. *27*, 469-480 (1931)
1.8   A.I. Kitaigorodsky: Tetrahedron *14*, 230-236 (1961)
1.9   A.I. Kitaigorodsky, K.V. Mirskaya: Crystallography *6*, 507-514 (1961)
1.10 A.I. Kitaigorodsky, K.V. Mirskaya: Crystallography *9*, 174-181 (1964)
1.11 A.Kitaigorodsky: J. Chim. Physique *63*, 9-16 (1966)
1.12 A.I. Kitaigorodsky: In *Sixth Molecular Crystal Symposium,* Schloß Elmau, Fed. Rep. of Germany (May 20-25, 1973) pp. 99-104
1.13 D.E. Williams: J. Chem. Phys. *45*, 3770-3778 (1966)
1.14 D.E. Williams: J. Chem. Phys. *47*, 4680-4684 (1967)
1.15 D.E. Williams: Acta Crystallogr. A *28*, 629-635 (1972)
1.16 F.A. Momany, G. Vanderkooi, H.A. Scheraga: Proc. Nat. Acad. Sci. USA *61*, 429-436 (1968)
1.17 A.M. Liquori, E. Giglio, L. Mazzarella: Nuovo Cimento B *55*, 476-480 (1968)
1.18 A. Warshel, S. Lifson: J. Chem. Phys. *53*, 582-594 (1970)
1.19 K.V. Mirskaya, I.E. Kozlova, V.F. Bereznitskaya: Phys. Status Solidi (b) *62*, 291 (1974)
1.20 G. Taddei, R. Righini, P. Manzelli: Acta Crystallogr. B *33*, 626-628 (1977)
1.21 J. Sworakowski: Sci. Papers of Wroclaw Tech. Univ. *6*, 56 (1974)
1.22 E.A. Silinsh, A.J. Jurgis: Izv. Akad. Nauk Latv. SSR, Ser. fiz. tekh. Nauk *2*, 21-25 (1977)
1.23 K. Mirsky, M.D. Cohen: J. Chem. Soc. Faraday Trans. II *72*, 2155-2163 (1976)
1.24 K. Mirsky, M.D. Cohen: Chem. Phys. Lett. *54*, 40-41 (1978)
1.25 S. Ramdas, J.M. Thomas, M.J. Goringe: J. Chem. Soc. Faraday Trans. II *73*, 551-561 (1977)
1.26 C.K. Ingold: *Structure and Mechanism in Organic Chemistry* (Cornell University Press, Ithaca, London 1969)
1.27 J.M. Robertson: Proc. Roy. Soc. A *140*, 79; *142*, 674 (1933)
1.28 A.M. Mathieson, J.M. Robertson, V.C. Sinclair: Acta Crystallogr. *3*, 245-250 (1950)
1.29 V.C. Sinclair, J.M. Robertson, A.M. Mathieson: Acta Crystallogr. *3*, 251-256 (1950)

1.30 D.W.J. Cruickshank: Acta Crystallogr. *9*, 915-923 (1956)
1.31 D.W.J. Cruickshank: Acta Crystallogr. *10*, 470-503 (1957)
1.32 J.M. Robertson: Rev. Mod. Phys. *30*, 155-158 (1958)
1.33 D.W.J. Cruickshank, R.A. Sparks: Proc. Roy. Soc. A *258*, 270-285 (1960)
1.34 D.W.J. Cruickshank: Tetrahedron *17*, 155-161 (1962)
1.35 G.J. Sloan: Mol. Cryst. *1*, 161-194 (1966)
1.36 G.J. Sloan: Mol. Cryst. *2*, 323-331 (1967)
1.37 G.J. Sloan, J.M. Thomas, J.O. Williams: Mol. Cryst. Liq. Cryst. *30*, 167-174 (1975)
1.38 F. Gutman, L.E. Lyons: *Organic Semiconductors* (Wiley, New York 1967)
1.39 L.E. Lyons: "Electron Transfer Across the Boundaries of Organic Solids", in *Physics and Chemistry of the Organic Solid State*, Vol. 1, ed. by D. Fox, M.M. Labes, A. Weissberger (Interscience, New York, London 1963)
1.40 J.H. Sharp, M. Smith: "Organic Semiconductors", in *Physical Chemistry*, Vol. 10 (Academic Press, New York 1970)
1.41 J.B. Birks: *Photophysics of Aromatic Molecules* (Wiley-Interscience, New York 1970)
1.42 M. Pope, H. Kallmann: Disc. Faraday Soc. *51*, 7-16 (1971)
1.43 M.S. Brodin, S.V. Marisova, A.F. Prihodko: "Exciton Spectrum of Anthracene Crystals", in *Excitons in Molecular Crystals* (in Russian), ed. by M.S. Brodin (Naukova Dumka, Kiev 1973) pp. 50-83
1.44 D.M. Hanson: Crit. Rev. Solid State Sci. *3*, 243-271 (1973)
1.45 N. Karl: "Organic Semiconductors", in *Festkörperprobleme*, Vol. XIV (Vieweg, Braunschweig 1974) pp. 261-290
1.46 J.B. Birks: "Photophysics of Aromatic Molecules", in *Organic Molecular Photophysics*, Vol. 2, ed. by J.B. Birks (Wiley-Interscience, New York 1975) pp. 409-613
1.47 U. Itoh: Res. Electrotechn. Lab. (Tokyo) *752*, 75 (1975)
1.48 R.G. Kepler: "Organic Molecular Crystals: Anthracene", in *Treatise on Solid State Chemistry*, Vol. 3, ed. by N.B. Hannay (Plenum Press, New York 1976) pp. 615-678
1.49 H. Meier: "Organic Semiconductors", in *Monographs in Modern Chemistry*, Vol. 2, ed. by H.E. Ebel (Verlag Chemie, Weinheim, Fed. Rep. of Germany 1974)
1.50 *Photonics of Organic Semiconductors* (Naukova Dumka, Kiev 1977) (in Russian)
1.51 C. Hamann, J. Heim: Organische Stoffe mit besonderen elektrischen Eigenschaften", in *Reinststoffprobleme V* (Academie-Verlag, Berlin 1977)
1.52 A.S. Davydov, E.F. Sheka: Phys. Status Solidi *11*, 877-890 (1965)
1.53 A.S. Davydov: *Theory of Molecular Excitons* (Nauka, Moscow 1968) (in Russian) [English transl.: (Plenum Press, New York 1971)]
1.54 O.H. Le Blanc: J. Chem. Phys. *36*, 1082-1087 (1962)
1.55 G.D. Thaxton, R.C. Jarnagin, M. Silver: J. Phys. Chem. *66*, 2461-2465 (1962)
1.56 J.L. Katz, S.A. Rice, S.I. Choi, J. Jortner: J. Chem. Phys. *39*, 1683-1697 (1963)
1.57 R. Silbey, J. Jortner, S.A. Rice, M.T. Vala: J. Chem. Phys. *42*, 733-737; *43*, 2925-2926 (1965)
1.58 R.M. Glaeser, R.S. Berry: J. Chem. Phys. *44*, 3797-3810 (1966)
1.59 M. Batley, L.J. Johnston, L.E. Lyons: Austral. J. Chem. *23*, 2397-2402 (1970)
1.60 G. Hug, R.S. Berry: J. Chem. Phys. *55*, 2516-2521 (1971)
1.61 A. Jurgis, E.A. Silinsh: Phys. Status Solidi (b) *53*, 735-743 (1972)
1.62 E.A. Silinsh, A. Jurgis, A.K. Gailis, L.F. Taure: *Electrical Properties of Organic Solids*, Summer School, Karpacz, Wroclaw, Sep. 1-7, 1974, pp. 40-68

1.63   E.A. Silinsh: Izv. Akad. Nauk Latv. SSR *6*, 40-56 (1975)
1.64   E.A. Silinsh, A.J. Jurgis: Izv. Akad. Nauk Latv. SSR, Ser. fiz. tekh.
        Nauk *1*, 73-82 (1977)
1.65   L.E. Lyons: J. Chem. Soc. 5001-5007 (1957)
1.66   E.A. Silinsh: Phys. Status Solidi (a) *3*, 817-828 (1970)
1.67   E.A. Silinsh, D.R. Balode, A.I. Belkind, A.K. Gailis, V.V. Grechov,
        A.J. Jurgis, L.F. Taure, S. Nešpurek: *Seventh Molecular Crystal Sym-
        posium*, Nikko, Japan (Sept. 8-12, 1975) pp. 145-150
1.68   S. Nešpurek, E.A. Silinsh: Phys. Status Solidi (a) *34*, 747-759 (1976)
1.69   E.A. Silinsh: Izv. Akad. Nauk Latv. SSR, Ser. fiz. tekh. Nauk *2*,
        26-33 (1977)
1.70   W. Helfrich, F.R. Lipsett: J. Chem. Phys. *43*, 4368-4376 (1965)
1.71   J.O. Williams, J.M. Thomas: "The Role of Structural Defects in the
        Luminescence of Organic Molecular Crystals", in *Surface and Defect
        Properties of Solids*, Vol. 2, ed. by M.W. Roberts, J.M. Thomas (Chem.
        Soc. Burlington House, London 1973) pp. 229-249
1.72   J.O. Williams, B.P. Clarke, J.M. Thomas, M.J. Shaw: Chem. Phys. Lett.
        38, 41-46 (1976)
1.73   V.A. Lisovenko, M.T. Shpak: Izv.AN SSSR, Ser. fiz. *39*, 2226-2230
        (1975)
1.74   V.A. Lisovenko, M.T. Shpak, B.G. Antonjuk: Chem. Phys. Lett. *42*,
        339-341 (1976)
1.75   W. Helfrich: "Space Charge Limited Currents in Organic Solids", in
        Physics and Chemistry of Organic Solid State, Vol. 3, ed. by D. Fox,
        M.M. Labes, A. Weissberger (Interscience, New York 1967) pp.1-58
1.76   M.A. Lampert, P. Mark: *Current Injection in Solids* (Academic Press,
        New York 1970)
1.77   J.M. Thomas, J.O. Williams: "Dislocations and the Reactivity of Organic
        Solids", in *Progress in Solid State Chemistry*, Vol. 6, ed. by H. Reiss
        and J.O. McCaldin (Pergamon Press, Oxford 1971) pp. 119-154
1.78   J.M. Thomas, E.L. Evans, J.O. Williams: Proc. Roy. Soc. London A *331*,
        417-427 (1972)
1.79   J.M. Thomas: *Electrical Properties of Organic Solids. Supplement.*
        Summer School, Karpacz, Wroclaw, Sept. 1-7, 1974, pp. 5-25
1.80   J.M. Thomas: Philos. Trans. Roy. Soc. London *277*, 251-286 (1974)
1.81   J.O. Williams: Sci. Prog. Oxf. *64*, 247-274 (1977)
1.82   A. Bondi: *Physical Properties of Molecular Crystals, Liquids and
        Glasses* (Wiley, New York 1968)
1.83   D.W.J. Cruickshank: Rev. Mod. Phys. *30*, 163 (1958)
1.83a   D.A. Dows, L. Hsu, S.S. Mitra, O. Brafman, M. Hayek, W.B. Daniels,
        R.K. Crawford: Chem. Phys. Lett. *22*, 595-599 (1973)
1.84   A. Hadni, B. Wyncke, G. Morlot, X. Gerbaux: J. Chem. Phys. *51*, 3514
        (1969)
1.85   S. Ramdas, J.O. Williams, J.M. Thomas, G.M. Parkinson, M.J. Goringe:
        Eighth Molecular Crystal Symposium, Santa Barbara, USA 1977, pp.
        280-285
1.86   G.M. Parkinson, M.J. Goringe, S. Ramdas, J.O. Williams, J.M. Thomas:
        J. Chem. Soc. Sect. D. 134-135 (1978)
1.87   D.P. Craig, J.F. Ogilvie, P.A. Reynolds: J. Chem. Soc. Faraday Trans.
        II *72*, 1603-1612 (1976)
1.88   J.O. Williams: J. Mater. Sci. Lett. *8*, 1361-1362 (1973)
1.89   W. Jones, J.M. Thomas, J.O. Williams: Philos. Mag. *32*, 1-11 (1975)
1.90   K.S. Sundarajan: Z. Kristallogr. *93*, 238-248 (1936)
1.91   M. Suzuki, T. Yokoyama, M. Ito: Spectrochim. Acta A *24*, 1091 (1968)
1.92   A.I. Belkind, V.V. Grechov: Phys. Status Solidi (a) *26*, 377-384 (1974)
1.93   S.C. Abrahams, J.M. Robertson, J.G. White: Acta Crystallogr. *2*, 238
        (1949)
1.94   D.W.J. Cruickshank: Acta Crystallogr. *10*, 504-508 (1957)

1.95   R.S. Becker, E. Chen: J. Chem. Phys. *45*, 2403-2409 (1966)
1.96   A. Terenin, F. Vilesov: Adv. Photochem. *2*, 385-423 (1964)
1.97   Y. Kamura, I. Shirotani, H. Inokuchi: Chem. Lett. 627-630 (1974)
1.98   Y. Maruyama, N. Iwasaki: Chem. Phys. Lett. *24*, 26-29 (1974)
1.99   E.A. Silinsh, A.I. Belkind, D.R. Balode, A.J. Biseniece, V.V. Gre-
       chov, L.F. Taure, M.V. Kurik, J.I. Vertzymacha, J. Bok: Phys. Status
       Solidi (a) *25*, 339-347 (1974)
1.100 W. Hofberger: Phys. Status Solidi (a) *30*, 271-278 (1975)
1.101 H. Müller, H. Baessler: Chem. Phys. Lett. *29*, 102-105 (1974)
1.102 W. Hofberger, H. Müller, H. Baessler: *Seventh Molecular Crystal Sym-
       posium*, Nikko, Japan (Sep. 8-12, 1975) pp. 51-52
1.103 I. Shirotani, Y. Kamura, H. Inokuchi: Mol. Cryst. Liq. Cryst. *28*,
       345-353 (1975)
1.104 D.R. Balode, A.I. Belkind, A.J. Biseniece, J.I. Vertzymacha, V.V. Gre-
       chov, J.V. Kalnach, M.V. Kurik, E.A. Silinsh, L.F. Taure: "Energy
       Structure and Electronic Microprocesses in Oriented and Quasi-amor-
       phous Layers of Tetracene and Pentacene", in *Organic Semiconductors*
       (in Russian), ed. by M.V. Kurik (Institute of Physics AN USSR, Kiev
       1976) pp. 15-20
1.105 J.M. Robertson, V.C. Sinclair, J. Trotter: Acta Crystallogr. *14*,
       697-704 (1961)
1.106 R.B. Campbell, J.M. Robertson, J. Trotter: Acta Crystallogr. *14*,
       705-711 (1961)
1.107 R.B. Campbell, J.M. Robertson, J. Trotter: Acta Crystallogr. *15*,
       289-290 (1962)
1.107a D.D. Kolendritskii, M.V. Kurik, Yu.P. Piryatinskii: Phys. Status
       Solidi (b) *91*, 741-751 (1979)
1.107b R. Jankowiak, J. Kalinowski, M. Konys, J. Bucheri: Chem. Phys. Lett.
       65, 549-553 (1979)
1.107c Y. Tomkiewicz, R.P. Groff, P. Avakian: J. Chem. Phys. *54*, 4504 (1971)
1.108 E. Clar: *Polycyclic Hydrocarbons,* Vol. 1 (Academic Press, New York
       1964)
1.109 B. Stevens: Spectrochim. Acta *18*, 439 (1962)
1.110 J. Tanaka: Bull. Chem. Soc. Jpn. *36*, 1237-1249 (1963)
1.111 Y. Nori, D.R. Kearns: Mol. Cryst. Liq. Cryst. *16*, 61-74 (1972)
1.112 J. Tanaka, T. Kishi, M. Tanaka: Bull. Chem. Soc. Jpn. *47*, 2376-2381
       (1974)
1.113 K. Fuke, K. Kava, T. Kajiwara, S. Nagakura: J. Mol. Spectrosc. *63*,
       98-107 (1976)
1.114 H. Tachikawa, L.R. Faulkner: Chem. Phys. Lett. *39*, 436-441 (1976)
1.115 E. von Freydorf, J. Kinder, M.E. Michel-Beyerle: Chem. Phys. *27*, 199-
       209 (1978)
1.116 T. Kobayashi: J. Chem. Phys. *68*, 3570-3574 (1978)
1.117 K.A. Nelson, D.D. Dlott, M.D. Fayer: Chem. Phys. Lett. *64*, 88-93
       (1979)
1.118 M.D. Cohen, R. Haberkorn, E. Huler, Z. Ludmer, M.E. Michel-Beyerle,
       D. Rabinovich, R. Sharon, A. Warshel, V. Yakhot: Chem. Phys. *27*,
       211-216 (1978)
1.119 V.V. Aleksandrov, A.I. Belkind, I.J. Muzikante, E.A. Silinsh, L.F.
       Taure: Fiz.Tverd.Tela *18*, 2410-2412 (1976)
1.120 J.M. Robertson, J.G. White: J. Chem. Soc. 358-368 (1947)
1.121 A. Camerman, J. Trotter: Acta Crystallogr. *18*, 636 (1965)
1.122 W. Jones, S. Ramdas, J.M. Thomas: Chem. Phys. Lett. *54*, 490-493 (1978)
1.123 D.M. Donaldson, J.M. Robertson, J.G. White: Proc. Roy. Soc. A *220*,
       311-321 (1953)
1.124 A.I. Kitaigorodsky: *Organic Crystalchemistry* (in Russian) (Izd. AN
       SSSR, Moscow 1955)
1.125 P. Petelenz: Phys. Status Solidi (b) *73*, 295-299 (1976)

1.126 V. Čápek: Czech. J. Phys. B *29*, 439-446 (1979)
1.127 D.D. Eley, H. Inokuchi, M.R. Willis: Discus. Faraday Soc. *28*, 54 (1959)
1.128 D.M. Burland: Phys. Rev. Lett. *33*, 833-835 (1974)
1.129 D.M. Burland, U. Konzelmann: J. Chem. Phys. *67*, 319 (1977)
1.130 T. Holstein: Ann. Phys. (N.Y.) *8*, 343 (1959)
1.131 J. Yamashita, T. Kurosana: Phys. Chem. Sol. *5*, 34 (1958)
1.132 J. Appel: Solid State Phys. *21*, 193-391 (1968)
1.133 L.B. Schein, C.B. Duke, A.R. McGhie: Phys. Rev. Lett. *40*, 197-200 (1978)
1.134 L.B. Schein: *Electrical and Related Properties of Organic Solids*, Papers of Intern. Conf., Karpacz, Poland (Sept. 18-23, 1978) pp. 273-277
1.135 K.H. Probst, N. Karl: Phys. Status Solidi (a) *27*, 499-523 (1975)
1.136 N. Karl: *Electrical and Related Properties of Organic Solids*, Papers of Intern. Conf., Karpacz, Poland (Sept. 18-23, 1978) pp. 43-50
1.136a L.B. Schein, A.R. McGhie: Phys. Rev. B *20*, 1631-1639 (1979)
1.136b L.B. Schein, A.R. McGhie: Chem. Phys. Lett. *62*, 356-359 (1979)
1.137 A. Madhukar, M. Post: Phys. Rev. Lett. *39*, 1424 (1977)
1.138 H. Haken, P. Reineker: Z. Phys. *249*, 253-268 (1972)
1.139 H. Haken: *Sixth Molecular Crystal Symposium*, Schloß Elmau, Fed. Rep. of Germany, May 20-25, 1973, pp. 13-15
1.140 H. Sumi: Solid State Commun. *28*, 309-312 (1978)
1.140a S. Efrima, H. Metiu: J. Chem. Phys. *69*, 5113 (1978); Chem. Phys. Lett. *60*, 226 (1979)
1.141 H. Sumi: Solid State Commun. *29*, 495-499 (1979)
1.142 L.B. Schein, P.J. Nigrey: Phys. Rev. B *18*, 2929-2930 (1978)
1.143 M. Silver, K. Resco: *Electrical and Related Properties of Organic Solids*, Papers of Intern. Conf., Karpacz, Poland (Sept. 18-23, 1978) Supplement
1.144 A.I. Belkind: *Photoelectron Emission from Organic Solid State* (Zinatne, Riga 1979) (in Russian)
1.145 S. Hino, N. Sato, H. Inokuchi: Chem. Phys. Lett. *37*, 494-496 (1976)
1.146 S. Hino, N. Sato, H. Inokuchi: J. Chem. Phys. *67*, 4139-4144 (1977)
1.147 S. Hino, N. Sato, H. Inokuchi: *Annual Review* (Inst. for Molecular Science, Okazaki, Japan 1978) pp. 68-69
1.148 L.Y. Bubnov, E.L. Frankevich: Phys. Status Solidi (b) *62*, 281-290 (1974)
1.149 A.I. Belkind, A.M. Brodsky, V.V. Grechov: Phys. Status Solidi (b) *85*, 456-472 (1978)
1.150 W. Pong, J.A. Smith: J. Appl. Phys. *44*, 174-176 (1973)
1.151 C.A. Burke, G.B. Birrell, G.H. Lesch, O.H. Griffith: Photochem. Photobiol. *19*, 29-34 (1974)
1.152 Y. Toyozawa: Progr. Theor. Phys. *12*, 421-443 (1954)
1.153 W. Siebrand: J. Chem. Phys. *41*, 3574-3581 (1964)
1.154 R.W. Munn, W. Siebrand: J. Chem. Phys. *52*, 47-63 (1970)
1.155 R.W. Munn, W. Siebrand: J. Chem. Phys. *52*, 6391-6406 (1970)
1.156 R.W. Munn, W. Siebrand: Disc. Faraday Soc. *51*, 17-23 (1971)
1.157 J. Vilfan: Phys. Status Solidi (b) *59*, 351-360 (1973)
1.158 P. Gosar, S. Choi: Phys. Rev. *150*, 529-538 (1966)
1.159 K. Seki, Y. Harada, K. Ohno, H. Inokuchi: Bull. Chem. Soc. Jpn. *47*, 1608-1610 (1974)
1.160 C.B. Duke: Surf. Sci. *70*, 674 (1978)
1.161 C.B. Duke: *Electrical and Related Properties of Organic Solids*, Papers of Intern. Conf., Karpacz, Poland (Sept. 18-23, 1978) pp. 9-11

Chapter 2

2.1 A.S. Davydov: *Theory of Light Absorption in Molecular Crystals* (in Russian) (AN USSR, Kiev 1951)
2.2 A.S. Davydov: *Solid State Theory* (in Russian) (Nauka, Moscow 1976)

2.3   J.I. Frenkel: *Collection of Selected Works* (in Russian) (AN SSSR, Moscow 1958)
2.4   V.M. Agranovich: *Exciton Theory* (in Russian) (Nauka, Moscow 1968)
2.5   V.M. Agranovich, M.D. Galanin: *Energy Transfer of Electron Excitation in Condensed Media* (in Russian) (Nauka, Moscow 1978)
2.6   D.S. McClure: Solid State Phys. *8*, 1 (1959)
2.7   H.C. Wolf: Solid State Phys. *9*, 1 (1959)
2.8   D.P. Craig, S.H. Walmsley: *Excitons in Molecular Crystals* (A. Benjamin, New York 1968)
2.9   S.A. Rice, J. Jortner: "Comments on the Theory of the Exciton of Molecular Crystals", in *Physics and Chemistry of the Organic Solid State*, Vol. 3 (Interscience, New York 1967)
2.10   H. Haken, G. Strobl: *Triplet State*, ed. by a. Zahlan (Cambridge University Press, Cambridge, 1967)
2.11   H. Haken, G. Strobl: Z. Phys. *262*, 135-148 (1973)
2.12   H. Haken, E. Schwarzer: Chem. Phys. Lett. *27*, 41 (1974)
2.13   V.M. Kenkre, R.S. Knox: Phys. Rev. B *9*, 5279-5290 (1974)
2.14   V.M. Kenkre: Phys. Rev. B *11*, 1741-1745 (1975)
2.15   V.M. Kenkre: Phys. Rev. B *12*, 2150-2159 (1975)
2.16   M. Grover, R. Silbey: J. Chem. Phys. *54*, 4843-4851 (1971)
2.17   R.W. Munn, R. Silbey: J. Chem. Phys. *68*, 2439-2450 (1978)
2.18   H. Sumi: J. Chem. Phys. *67*, 2943 (1977)
2.19   M.V. Kurik, L.I. Tsikora: Phys. Status Solidi (b) *66*, 695-702 (1974)
2.20   M.V. Kurik, Y.P. Piryatinsky: Fiz. Tverd. Tela *13*, 2877-2882 (1971)
2.21   V. Ern, A. Suna, Y. Tomkiewicz, P. Avakian, R.P. Groff: Phys. Rev. B *5*, 3222 (1972)
2.22   A.J. Campillo, S.L. Shapiro: "Picosecond Relaxation Measurements in Biology", in *Ultrashort Light Pulses. Picosecond Techniques and Applications*, ed by S.L. Shapiro, Topics in Applied Physics, Vol. 18 (Springer, Berlin, Heidelberg, New York 1977)
2.23   P. Reineker: Z. Naturforsch. a *29*, 282 (1974)
2.24   A.A. Ovchinnikov, N.S. Erikhman: Zh. Eksp. Teor. Fiz. *67*, 1474 (1974)
2.25   I.B. Rips: Teor. Mat. Fiz. *20*, 275 (1979)
2.26   R. Kühne, P. Reineker: Phys. Status Solidi B *89*, 131-141 (1978)
2.27   M. Grover, R. Silbey: J. Chem. Phys. *52*, 2099-2108 (1970)
2.28   J.A. Reissland: *The Physics of Phonons* (Wiley, New York 1973)
2.29   S.D. Druger: "Theory of Charge Transport Processes in Organic Molecular Solids", in *Organic Molecular Photophysics*, Vol. 2, ed. by J.B. Birks (Wiley, Interscience, New York 1975) pp. 313-394
2.30   W.H. McCrea: Interdiscipl. Sci. Rev. *3*, 267-274 (1978)
2.31   L.E. Lyons, J.C. Mackie: Proc. Chem. Soc. 71-77 (1962)
2.32   J.C. Mackie: Ph. D. Thesis, University of Sydney (1963)
2.33   G. Vaubel, H. Baessler: Phys. Status Solidi *26*, 599-606 (1968)
2.34   H. Baessler, H. Killesreiter: Mol. Cryst. Liq. Cryst. *24*, 21-31 (1973)
2.35   J. Dresner: J. Chem. Phys. *52*, 6343-6347 (1970)
2.36   R.S. Knox: *Theory of Excitons* (Academic Press, New York 1963)
2.37   S.I. Choi, J. Jortner, S. Rice, R. Silbey: J. Chem. Phys. *41*, 3294-3306 (1964)
2.38   M. Pope, J. Burgos: Mol. Cryst. *1*, 395 (1966)
2.39   E.A. Silinsh, L.F. Taure: "Energy Level Spectra of Ionized States in Molecular Crystals", in *Photonics of Organic Semiconductors* (in Russian) (Naukova Dumka, Kiev 1977) pp. 3-32
2.40   H. Fröhlich, H. Pelzer, S. Zienau: Philos. Mag. *41*, 221-242 (1950)
2.41   H. Haken, W. Schottky: Z. Phys. Chem. *16*, 218-244 (1958)
2.42   S. Wang, C.K. Mahutte, M. Matsura: Phys. Status Solidi (b) *51*, 11-55 (1972)
2.43   V.A. Bendersky: Zh. Strukt. Khim. *4*, 415-423 (1963)
2.44   N.F. Mott, M.J. Littleton: Trans. Faraday Soc. *34*, 485-491 (1938)

2.45   N.F. Mott, R.W. Gurney: *Electronic Processes in Ionic Crystals* (Univ. Press, London-Oxford 1940, The Clarendon Press, Oxford 1948, 2nd ed., reprinted by Dover Publ. Inc., New York 1964)

2.46   E.S. Rittner, R.A. Hunter, F.K. Du Prê: J. Chem. Phys. *17*, 198-208 (1949)

2.47   S.D. Druger, R.S. Knox: J. Chem. Phys. *50*, 3143-3153 (1969)

2.48   W.B. Fowler: Phys. Rev. *151*, 657-667 (1966)

2.49   V. Čápek: Czech. J. Phys. B *28*, 567-570 (1978)

2.50   G.L. Hug: *"Ionic States in Molecular Crystals"*; Dissertation, University of Chicago, Chicago, Ill. (1970)

2.51   A. Schweig: Chem. Phys. Lett. *1*, 195-199 (1967)

2.52   C.G. Le Fevre, R.J.W. Le Fevre: Rev. Pure Appl. Chem. *5*, 261-318 (1955)

2.53   R.J.W. Le Fevre, K.M. Sundaram: J. Chem. Soc., 4442-4446 (1963)

2.54   K. Higasi, H. Baba, A. Rembaum: *Quantum Organic Chemistry* (Interscience Publ., New York 1965)

2.55   V. Čápek: Czech. J. Phys. B *28*, 773 (1978)

2.56   P.J. Bounds, R.W. Munn: Chem. Phys. *44*, 103-112 (1979)

2.57   P.G. Cummins, D.J. Dunmur, R.W. Munn: Chem. Phys. Lett. *22*, 519-522 (1973)

2.58   R.J.W. Le Fevre: Rev. Pure Appl. Chem. *20*, 67-79 (1970)

2.59   R.J.W. Le Fevre, L. Radom: J. Chem. Soc. Sect. B, 1295-1298 (1967)

2.60   R.J.W. Le Fevre, L. Radom, G.L.D. Ritchie: J. Chem. Soc. Sect. B, 775-778 (1968)

2.61   C.L. Cheng, D.S.N. Murthy, G.L. D. Ritchie: Austral. J. Chem. *25*, 1301-1305 (1972)

2.62   M.F. Vuks: Opt. Spektrosk. *20*, 644-651 (1966)

2.63   I.V. Obreimov, A.F. Prihodko, I.V. Rodnikova: Zh. Eksp. Teor. Fiz. *18*, 409-418 (1948)

2.64   I.V. Obreimov: Tr. Gos. Opt. Inst. *24*, 5-11 (1957)

2.65   R. Mathies, A.C. Albrecht: J. Chem. Phys. *60*, 2500-2508 (1974)

2.66   I. Nakada: J. Phys. Soc. Jpn. *17*, 113-118 (1962)

2.67   N.V. Cohan, C.A. Coulson, J.B. Jamieson: Trans. Faraday Soc. *53*, 582-588 (1957)

2.68   S. Takashima: Biopolymers *6*, 1437-1452 (1968)

2.69   S. Takashima: Biopolymers *8*, 199-216 (1969)

2.70   D.A. Dunmur: Mol. Phys. *23*, 109-115 (1972)

2.71   A.Jr. Streitwieser: *Molecular Orbital Theory* (Wiley, New York 1962)

2.72   B. Pullman, A. Pullman: *Quantum Biochemistry* (Interscience, New York 1963)

2.73   J.A. Pople: Trans. Faraday Soc. *49*, 1375-1385 (1953)

2.74   R. Pariser, R.G. Parr: J. Chem. Phys. *21*, 466-471 (1953)

2.75   R. Pariser, R.G. Parr: J. Chem. Phys. *21*, 767-776 (1953)

2.76   A. Brickstock, J.A. Pople: Trans. Faraday Soc. *50*, 901-911 (1954)

2.77   J.A. Pople: J. Phys. Chem. *61*, 6-10 (1957)

2.78   R. Hoffmann: J. Chem. Phys. *39*, 1397-1412 (1963)

2.79   L. Libit, R. Hoffmann: J. Amer. Chem. Soc. *96*, 1370-1383 (1974)

2.80   J.A. Pople, D.P. Santry, G.A. Segal: J. Chem. Phys. *43*, S129-S135 (1965)

2.81   J.A. Pople, D.L. Beveridge: *Approximate Molecular Orbital Theory* (McGraw-Hill, New York 1970)

2.82   R.J. Buenker, S.D. Peyerimhoff: Chem. Phys. Lett. *3*, 37-42 (1969)

2.83   T. Koopmans: Physica *1*, 104-113 (1934)

2.84   J.R. Hoyland, L. Goodman: J. Chem. Phys. *36*, 12-24 (1962)

2.85   K. Ohno, T. Hirooka, Y. Harada, H. Inokuchi: Bull. Chem. Soc. Jpn. *46*, 2353-2355 (1976)

2.86   J.H. Eland: Int. J. Mass. Spectr. Ion Phys. *9*, 214-219 (1972)

2.87   B.A. Harrison: *Solid State Theory* (McGraw-Hill, New York 1970)

2.88   M.J.S. Dewar, J.A. Hashmall, C.G. Venier: J. Amer. Chem. Soc. *90*, 1953 (1968)

2.89   M.J.S. Dewar: *The Molecular Orbital Theory of Organic Chemistry*
        (McGraw-Hill, New York 1969)
2.90   D.W. Turner: Philos. Trans. Roy. Soc. London A *268*, 7-31 (1970)
2.91   D.E. Freeman, I.G. Ross: Spectrochim. Acta *16*, 1393-1408 (1960)
2.92   D.B. Scully, D.H. Whiffen: Spectrochim. Acta *16*, 1409-1415 (1960)
2.93   C.L. Dodson, J.F. Graham: J. Phys. Chem. 77, 2903-2906 (1973)
2.94   M.A. Jeljashevich: *Atomic and Molecular Spectroscopy* (in Russian)
        (Izd.Fiz. Mat. Lit., Moscow 1962)
2.95   I.J. Kachkurova: Zh. Prikl. Spektr. *22*, 689-695 (1975)
2.96   I.V. Aleksandrov, J.S. Bobovich, V.G. Maslov, A.N. Sidorov: Pisma
        Zh. Eksp. Teor. Fiz. *17*, 306-309 (1973)
2.97   I.V. Aleksandrov, J.S. Bobovich, V.G. Maslov, A.N. Sidorov: Opt. Spek-
        trosk. *35*, 264-269 (1973)
2.98   A.I. Belkind, J. Bok: Phys. Status Solidi A *22*, K37-K40 (1974)
2.99   N. Karl, H. Feederle: Phys. Status Solidi A *34*, 497-503 (1976)
2.100   P. Schlotter, H. Baessler: Chem. Phys. *19*, 353-360 (1977)
2.101   L.M. Blinov: "Stark Spectroscopy of Molecular Crystals", in *Modulation
        Spectroscopy of Semiconductors and Dielectrics* (in Russian) (Tbilisi
        1975) pp. 79-85
2.102   S.C. Abbi, D.M. Hanson: J. Chem. Phys. *60*, 319-320 (1974)
2.103   L.M. Blinov, N.A. Kirichenko: Opt. Spektrosk. *37*, 897-902 (1974)
2.104   H. Killesreiter, H. Baessler: Phys. Status Solidi B *51*, 657-668 (1972)
2.105   H. Baessler, H. Killesreiter: Phys. Status Solidi B *53*, 183-192 (1972)
2.106   W. Spannring, H. Baessler: Chem. Phys. *25*, 325-331 (1977)
2.107   I.S. Kaulach, E.A. Silinsh: Phys. Status Solidi B *75*, 247-254 (1976)
2.108   T. Kitagawa: J. Mol. Spectrosc. *26*, 1-23 (1968)
2.109   M. Pope, J. Burgos: Mol. Cryst. *3*, 215-226 (1967)
2.110   N. Geacintov, M. Pope: J. Chem. Phys. *50*, 814-821 (1969)
2.111   U. Fano: Phys. Rev. *124*, 1866-1878 (1961)
2.112   L. Onsager: Phys. Rev. *54*, 554-557 (1938)
2.113   R.C. Hughes: J. Chem. Phys. *55*, 5442-5447 (1971)
2.114   R.C. Hughes: "Geminate Recombination of Charge Carriers in Organic
        Solids", in *Annual Report, Conference on Electrical Insulation and
        Dielectric Phenomena* (National Academy of Sciences, Washington D. C.
        1971) pp. 8-16
2.115   R.H. Batt, C.L. Braun, J.F. Hornig: J. Chem. Phys. *49*, 167-168 (1968)
2.116   R.H. Batt, C.L. Braun, J.F. Hornig: Appl. Opt., Suppl. *3*, 20-28 (1969)
2.117   R.R. Chance, C.L. Braun: J. Chem. Phys. *59*, 2269-2272 (1973)
2.118   R.R. Chance, C.L. Braun: J. Chem. Phys. *64*, 3573-3581 (1976)
2.118a B.M. Borsenberger, A.I. Ateya: J. Appl. Phys. *50*, 909-913 (1979)
2.118b C.S. Ryan, J.B. Webb, D.F. Williams: Mol. Cryst. Liq. Cryst. *56*,
        (Letters), 69-74 (1979)
2.119   L.E. Lyons, K.A. Milne: *Seventh Molecular Crystal Symposium*, Nikko,
        Japan (Sept. 8-12, 1975) pp. 125-128
2.120   L.E. Lyons, K.A. Milne: J. Chem. Phys. *65*, 1474-1484 (1976)
2.121   A.K. Gailis, V.A. Kolesnikov, E.A. Silinsh: Izv. Akad. Nauk Latv. SSR,
        Ser. fiz. tekh. Nauk *1*, 28-34 (1978)
2.122   E.A. Silinsh, A.K. Gailis, V.A. Kolesnikov, I.J. Muzikante: *Electrical
        and Related Properties of Organic Solids*, Papers of Intern. Conf.,
        Karpacz, Poland (Sept. 18-23, 1978) p. 278
2.123   P. Melz: J. Chem. Phys. *57*, 1694-1699 (1972)
2.124   K. Lochner, B. Reimer, H. Baessler: Phys. Status Solidi B *76*, 533-540
        (1976)
2.125   D.M. Pai, R.C. Enck: Phys. Rev. B *11*, 5163-5174 (1975)
2.126   G. Klein, R. Voltz: Int. J. Radiat. Chem. *7*, 155-174 (1975)
2.127   J.C. Knights, E.A. Davis: J. Phys. Chem. Sol. *35*, 543-554 (1974)
2.127a L.B. Schein, R.W. Anderson, R.C. Enck, A.R. McGhie: J. Chem. Phys. *71*,
        3189-3193 (1979)

2.128 K.M. Hong, J. Noolandi: J. Chem. Phys. *69*, 5026-5039 (1978)
2.129 R.G. Kepler, F.N. Coppage: Phys. Rev. *151*, 610 (1966)
2.130 J.I. Vertzymacha, M.V. Kurik, E.A. Silinsh, L.F. Taure: Izv. Akad. Nauk Latv. SSR, Ser. fiz. tekh. Nauk *2*, 47-53 (1977)
2.131 T. Hirooka, M. Kochi, J. Aihara, H. Inokuchi, Y. Harada: Bull. Chem. Soc. Jpn. *42*, 1481-1486 (1969)
2.132 M. Kochi, Y. Harada, T. Hirooka, H. Inokuchi: Bull. Chem. Soc. Jpn. *43*, 2690-2702 (1970)
2.133 A.I. Belkind, R.I. Kalendarev: Phys. Status Solidi A *14*, 681-688 (1972)
2.134 E.A. Silinsh, L.F. Taure, D.R. Balode: "Electrophysical and Photoelectrical Properties of Thin Films of 1,3-indandione Compounds", in *Structure and Tautomerism of β-Dicarbonil Compounds* (Zinatne, Riga 1976)
2.135 A.I. Belkind, S.B. Aleksandrov: Phys. Status Solidi A *9*, 105 (1972)
2.136 A.I. Belkind, S.B. Aleksandrov, V.V. Aleksandrov, V.V. Grechov: *Electrical Properties of Organic Solids*, Summer School, Karpacz, Wroclaw (Sept. 1-7, 1974) pp. 8-24
2.137 R. Boschi, J.N. Murell, W. Schmidt: Faraday Disc. Chem. Soc. *54*, 116-126 (1972)
2.138 Y. Kamura, K. Seki, H. Inokuchi: Chem. Phys. Lett. *30*, 35-38 (1975)
2.139 N.E. Geacintov, J. Burgos, M. Pope, C. Strom: Chem. Phys. Lett. *11*, 504 (1971)
2.140 V.V. Aleksandrov, A.I. Belkind, I.S. Kaulach, I.J. Muzikante, E.A. Silinsh, L.F. Taure: "Photoconductivity Mechanisms and Energy Level Spectra of Perylene", in *Organic Semiconductors* (in Russian), ed. by M.V. Kurik (Institute of Physics AN USSR, Kiev 1976) pp. 20-24
2.141 A.K. Gailis, E.A. Silinsh: Izv. Akad. Nauk Latv. SSR, Ser. fiz. tekh. Nauk *4*, 29-38 (1973)
2.142 D.R. Balode, E.A. Silinsh, L.F. Taure: Izv. Akad. Nauk Latv. SSR, Ser. fiz. tekh. Nauk *1*, 35-45 (1978)
2.143 M.J.S. Dewar, D.W. Goodman: J. Chem. Soc. Faraday Trans. II *68*, 1784-1788 (1972)
2.144 J.G. Angus, G.C. Morris: J. Mol. Spectrosc. *21*, 310-324 (1966)
2.145 G. Briegleb, Z. Czekalla: Z. Electrochem. *63*, 6-12 (1959)
2.146 R. Foster: Nature *183*, 1253-1255 (1959)
2.147 L.E. Lyons, G.C. Morris, L.J. Warren: J. Phys. Chem. *72*, 3677-3678 (1968)
2.148 L.E. Lyons, G.C. Morris: J. Chem. Soc. 5192-5199 (1960)
2.149 M. Fujihira, T. Hirooka, H. Inokuchi: Chem. Phys. Lett. *19*, 584-587 (1973)
2.150 A.P. Marchetti, D.R. Kearns: Mol. Cryst. Liq. Cryst. *6*, 299-317 (1970)
2.151 N. Geacintov, M. Pope: J. Chem. Phys. *45*, 3884-3885 (1966)
2.152 R.F. Chaiken, D.R. Kearns: J. Chem. Phys. *45*, 3966-3976 (1966)
2.153 V.V. Grechov, J.V. Kalnach, A.I. Belkind: Izv. Akad. Nauk Latv. SSR, Ser. fiz. tekh. Nauk *5*, 64-75 (1977)
2.154 M. Batley, L.E. Lyons: Austral. J. Chem. *19*, 345-350 (1966)

Chapter 3

3.1   D.R. Balode, A.K. Gailis, E.A. Silinsh, L.F. Taure: "Investigation of Thin Layers of Organic Semiconductors by SCLC Method", in *Semiconductors and Their Application in Electrotechnique*, Vol. 5 (in Russian) (Zinatne, Riga 1971) pp. 197-220
3.2   P. Jansen, W. Helfrich, N. Riehl: Phys. Status Solidi *7*, 851-861 (1964)
3.3   A. Sussman: J. Appl. Phys. *38*, 2738-2752 (1967)
3.4   H. Baessler, G. Hermann, N. Riehl, G. Vaubel: J. Phys. Chem. Sol. *30*, 1579-1585 (1969)

3.5   J. Sworakowski: Mol. Cryst. Liq. Cryst. *11*, 1-11 (1970)
3.6   J. Sworakowski: Mol. Cryst. Liq. Cryst. *19*, 259-268 (1973)
3.7   J.M. Thomas, J.O. Williams, G.A. Cox: Trans. Faraday Soc. *64*,
        2496-2504 (1968)
3.8   J.M. Thomas: Israel J. Chem. *10*, 573-580 (1972)
3.9   J.M. Thomas: *Sixth Molecular Crystal Symposium:* Schloß Elmau, Fed. Rep.
        of Germany (May 20-25, 1973) pp. 124-127
3.10   J.M. Thomas, J.O. Williams, W. Jones: *Reactivity of Solids*, Proc. Se-
        venth Intern. Symp. on the Reactivity of Solids, Bristol (1972) pp.
        515-524
3.11   E.A. Silinsh, A.J. Jurgis: Izv. Akad. Nauk Latv. SSR, Ser. fiz. tekh.
        Nauk *1*, 36-39 (1977)
3.12   J. Sworakowski: Politech. Wroclaw. Komunikati N *78*, 1-13 (1975)
3.13   J. Sworakowski: Mol. Cryst. Liq. Cryst. *33*, 83-89 (1976)
3.14   J.H. Holloman, R. Maurer, F. Seitz: *Imperfection in Nearly Perfect
        Crystals*, ed. by W. Schockley (Wiley, New York 1952)
3.15   H.G. Van Bueren: *Imperfections in Crystals* (North-Holland, Amsterdam
        1960)
3.16   J.N. Sherwood: "Diffusion in Molecular Solids", in *Surface and Defect
        Properties of Solids*, Vol. 2 (Chem. Soc., Burlington House, London
        1973) pp. 250-268
3.17   A.V. Chadwick, J.N. Sherwood: "Point Defects in Molecular Solids", in
        Point Defects in Solids, Vol. 2, ed. by J.H. Crawford Jr., L.M. Slif-
        kin (Plenum Press, New York 1975)
3.18   W.F. Harris: "The Geometry of Disclinations in Crystals", in *Surface
        and Defect Properties of Solids*, Vol. 3 (Chem. Soc., Burlington
        House, London 1974)
3.19   W.F. Harris: Sci. Amer. *12*, 130-145 (1977)
3.20   A. Kelly. G.W. Groves: *Crystallography and Crystal Defects* (Longman,
        London 1970)
3.21   R.A. Laudise: *The Growth of Single Crystals* (Prentice-Hall, Engle-
        wood Cliffs, N. J. 1970)
3.22   H.F. Mataré: *Defect Electronics in Semiconductors* (Wiley Interscience,
        New York 1971)
3.23   J.N. Sherwood: *IV International Symposium on Organic Solid State*, Bor-
        deaux, France (July 16-18, 1975) p. 20
3.24   R. Fox, J.N. Sherwood: Trans. Faraday Soc. *67*, 3364-3371 (1971)
3.25   J.N. Sherwood, D.J. White: Philos. Mag. *16*, 975-980 (1967)
3.26   S. Ramdas, J.M. Thomas: J. Chem. Soc. Faraday Trans. II *72*, 1251-1258
        (1976)
3.27   P.E. Schipper, S.H. Wamsley: Proc. Roy. Soc. A *348*, 203-219 (1976)
3.28   J.O. Williams, D. Donati, J.M. Thomas: J. Chem. Soc. Faraday Trans.
        II *73*, 1169-1177 (1977)
3.29   J.N. Sherwood, D.J. White: Philos. Mag. *15*, 745-753 (1967)
3.30   J.N. Sherwood, S.J. Thomson: Trans. Faraday Soc. *56*, 1443-1451 (1960)
3.31   C.H. Lee, H.K. Kevorkian, P.J. Reucroft, M.M. Labes: J. Chem. Phys. *42*,
        1406-1410 (1965)
3.32   P.J. Reucroft, H.K. Kevorkian, M.M. Labes: J. Chem. Phys. *44*, 4416-
        -4420 (1966)
3.33   G. Burns, J.N. Sherwood: Mol. Cryst. Liq. Cryst. *18*, 91-94 (1972)
3.34   J.P. Hirth, J. Lothe: *Theory of Dislocations* (McGraw-Hill, New York
        1968)
3.35   W.T. Read Jr.: *Dislocations in Crystals* (McGraw-Hill, New York 1953)
3.36   W.T. Read Jr.: Philos. Mag. *46*, 111-131 (1955)
3.37   W. Bardsley: "The Electrical Effects of Dislocations in Semiconduc-
        tors", in *Progress in Semiconductors*, Vol. 4 (Heywood, London 1960)
        pp. 155-203
3.38   R.M. Broudy: Adv. Phys. *12*, 135-184 (1963)

3.39   A.H. Cottrell: *Theory of Crystal Dislocation* (Blackie and Son, London 1965)
3.40   R.E. Peierls: Proc. Phys. Soc. *52*, 34-37 (1940)
3.41   A.S. Nowick, B.C. Berry: *Anelastic Relaxation in Crystalline Solids* (Academic Press, New York 1972)
3.42   T. Danno, H. Inokuchi: Bull. Chem. Soc. Jpn. *41*, 1783-1787 (1968)
3.43   G.K. Afanas'eva, R.M. Miasnikova: Crystallography *15*, 189-190 (1970)
3.44   H.B. Huntington, J.E. Dickey, R. Thomson: Phys. Rev. *100*, 1117-1128 (1955)
3.45   R.M. Cotterill, M. Doyama: Phys. Rev. *145*, 465-478 (1966)
3.46   A. Englert, H. Tompa: J. Phys. Chem. Sol. *21*, 306-309 (1961)
3.47   S. Amelinckx, W. Dekeyser: "The Structure and Properties of Grain Boundaries", in *Solid State Physics*, Vol. 8 (Academic Press, New York 1959)
3.48   P.M. Robinson, H.G. Scott: Phys. Status Solidi *20*, 461-471 (1967)
3.49   M. August: *Theorie statischer Versetzungen* (Teubner, Leipzig 1966)
3.50   B.M. Strunin: Fiz. Tverd. Tela *9*, 803-812 (1967)
3.51   A.A. Predvoditelev: "Investigation of Dislocation Ensembles", in *Problems of Modern Crystallography* (in Russian) (Nauka, Moscow 1975)
3.52   J. Friedel: Philos. Mag. *46*, 1169-1186 (1965)
3.53   J.O. Williams, J.M. Thomas: Trans. Faraday Soc. *63*, 1710-1729 (1967)
3.54   S. Amelinckx, P. Delavignette: "Dislocations in Layer Structures", in *Direct Observations of Imperfections in Crystals*, ed. by J.B. Newkirk, J.H. Wernick (Wiley, New York 1962)
3.55   P.M. Robinson, H.G. Scott: Mol. Cryst. Liq. Cryst. *11*, 13-23 (1970)
3.56   P.M. Robinson, H.G. Scott: Acta Metall. *15*, 1230-1231 (1967)
3.57   D. Michell, P.M. Robinson, A.P. Smith: Phys. Status Solidi *26*, K93-K95 (1968)
3.58   K. Kojima: Phys. Status Solidi, A *51*, 71-78 (1979)
3.59   W. Jones, J.M. Thomas, J.O. Williams, L.W. Hobbs: J. Chem. Soc. Faraday Trans. II *71*, 138-145 (1975)
3.60   J. Tanaka, T. Koda, S. Shionoya, S. Minomura: Bull. Chem. Soc. Jpn. *38*, 1559-1560 (1965)
3.61   H. Müller, H. Baessler: Chem. Phys. Lett. *36*, 312-315 (1975)
3.62   H. Müller, H. Baessler, G. Vaubel: Chem. Phys. Lett. *29*, 102-105 (1974)
3.63   R. Eiermann, W. Hofberger, H. Baessler: J. Non-Cryst. Solids *28*, 415-428 (1978)
3.64   J. Hoffmann, K.P. Seefeld, W. Hofberger, H. Baessler: Molec. Phys. *37*, 973-979 (1979)
3.65   J.P. Hirth: J. Appl. Phys. *32*, 700-711 (1961)
3.66   Yu.V. Mnyukh, N.N. Petropavlov: J. Phys. Chem. Sol. *33*, 2079-2087 (1973)
3.67   Yu.V. Mnyukh, N.A. Panfilova: J. Phys. Chem. Sol. *34*, 159-170 (1973)
3.68   P.M. Robinson, H.G. Scott: J. Cryst. Growth *1*, 187-194 (1967)
3.69   A.G. McGhie, P.J. Reucroft, M.M. Labes: J. Chem. Phys. *45*, 3163 (1966)
3.70   N.T. Corke, A.A. Kawada, J.N. Sherwood: Nature *213*, 62-63 (1967)
3.71   J.W. Mitchell: "Direct Observation of Dislocations in Crystals by Optical and Electron Microscopy", in *Direct Observation of Imperfections in Crystals*, ed. by J.B. Newkirk, J.H. Wernick (Wiley, New York 1962) pp. 3-27
3.72   M.D. Cohen, I. Ron, G.M. Schmidt, J.M. Thomas: Nature *224*, 167-168 (1969)
3.73   G. Nomarski, A.R. Weil: Rev. Met. *52*, 121-134 (1955)
3.74   R.D. Heidenreich: *Fundamentals of Transmission Electron Microscopy* (Wiley, New York 1964)
3.75   J.W. Menter: Proc. Roy. Soc. A *236*, 119-125 (1956)
3.76   W. Jones, J.M. Thomas, J.O. Williams, M.J. Goringe, L. Hobbs: *IV International Symposium on Organic Solid State*, Bordeaux, France (July 16-18, 1975) p. 21
3.77   P.B. Hirsch, A. Howie, R.B. Nickolson, D.W. Pashley: *Electron Microscopy of Thin Crystals* (Butterworth, London 1965)

3.78   J.I. Goldstein, H. Yakowitz (eds.): *Practical Scanning Electron Microscopy* (Plenum Press, New York 1975)
3.79   T.E. Everhart, T.L. Hayes: Sci. Amer. *226*, 55-69 (1972)
3.80   G. Bak, A. Lipinski, W. Mycielski: Thin Solid Films *56*, 343-348 (1979)
3.81   T.N. Rhodin, D.S.Y. Tong: Phys. Today *28*, 23-28, 32 (1975)
3.82   L.E. Firment, G.A. Somorjai: Surf. Sci. *55*, 413-426 (1976)
3.83   J.C. Buchholz, G.A. Somorjai: J. Chem. Phys. *66*, 573-580 (1977)
3.84   W.W. Webb: "X-Ray Diffraction Topography", in *Direct Observation of Imperfections in Crystals,* ed. by J.B. Newkirk, J.H. Wernick (Wiley, New York 1962)
3.85   P. Gay, P.B. Hirsch, A. Kelley: Acta Metall. *1*, 315-319 (1953)
3.86   Z.G. Pinsker: *Dynamic Theory of Scattering of X-Rays in Crystals,* Springer Series in Solid-State Sciences, Vol. 3 (Springer, Berlin, Heidelberg, New York 1978)
3.87   L.G. Schulz: J. Metals, AIME Trans. *200*, 1082-1083 (1954)
3.88   *Direct Observation of Imperfections in Crystals,* ed. by J.B. Newkirk, J.H. Wernick (Wiley, New York 1962)
3.89   A.R. Lang: Acta Crystallogr. *12*, 249-250 (1959)
3.90   A.E. Jenkinson, A.R. Lang: "X-Ray Diffraction Topographic Studies of Dislocations in Floating Zone-grown Si", in *Direct Observation of Imperfections in Crystals,* ed. by J.B. Newkirk, J.H. Wernick (Wiley, New York 1962) pp. 471-495
3.91   W.L. Bond, J. Andrus: Amer. Mineralogist *37*, 622-632 (1952)
3.92   U. Bonse, M. Hart: Z. Phys. *190*, 455-467 (1966); see also U. Bonse, W. Graeff: "X-Ray and Neutron Interferometry" in *X-Ray Optics,* ed. by H.J. Queisser, Topics in Applied Physics, Vol. 22 (Springer,Berlin, Heidelberg, New York 1977)
3.93   W.L. Bond: Acta Crystallogr. *13*, 814-818 (1960)
3.94   R.L. Barns: Mater. Res. Bull. *2*, 273-282 (1967)
3.95   D. Goode, Y. Lupien, W. Siebrand, D.F. Williams, J.M. Thomas, J.O. Williams: Chem. Phys. Lett. *25*, 308-311 (1974)
3.96   H.G. Drickamer, C.W. Frank: Ann. Rev. Phys. Chem. *23*, 39-64 (1972)
3.97   V.J. Litvinenko, A.I. Pushareva, O.I. Kozhuhova: Crystallography *21*, 850-852 (1976)
3.98   J.N. Sherwood: Bull. Soc. Sci. Bretagne, Sci. math., phys. et nat. *48*, 11 (1973)
3.99   G.F. Reynolds: "Crystal Growth", in *Physics and Chemistry of the Organic Solid State,* Vol. 1, ed. by D. Fox, M.M. Labes, A. Weissberger (Interscience, New York 1963)

Chapter 4

4.1   C.E. Swenberg, N.E. Geacintov: "Exciton Interactions in Organic Solids", in *Organic Molecular Photophysics,* ed. by J.B. Birks (Wiley, New York 1973) pp. 489-564
4.2   N.E. Geacintov, C.E. Swenberg: "Exciton Interactions in Organic Solids", in *Organic Molecular Photophysics,* Vol. 2, ed. by J.B. Birks (Wiley, Interscience, New York 1975) pp. 395-408
4.3   M. Schott: *Electrical Properties of Organic Solids.* Supplement. Sci. Papers of Wroclaw Tech. Univ. (Wroclaw 1976) pp. 27-64
4.4   E.I. Rashba: Opt. Spektrosk. *2*, 568-577 (1957)
4.5   E.I. Rashba: Fiz. Tverd. Tela *4*, 3301-3320 (1962)
4.6   R.E. Merrifield: J. Chem. Phys. *38*, 920-924 (1963)
4.7   V.I. Sugakov: Opt. Spektrosk. *21*, 574-582 (1966)
4.8   D.P. Craig, M.R. Philpott: Proc. Roy. Soc. A *293*, 213-234 (1966)
4.9   R. Silbey: Chem. Phys. Lett. *14*, 609-611 (1972)
4.10   N.N. Ostapenko, V.I. Sugakov, M.T. Shpak: "Localized Excitons in Molecular Crystals", in *Excitons in Molecular Crystals* (in Russian)

(Naukova Dumka, Kiev 1973) pp. 92-140

4.11   P.E. Schipper: Mol. Cryst. Liq. Cryst. *28*, 401-421 (1974)
4.12   S.A. Rice, J. Jortner: "Possible Uses of High Pressure Techniques for
       the Study of Electronic States of Molecular Crystals", in *Physics of
       Solids at High Pressure*, ed. by C.T. Tomizuka, R.M. Emrick (Academic
       Press, New York 1965) pp. 63-168
4.13   H.W. Offen: "Absorption and Luminescence of Aromatic Molecules at High
       Pressure", in *Organic Molecular Photophysics*, Vol. 1, ed. by J.B. Birks
       (Wiley, London 1973) pp. 103-151
4.14   S. Wiederhorn, H.G. Drickamer: J. Phys. Chem. Sol. *9*, 330-334 (1959)
4.15   R.B. Aust, W.H. Bentley, H.G. Drickamer: J. Chem. Phys.*41*, 1856-1864
       (1964)
4.16   D.C. Fischer, H.G. Drickamer: J. Chem. Phys. *54*, 4825-4837 (1971)
4.17   H. Ohigashi, I. Shirotani, H. Inokuchi, S. Minomura: J. Chem. Phys. *43*,
       314-315 (1965)
4.18   I. Shirotani, K. Kawada, H. Inokuchi: Bull. Chem. Soc. Jpn. *43*, 2381-
       -2385 (1970)
4.19   S. Arnold, W.B. Whitten, A.C Damask: J. Chem. Phys. *61*, 5162-5166
       (1974)
4.20   W.B. Whitten, S. Arnold: Phys. Status Solidi B *74*, 401-407 (1976)
4.21   R. Schnaithmann, H.C. Wolf: Z. Naturforsch. *20a*, 76-81 (1965)
4.22   F.R. Lipsett, G. MacPherson: Canad. J. Phys. *44*, 1485-1515 (1966)
4.23   N.J. Bridge, D. Vincent: J. Chem. Soc. Faraday Trans. II *9*, 1522-1535
       (1972)
4.24   L.E. Lyons, L.J. Warren: *Proc. of the 5th Molecular Crystal Symposium*,
       Philadelphia, USA, 1970, p. 115
4.25   V.A. Lisovenko, I.K. Khutornaya, M.T. Shpak: Phys. Status Solidi A *42*,
       433-437 (1977)
4.26   J.O. Williams, Z. Zboinski: J. Chem. Soc. Faraday Trans. II *74*, 618-
       -629 (1978)
4.27   A.S. Gaievskii, V.A. Lisovenko, A.N. Faidysh, L.B. Yankovskaya: Phys.
       Status Solidi (b) *92*, 31-37 (1979)
4.28   J.O. Williams, B.P. Clarke: J. Chem. Soc. Faraday Trans. II *73*, 1371-
       1384 (1977)
4.29   N.I. Ostapenko, A.R. Skrishevsky, A.I. Faidish, M.T. Shpak: "The In-
       fluence of Structural Defects on Temperature Dependent Intensity of
       Luminescence in Naphthalene Crystals", in *Problems of Molecular Crys-
       tals Spectroscopy* (in Russian) (Naukova Dumka, Kiev 1976) pp. 158-165

Chapter 5

5.1   A.J. Jurgis, E.A. Silinsh: "Calculations of Electronic Polarization
       Energy Increase in Defective Regions of Lattice Compression", in *Or-
       ganic Semiconductors* (in Russian) (Institute of Physics AN USSR, Kiev
       1976) pp. 10-15
5.2   J. Sworakowski: *Electrical Properties of Organic Solids*, Summer School,
       Karpacz, Wroclaw (Sept. 1-7, 1974) pp. 191-199
5.3   G.M. Parkinson, J.M. Thomas, J.O. Williams: J. Phys. C *7*, L310-L313
       (1974)
5.4   A. Samoč, M. Samoč, J. Sworakowski, J.O. Williams: *Seventh Molecular
       Crystal Symposium*, Nikko, Japan (Sept. 8-12, 1975) pp. 141-144
5.5   A. Samoč, M. Samoč, J. Sworakowski, J.O. Williams, J.M. Thomas: Phys.
       Status Solidi A *37*, 271-278 (1976)
5.6   E.A. Silinsh: "On the Physical Nature of Traps in Molecular Crystals",
       in *Semiconductors and Their Applications in Electrotechnique* (in Rus-
       sian), Vol. 5 (Zinatne, Riga 1971) pp. 179-196
5.7   D.I. Hudson: *Statistics*, Lectures on Elementary Statistics and Proba-
       bility, Geneva (1964)

5.8   G.L. Squires: *Practical Physics* (McGraw-Hill, London 1968)
5.9   L.D. Landau, E.M. Lifshic: *Statistical Physics* (in Russian), Vol. 1 (Nauka, Moscow 1976)
5.10 A.A. Predvoditelev, N.H. Rakova, Nan Hyn-bin: Fiz. Tverd. Tela *9*, 300-308 (1967)
5.11 A.D. Mishkis: *Lectures on Advanced Mathematics* (in Russian) (Nauka, Moscow 1967)
5.12 F. Reif: *Statistical Physics*. Berkley Physics Course, Vol. 5 (McGraw-Hill, New York 1967)
5.13 A. Rose: Phys. Rev. *97*, 1538-1544 (1955)
5.14 A. Rose: *Concepts in Photoconductivity and Allied Problems* (Interscience, New York 1963)
5.15 M.A. Lampert: Phys. Rev. *103*, 1648-1656 (1956)
5.16 M.A. Lampert: Rep. Progr. Phys. *27*, 329-367 (1964)
5.17 *Problems of Thin-Layer Electronics* (in Russian) (Nauka, Moscow 1966)
5.18 E.A. Silinsh, L.F. Taure: Izv. Akad. Nauk Latv. SSR, Ser. fiz. tekh. Nauk *1*, 18-27 (1970)
5.19 K.J. Hall, J.S. Bonham, L.E. Lyons: Aust. J. Chem. *31*, 1661-1677 (1978)
5.20 J.S. Bonham: Aust. J. Chem. *31*, 2117-2130 (1978)
5.21 H.P.D. Lanyon: Phys. Rev. *130*, 134-143 (1963)
5.22 J.S. Bonham, D.H. Jarvis: Aust. J. Chem. *30*, 705-720 (1977)
5.23 J.S. Bonham, D.H. Jarvis: Aust. J. Chem. *31*, 2103-2115 (1978)
5.24 J.S. Bonham: Aust. J. Chem. *31*, 2291-2294 (1978)
5.25 E.A. Silinsh, L.F. Taure: *IV Molecular Crystal Symposium*, Netherlands (1968)
5.26 E.A. Silinsh, L.F. Taure: Phys. Status Solidi *32*, 847-852 (1969)
5.27 L.F. Taure: Ph. D. Thesis, Riga (1975)
5.28 P.N. Murgetroyd: J. Phys. D *3*, 151-156 (1970)
5.29 P.A. Raykerus: Radiotekh. Elektron. *16*, 609-613 (1971)
5.30 A.K. Gailis, E.A. Silinsh: Izv. Akad. Nauk.Latv. SSR, Ser. fiz. tekh. Nauk *5*, 13-21 (1972)
5.31 P. Mark, W. Helfrich: J. Appl. Phys. *33*, 205-215 (1962)
5.32 W. Helfrich, P. Mark: Z. Phys. *171*, 527-536 (1963)
5.33 W. Helfrich: Phys. Status Solidi *7*, 263-268 (1964)
5.34 H. Bauser, H.H. Ruf: Phys. Status Solidi *32*, 135-149 (1969)
5.35 G.H. Heilmeier, G. Warfield: J. Chem. Phys. *38*, 163-168 (1963)
5.36 A. Many, J. Levinson, L. Teucher: Mol. Cryst. *5*, 273-294 (1969)
5.37 Z. Burstein, A. Many: Mol. Cryst. Liq. Cryst. *25*, 31-44 (1974)
5.38 G.P. Owen, J. Sworakowski, J.M. Thomas, D.F. Williams, J.O. Williams: J. Chem. Soc. Faraday Trans. II *70*, 853-861 (1974)
5.39 P.J. Reucroft, F.D. Mullins: J. Phys. Chem. Sol. *35*, 347-353 (1974)
5.40 S. Nešpurek, J. Sworakowski, J.O. Williams: J. Phys. C *9*, 2073-2080 (1976)
5.41 G.P. Owen, A. Charlesby: J. Phys. C *7*, L400-L402 (1974)
5.42 E.A. Silinsh, I.J. Muzikante, L.F. Taure: *Electrical and Related Properties of Organic Solids,* Papers of Intern. Conf., Karpacz, Poland (Sept. 18-23, 1978) pp. 71-79
5.43 N. Croitoru, S. Grigorescu: Rev. Roum. Phys. *15*, 465-472 (1970)
5.44 S. Nešpurek, P. Smejtek: Czech. J. Phys. B *22*, 160-175 (1972)
5.45 S. Nešpurek: Czech. J. Phys. B *24*, 660-670 (1974)
5.46 J. Bonham: Aust. J. Chem. *26*, 927-939 (1973)
5.47 M. Gamoudi, N. Rosenburg, G. Guillard, M. Maitrot, G. Mesnard: J. Phys. C *7*, 1149-1159 (1974)
5.48 M. Pope, H. Kallmann: J. Israel. Chem. Soc. *10*, 269-286 (1972)
5.49 S. Nešpurek, J. Sworakowski: Phys. Status Solidi A *41*, 619-627 (1977
5.50 S. Nešpurek, J. Sworakowski: Phys. Status Solidi A *49*, K149-K152 (1978)
5.51 S. Nešpurek, J. Obrada, J. Sworakowski: Phys. Status Solidi A *46*, 273--280 (1978)

5.52   M.A. Nicolet: J. Appl. Phys. *37*, 4224 (1966)
5.53   M.A. Nicolet, V. Rodrignez, D. Stolfa: Surf. Sci. *10*, 146 (1968)
5.54   J. Sworakowski: J. Appl. Phys. *41*, 292 (1970)
5.55   A.I. Rozental, L.G. Paritskii: Sov. Phys.-Semicond. *5*, 2100 (1972)
5.56   W. Hwang, K.C. Kao: Solid-State Electron. *15*, 523-529 (1972)
5.57   P. Delannoy, M. Schott, J. Berrehar: Phys. Status Solidi A *32*, 577
       (1975)
5.58   W. Hwang, K.C. Kao: Solid-State Electron. *19*, 1045 (1976)
5.59   K.H. Nicholas, J. Woods: Brit. J. Appl. Phys. *15*, 783-795 (1964)
5.60   G.A. Dussel, R.H. Bube: Phys. Rev. *155*, 764-779 (1967)
5.61   M.M. Perlman: J. Electrochem. Soc. *119*, 892-898 (1972)
5.62   M.C. Driver, G.T. Wright: Proc. Phys. Soc. *81*, 141-147 (1963)
5.63   K. Kirov, V. Zhelev: Phys. Status Solidi *8*, 431-440 (1965)
5.64   P. Devaux, M. Schott: Phys. Status Solidi *20*, 301-309 (1967)
5.65   J.G. Simmons, G.W. Taylor, M.C. Tam: Phys. Rev. B *7*, 3714-3719 (1973)
5.66   A. Samoč, M. Samoč, J. Sworakowski: Phys. Status Solidi A *36*, 735-
       -745 (1976)
5.66a S. Maeta, K. Sakaguchi: Jpn. J. Appl. Phys. *18*, 1983-1985 (1979)
5.67   R. Eiermann, W. Hofberger: *Electrical and Related Properties of Organ-
       ic Solids,* Papers of Intern. Conf., Karpacz, Poland (Sept. 18-23, 1978)
       p. 103
5.67a T. Sakurai: Jpn. J. Appl. Phys. *18*, 2083-2084 (1979)
5.68   A.N. Blagodarov, E.L. Lutsenko, L.D. Rosenshtein: Phys. Status Solidi
       A *5*, 333-340 (1971)
5.69   R.A. Creswell, M.M. Perlman: J. Appl. Phys. *41*, 2365-2375 (1970)
5.70   J.A. Gorohovatskii, A.N. Rodionov, V.I. Zhdanok: Izv. Akad. Nauk Latv.
       SSR, Ser. fiz. tekh. Nauk *2*, 21-28 (1976)
5.71   J.G. Simmons, M.C. Tam: Phys. Rev. B *7*, 3706-3713 (1973)
5.72   H. Gobrecht, D. Hofmann: J. Phys. Chem. Sol. *27*, 509-522 (1966)
5.73   W. Helfrich, N. Riehl, P. Thoma: Phys. Lett. *10*, 31-32 (1964)
5.74   N. Shiomi: J. Phys. Soc. Jpn. *21*, 907-916 (1966)
5.75   O. Jelinek, I.A. Tale: J. Lumin. *10*, 371-379 (1975)
5.76   L.D. Rosenstein, V.V. Sinitskii, Yu.A. Vidadi: Fiz. Tekh. Polupr. *3*,
       118-120 (1969)
5.77   V.V. Sinitskii, Yu.A. Vidadi, L.D. Rosenstein: Phys. Status Solidi
       A *5*, 327-331 (1971)
5.78   M.I. Gugeshashvili, I.A. Eligulashvili, G.A. Nakashidze: Fiz. Tekh.
       Polupr. *2*, 144-146 (1968)
5.79   E.L. Frankevich, I.A. Sokolik: Solid State Commun. *8*, 251-253 (1970)
5.80   E.L. Frankevich, I.A. Sokolik, L.V. Lukin: Phys. Status Solidi B *54*,
       61-65 (1972)
5.81   N. Wakayama, D.F. Williams: J. Chem. Phys. *57*, 1770-1779 (1972)
5.82   J. Kalinowski, J. Godlewski: Phys. Status Solidi A *20*, 403-410 (1973)
5.83   J.D. Brodribb, D. O'Colmain, D.M. Hughes: J. Phys. D *9*, 253-263 (1976)
5.84   R. Pethig: "Some Dielectric and Electronic Properties of Biomacromole-
       cules ", in *Dielectric and Related Molecular Processes.* Specialist
       Periodical Report of the Chemical Society, London, Vol. 3 (1977)
5.85   S. Bone, R. Pethig: J. Chem. Soc. Faraday Trans. I *74*, 720-726 (1978)
5.86   A.J. Twarowski, A.C. Albrecht: J. Chem. Phys. *70*, 2255-2261 (1979)
5.87   I.S. Kaulach, E.A. Silinsh: Izv. Nauk Latv. SSR, Ser. fiz. tekh. Nauk
       3, 34-40 (1976)
5.88   A. Campos, J.A. Giacometti, M. Silver: Appl. Phys. Lett. *34*, 226-228
       (1979)
5.89   W. Arden, L.M. Peter, G. Vaubel: J. Lumin. *9*, 257-266 (1974)
5.90   Y. Maruyama, K. Machida, N. Iwasaki, S. Iwashima: Chem. Lett. 911-914
       (1975)
5.91   Y. Kamura, I. Shirotani, K. Ohno, K. Seki, H. Inokuchi: Bull. Chem.
       Soc. Jpn. *49*, 418-422 (1976)

5.92   S.D. Druger: "Theory of Charge Transport Properties in Organic Molecular Solids", in *Organic Molecular Photophysics* (Wiley, Interscience, New York 1975) pp. 313-394
5.93   R.W. Munn: Mol. Cryst. Liq. Cryst. *31*, 105-113 (1975)
5.94   D.C. Hoesterey, G.M. Letson: J. Phys. Chem. Sol. *24*, 1609 (1963)

Additional References with Titles

K. Kato, C.L. Braun: The photoconduction threshold in anthracene single crystals, J. Chem. Phys. *72*, 172-176 (1980)

R.W. Munn, W. Siebrand: Charge-carrier drift mobility analysis for naphthalene, J. Chem. Phys. *72*, 3428-3430 (1980)

E.I.P. Walker, A.P. Marchetti, R.H. Young: Off-diagonal mobility components for electrons and holes in naphthalene, J. Chem. Phys. *72*, 3426-3428 (1980)

D.M. Lubman, R. Naaman, R.N. Zare: Multiphoton ionization of azulene and naphthalene, J. Chem. Phys. *72*, 3034-3040 (1980)

D.P. Craig, L.A. Dissado, S.H. Walmsley: Exciton-phonon coupling in molecular crystal trapping processes, Chem. Phys. *46*, 87-105 (1980)

J. Klafter, J. Jortner: Some features of two-particle exciton-phonon excitations in molecular crystals, Chem. Phys. *47*, 25-48 (1980)

V. Capek, I. Rips: Towards fading of memory in the generalized master equation theories of excitation transfer, Phys. Status Solidi (b) *97*, K93-K97 (1980)

R. Kopelman, P. Argyrakis: Diffusive and percolative lattice migrations: Excitons, J. Chem. Phys. *72*, 3053-3060 (1980)

H. Seki, U. Itoh: Ultra thin films of anthracene on fused quartz and sapphire, J. Chem. Phys. *72*, 2166-2178 (1980)

J.B. Web, D.F. Williams, J. Noolandi: Observation of dispersive transport in single crystal anthracene, Sol. State Commun. *31*, 905-907 (1979)

J.S. Bonham: On the theory of space-charge-limited current with diffusion for an exponential trap distribution, Phys. Status Solidi (a) *55*, 61-65 (1979)

S. Maeta, K. Sakaguchi: On the determination of trap depth from thermally stimulated currents, Jpn. J. Appl. Phys. *19*, 519-526 (1980)

M. Pope, C.E. Swenberg: *Electronic Processes in Organic Crystals* (Oxford University Press 1980) (to be published)

K.C. Kao, W. Hwang: *Electrical Transport in Solids – With Particular Reference to Organic Semiconductors* (Pergamon Press, Oxford 1980)

.N. Karl: "High Purity Organic Molecular Crystals", in *Organic Crystals. Germanates. Semi-Conductors*, ed. by H.C. Freyhardt, Springer Series in Crystals/Growth, Properties, Applications (Springer, Berlin, Heidelberg, New York, in preparation)

Subject Index

Neutron Diffraction

Editor: H. Dachs

1978. 138 figures, 32 tables. XIII, 357 pages
(Topics in Current Physics, Volume 6)
ISBN 3-540-08710-9

Contents:
H. Dachs: Principles of Neutron Diffraction. -
J. B. Hayter: Polarized Neutrons. - *P. Coppens:*
Combining X-Ray and Neutron Diffraction:
The Study of Charge Density Distributions in
Solids. - *W. Prandl:* The Determination of
Magnetic Structures. - *W. Schmatz:* Disordered Structures. - *P.-A. Lindgård:* Phase
Transitions and Critical Phenomena. -
G. Zaccái: Application of Neutron Diffraction
to Biological Problems. - *P. Chieux:* Liquid
Structure Investigation by Neutron Scattering. - *H. Rauch, D. Petrascheck:* Dynamical
Neutron Diffraction and Its Application.

Electron Spectroscopy for Surface Analysis

Editor: H. Ibach

1977. 123 figures, 5 tables. XI, 255 pages
(Topics in Current Physics, Volume 4)
ISBN 3-540-08078-3

Contents:
H. Ibach: Introduction. - *D. Roy, J. D. Carette:*
Design of Electron Spectrometers for Surface
Analysis. - *J. Kirschner:* Electron-Excited Core
Level Spectroscopies. - *M. Henzler:* Electron
Diffraction and Surface Defect Structure. -
B. Feuerbacher, B. Fitton: Photoemission
Spectroscopy. - *H. Froitzheim:* Electron
Energy Loss Spectroscopy.

Positrons in Solids

Editor: P. Hautojärvi

1979. 66 figures, 25 tables. XIII, 255 pages
(Topics in Current Physics, Volume 12)
ISBN 3-540-09271-4

Contents:
P. Hautojärvi, A. Vehanen: Introduction to
Positron Annihilation. - *P. E. Mijnarends:*
Electron Momentum Densities in Metals and
Alloys. - *R. N. West:* Positron Studies of Lattice
Defects in Metals. - *R. M. Nieminen, M. J.
Manninen:* Positrons in Imperfect Solids:
Theory. - *A. Dupasquier:* Positrons in Ionic
Solids.

Excitons

Editor: K. Cho

1979. 118 figures, 8 tables. XI, 274 pages
(Topics in Current Physics, Volume 14)
ISBN 3-540-09567-5

Contents:
K. Cho: Introduction. - *K. Cho:* Internal Structure of Excitons. - *J. P. Dean, D. C. Herbert:*
Bound Excitons in Seminconductors. - *B. Fischer, J. Lagois:* Surface Exciton Polaritons. -
P. Y. Yu: Study of Excitons and Exciton. -
Phonon Interactions by Resonant Raman and
Brillouin Spectroscopies.

Springer-Verlag
Berlin
Heidelberg
New York

Excitons at High Density

Editors: H. Haken, S. Nikitine

1975. 120 figures. IV, 303 pages
(Springer Tracts in Modern Physics,
Volume 73)
ISBN 3-540-06943-7

Contents:
Biexcitons. – Electron-Hole Droplets. – Biexcitons and Droplets. – Special Optical Properties of Excitons at High Density. – Laser Action of Excitons. – Excitonic Polaritons at Higher Densities.

X-Ray Optics

Applications to Solids

Editor: H.-J. Queisser
1977. 133 figures, 14 tables. XI. 227 pages
(Topics in Applied Physics, Volume 22)
ISBN 3-540-08462-2

Contents:
H. J. Queisser: Introduction: Structure and Structuring of Solids. – *M. Yoshimatsu, S. Kozaki:* High Brillance X-Ray Sources. – *E. Spiller, R. Feder:* X-Ray Lithography. – *U. Bonse, W. Graeff:* X-Ray and Neutron Interferometry. – *A. Authier:* Section Topography. – *W. Hartmann:* Live Topography.

Springer-Verlag
Berlin
Heidelberg
New York

G. Leibfried, N. Breuer

Point Defetcs in Metals I

Introduction to the Theory

1978. 138 figures, 22 tables. XIV, 342 pages
(Springer Tracts in Modern Physics,
Volume 81)
ISBN 3-540-08375-8

Contents:
Introduction and Survey. – Harmonic Approximation and Linear Response (Green's Function) of an Arbitrary System. – Lattice Theory. – Continuum Theory. – Transition from Lattice to Continuum Theory. – Statics and Dynamics of Simple Single Point Defects. – Scattering of Neutrons and X-Rays by Crystals. – Probability, Distributions and Statistics. – Properties of Crystals with Defects in Small Concentration. – Appendix.

H. Bilz, W. Kress

Phonon Dispersion Relations in Insulators

1979. 162 figures in 271 sepparate Illustrations
VIII, 241 pages
(Springer Series in Solid-State Sciences,
Volume 10)
ISBN 3-540-09399-0

Contents:
Summary of Theory of Phonons: Introduction. Phonon Dispersion Relations and Phonon Models. – Phonon Atlas of Dispersion Curves and Densities of States: Rare-Gas Crystals. Alkali Halides (Rock Salt Structure). Metal Oxides (Rock Salt Structure). Transition Metal Compounds (Rock Salt Structure). Other Cubic Crystals (Rock Salt Structure). Cesium Chloride Structure Crystals. Diamond Structure Crystals. Zinc-Blende Structure Crystals. Wurtzite Structure Crystals. Fluorite Structure Crystals. Rutile Structure Crystals. ABO_3 and ABX_3 Crystals. Layered Structure Crystals. Other Low-Symmetry Crystals. Molecular Crystals. – Mixed Crystals. Organic Crystals. – References. – Subject Index.